# Mobile IPv6

## MOBILITY IN A WIRELESS INTERNET

Hesham Soliman



Addison-Wesley

Boston • San Francisco • New York • Toronto • Montreal
London • Munich • Paris • Madrid
Cape Town • Sydney • Tokyo • Singapore • Mexico City

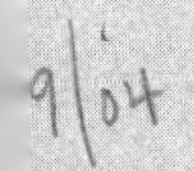

The publisher offers discounts on this book when ordered in quantity for bulk purchases and special sales. For more information, please contact:

U.S. Corporate and Government Sales
(800) 382-3419
corpsales@pearsontechgroup.com

For sales outside of the U.S., please contact:

International Sales
(317) 581-3793
international@pearsontechgroup.com

Visit Addison-Wesley on the Web: www.awprofessional.com

Library of Congress Cataloging-in-Publication Data
Soliman, Hesham.
Mobile IPv6 : mobility in a wireless Internet/Hesham Soliman.
p. cm.
Includes bibliographical references and index.
ISBN: 0-201-78897-7
1. TCP/IP (Computer network protocol). 2. Wireless Internet. 3. Mobile computing.
I. Title.

TK5105.585 2004 200413511646



ISBN: 0-201-78897-7

Text printed on recycled paper

First printing, April 2004

*To the memory of my father*
*and to my mother,*
*For their unconditional love and support throughout my life*

# CONTENTS

*Foreword* xv

*Preface* xvii

*Acknowledgments* xxiii

# PART ONE INTRODUCTION 1

## ONE Introduction 3

*1.1 The Internet Protocol Suite* 3

1.1.1 Networking with the Internet Protocol Suite 5

*1.2 IP Addresses* 7

*1.3 The Domain Name System* 7

*1.4 Host-to-Host Communication* 9

1.4.1 Encapsulation 10

1.4.2 Demultiplexing 11

*1.5 Routing in the Internet* 13

*1.6 Client-Server Versus Peer-to-Peer Communication* 17

*1.7 The Need for IPv6* 18

*1.8 What Is IP Mobility?* 19

1.8.1 How Important Is Mobility? 21

1.8.2 Where Do I Need Mobility Management: Layer 2, Layer 3, or Upper Layers? 21

1.8.3 Mobile IPv6: Main Requirements 23

*1.9 Summary* 24

## TWO An IPv6 Primer 25

*2.1 The IPv6 Protocol 26*

2.1.1 Why Doesn't the IPv6 Header Contain a Checksum Field? 27

2.1.2 Do We Need a Larger Payload Length Field? 28

2.1.3 The Flow Label 28

*2.2 IPv6 Extension Headers 29*

2.2.1 The Hop-by-Hop Options Header 30

2.2.2 The Routing Header 31

2.2.3 The Fragmentation Header 35

2.2.4 IP Layer Security 39

2.2.5 The Destination Options Header 43

2.2.6 Ordering of the Extension Headers 43

*2.3 ICMPv6 44*

2.3.1 ICMPv6 Error Messages 45

2.3.2 ICMPv6 Informational Messages 47

*2.4 Tunneling 48*

2.4.1 What Happens to other Fields in the Tunnel Header? 50

2.4.2 How Many Times Can an IPv6 Packet Be Tunneled? 50

*2.5 IPv6 Addresses 51*

2.5.1 Textual Representation of IPv6 Addresses 52

2.5.2 Unicast Addresses 53

2.5.3 Multicast Addresses 58

2.5.4 Anycast Addresses 60

2.5.5 The Unspecified Address 61

2.5.6 IPv6 Addresses Containing IPv4 Addresses 61

*2.6 Neighbor Discovery 61*

2.6.1 Why Does a Node Need to Discover a Neighbor? 62

*2.7 Stateless Address Autoconfiguration 73*

2.7.1 Ingress Filtering 74

*2.8 A Communication Example 74*

*2.9 Summary 77*

# PART TWO MOBILE IPv6 79


## THREE Mobile IPv6 81

*3.1 Mobile IPv6 Terminology 82*

*3.2 Overview of Mobile IPv6 83*

3.2.1 Binding Updates and Acknowledgments 88

3.2.2 Refreshing Bindings 93

3.2.3 Why Reverse Tunneling? 93

3.2.4 Movement Detection 93

3.2.5 Returning Home 97

3.2.6 Source Address Selection in Mobile Nodes 99

3.2.7 Dynamic Home Agent Discovery 99

3.2.8 Challenges Associated with Plug and Play for Mobile Nodes 103

3.2.9 Can a Mobile Node Have More than One Home Agent? 105

3.2.10 Virtual Home Links 106

*3.3 Route Optimization 106*

3.3.1 Sending Route Optimized Packets to Correspondent Nodes 109

3.3.2 Receiving Route Optimized Packets from Correspondent Nodes 111

3.3.3 Acknowledging Binding Updates Sent to Correspondent Nodes 112

3.3.4 What if the Correspondent Node Failed? 112

3.3.5 Why Not IP in IP Tunneling for Route Optimization? 113

*3.4 What if the Mobile Node Failed? 114*

*3.5 Site-Local Addresses and Mobile IPv6 115*

*3.6 A Communication Example 116*

*3.7 Summary 118*

## FOUR Introduction to Security 121

*4.1 What Is Security and Why Is It Needed? 121*

*4.2 Authentication 122*

*4.3 Authorization 123*

*4.4 Confidentiality, Integrity Checks, Nonrepudiation, and Replay Attacks 123*

*4.5 Cryptography 124*

4.5.1 Encryption Algorithms and Keys 126
4.5.2 Secret Key Encryption 127
4.5.3 Public Key Cryptography 130
4.5.4 Hash Functions, Message Digests, and Message Authentication Codes 137
4.5.5 Nonces and Cookies 139
4.5.6 Establishing a Security Association 140
4.5.7 Cryptographically Generated Addresses 142
4.5.8 Firewalls and Application Level Gateways 145

*4.6 Summary 146*

## FIVE Securing Mobile IPv6 Signaling 149

*5.1 Why Do We Need to Secure Mobile IPv6? 149*

5.1.1 Using Binding Updates to Launch Attacks 150
5.1.2 Attacks Using the Routing Header and Home Address Option 153
5.1.3 MITM Attacks on MPS/MPA 154

*5.2 Requirements for Mobile IPv6 Security 154*

5.2.1 Securing Communication Between Mobile and Correspondent Nodes 155
5.2.2 Securing Messages to the Home Agent 157
5.2.3 Assumptions about Mobile IPv6 Security 157

*5.3 Mobile IPv6 Security 157*

5.3.1 Securing Binding Updates to the Home Agent 158
5.3.2 Securing Mobile Prefix Solicitations and Advertisements 162

5.3.3 Manual Versus Dynamic SAs Between the Mobile Node and Its Home Agent Configuration 162
5.3.4 Securing Binding Updates to Correspondent Nodes 165
5.3.5 Preventing Attacks Using Home Address Options and Routing Headers 184

*5.4 Future Mechanisms for Authenticating Binding Updates 185*
5.4.1 Alternative 1: Using a Cryptographically Generated Home Address 186
5.4.2 Alternative 2: Using Cryptographically Generated Home and Care-of Addresses 188
5.4.3 Other Improvements Gained from CGAs 188

*5.5 Summary 189*

PART THREE HANDOVER OPTIMIZATIONS FOR WIRELESS NETWORKS 79

▼ SIX Evaluating Mobile IPv6 Handovers 195

*6.1 Layer 2 Versus Layer 3 Handovers 196*
6.1.1 Where Is Layer 2 Terminated? 198
6.1.2 Two Different Categories of Wireless Links 198
6.1.3 Make-Before-Break Versus Break-Before-Make Handovers 201

*6.2 How Long Does a Mobile IPv6 Handover Take? 202*
6.2.1 Reducing Neighbor Discovery and DAD Delays 205

*6.3 Handover Impacts on TCP and UDP Traffic 207*
6.3.1 How Does TCP Work? 207
6.3.2 Mobility Impacts on TCP 215
6.3.3 What About UDP? 217

*6.4 Summary 218*

## ▼ SEVEN Mobile IPv6: Handover Optimizations and Extensions 221

*7.1 Fast Handovers for Mobile IPv6 221*

7.1.1 Anticipation and Handover Initiation 222
7.1.2 Updating the Current Access Router 224
7.1.3 Moving to a New Link 226
7.1.4 Failure Cases 227
7.1.5 The Cost of Anticipation 228
7.1.6 Security Issues 235
7.1.7 Can We Use CGAs to Secure Fast Handover Signaling? 236
7.1.8 An Alternate Approach to Fast Handovers 237

*7.2 Hierarchical Mobile IPv6 (HMIPv6) 239*

7.2.1 HMIPv6 Overview 242
7.2.2 MAP Discovery 245
7.2.3 Deploying HMIPv6 248
7.2.4 Location Privacy 251
7.2.5 Local Mobility Without Updating Correspondent Nodes 252
7.2.6 Securing Binding Updates Between a Mobile Node and a MAP 252

*7.3 Combining Fast Handovers and HMIPv6 253*

*7.4 Flow Movement in Mobile IPv6 259*

*7.5 Summary 263*

## ▼ EIGHT Current and Future Work on IPv6 Mobility 265

*8.1 AAA as an Enabler for Mobility 265*

*8.2 Achieving Seamless Mobility 268*

8.2.1 Link-Layer Agnostic Interface to the IP Layer 268
8.2.2 Context Transfer 269
8.2.3 Candidate Access Router Discovery (CARD) 270

*8.3 Network Mobility 270*

*8.4 Summary 274*

# PART FOUR IPv6 AND MOVILE IPv6 DEPLOYMENT 277

## NINE IPv6 in an IPv4 Internet: Migration and Coexistence 279


*9.1 How and When Will IPv6 Be Deployed? 280*

*9.2 What Are the Problems? 280*

*9.3 Tunneling 281*

9.3.1 Configured Tunnels 282

9.3.2 6-to-4 Tunneling 283

9.3.3 Routing Protocols-Based Tunnel End Point Discovery 286

9.3.4 Intrasite Automatic Addressing Protocol (ISATAP) 286

*9.4 Translation 288*

9.4.1 Stateless IP ICMP Translator (SIIT) 289

9.4.2 Network Address Translator and Protocol Translator (NAT-PT) 292

*9.5 Other Deployment Scenarios and Considerations 293*

9.5.1 IPv6-Only Networks 294

9.5.2 Mobility Considerations 294

*9.6 Summary 296*


## TEN A Case Study: IPv6 in 3GPP Networks 299


*10.1 3GPP Background 300*

*10.2 3GPP UMTS Network Architecture 301*

10.2.1 Packet-Switched Core Network 302

10.2.2 Circuit-Switched Core Network 305

*10.3 UTRAN Architecture 305*

10.3.1 Wideband Code Division Multiple Access 306

10.3.2 Power Control and Handovers 308

*10.4 UMTS Core Network 311*

10.4.1 PDP Context Activation 312

10.4.2 Mobility Management in the Core Network 315

*10.5 IPv6 in UMTS 315*
10.5.1 Address Configuration 316
10.5.2 Transition and Coexistence 317
10.5.3 IPv6 Mobility 320
*10.6 Summary 325*

Index 327

# FOREWORD

Mobile IPv6 is a network layer protocol for enabling mobility in IPv6 networks. IP mobility technology has gained a significant amount of traction over the last few years, driven by a number of factors. Among them are the emergence of 3G wireless networks that support packet data services; deployment of high-speed wireless networks such as UMTS, cdma2000 and GPRS/EDGE; devices that are multifunctional and capable of services that go beyond just voice or text messaging (SMS); and the increasing dependence of society on information and the need to access it from any place and any time. The number of notebook computers sold has exceeded that of desktop machines in the first half of 2003, and the availability of 802.11 wireless LANs in homes and public places such as airports, hotels, and coffee shops has made untethered wireless computing much more attractive to a very large set of users. But what does all this have to do with Mobile IPv6? The simple answer is that Mobile IPv6 is a protocol solution for achieving seamless mobility across heterogeneous access network types.

IPv6 networks are beginning to be deployed in some parts of the world, and it is anticipated that the IPv4 address shortage issues will accelerate the deployment of IPv6 networks even faster in rapidly developing countries in Asia-Pacific like China and India, Latin America as well as developed countries like Japan, S. Korea and many countries in Western Europe. Mobile IPv6 enables mobility of the devices or hosts in these next-generation IP networks. Mobility for IP devices has not been a major driver in the past. However, with the inclusion of IP stacks in PDAs, mobile phones, and various forms of notebook and tablet PCs, mobility is becoming an increasingly critical need. Mobile IPv6 is an integral part of the IPv6 protocol stack and hence is expected to be a standard component of every IPv6 device in the near future. With this capability built in, user experience, which has been primarily limited to being nomadic in nature as far as mobility of IP devices is concerned, will shift to one that is seamless in nature and unnoticed.

Mobile IPv6 has a long history in the IETF. Work on the protocol was started in the mid 1990s. It was approved for publication as an RFC only in July 2003. A lot of work has gone into this specification over the years. The protocol has evolved significantly since 2001 due to the discovery of a flaw in the assumptions of security for route optimization. Hesham Soliman, the author of this book, has been a major contributor to the development of the Mobile IPv6 protocol in the IETF and brings a vast amount of experience as well as in-depth knowledge to this book. He has been involved in all the key design decisions and discussions that have taken place with respect to

Mobile IPv6 since 2000. Hesham is also a recognized expert on IPv6. An indepth understanding of IPv6 is absolutely essential to grasp the nuances of Mobile IPv6 and the author's expertise in this domain brings added value to this book. Enough issues have been raised, discussed, and closed on this protocol and enough text has been generated to fill a book by itself.

A number of implementations of Mobile IPv6 have been done, and many more are now in various phases of development. This book serves as an excellent reference source for implementors, not only for understanding the reasons why certain features of the protocol have been designed in a specific way but also for clarifying many of the ambiguities that arise from a direct reading of the specification published as an RFC. Many of the IPv6 features, such as stateless address allocation and Neighbor Discovery, and how they are used in conjunction with Mobile IPv6 are explained in detail. Enhancements to Mobile IPv6 with solutions such as hierarchical mobility and dealing with handoffs in a rapid manner provide value-added material to the understanding of the basic protocol.

I believe this book provides an insightful understanding for anyone interested in Mobile IPv6. The timing of the book could not be any better, since this is the first effort at capturing all the details of the protocol. Mobile IPv6 at the 10,000-feet level may seem simple enough, but when you delve into the details, you will appreciate all the finer complexities of achieving mobility with security. This attention to detail can only come from one who has participated in the standards efforts associated with the protocol and from doing actual implementation. Hesham brings both of these qualities to this book without a doubt.

*Basavaraj Patil*
*Co-chair of the Mobile IP WG in IETF*
*Senior Technology Specialist, Nokia Networks*

# PREFACE

This book focuses primarily on the current Mobile IPv6 standard, which was finalized in 2003 by the Internet Engineering Task Force (IETF). It also covers current proposed optimizations and applications. The aim of this book is to explain in detail the current Mobile IPv6 specification and, more importantly, the assumptions that led to its final design. These assumptions are extremely important to developers and researchers in the field of IPv6 mobility. The topic of IPv6 mobility management has gained a lot of momentum over the last decade, especially with the growth of the wireless industry and the more recent convergence between wireless telecommunication networks (supporting over 1.5 billion devices) and the Internet.

Research on packet switching started back in the 1960s and continued through the 1980s, producing the Internet as we know it today. The predominant Internet Protocol (IP) supported on the Internet today is the IP version 4 (IPv4). When IPv4 was first designed, the Internet was not expected to grow as fast as it did or to be used for the vast and diverse number of applications used today. In 1992, the IAB (known then as the Internet Activities Board and now called the Internet Architecture Board) forecasted the expected shortage in IP addresses in the near future and the imminent need for a new version of IP. A Next Generation Directorate was formed in IETF to choose the new protocol. In 1994, the Next Generation Directorate made its decision. The new protocol was named IP version 6 (IPv6), since version number 5 was already taken by another experimental protocol.

Mobile IPv6 provides mobility support for IPv6, allowing devices to move within the Internet topology while being reachable and maintaining ongoing connections. To understand Mobile IPv6 and the assumptions behind the final specification, IPv6 needs to be understood in some detail. In addition, the fundamental principles of Internet routing and naming services must be understood. In this regard, this book is self-contained; it starts by describing the Internet protocol suite, commonly called the TCP/IP protocol stack, then describes the core IPv6 specifications in detail before getting into the details of Mobile IPv6 and its optimizations. Finally, the use of IPv6 and Mobile IPv6 are presented in the context of a third-generation cellular system based on the third Generation Partnership Project (3GPP) standards.

## Organization of This Book

Mobility management is a very large area that impacts many parts of the Internet protocol stack. In order for you to gain the maximum benefit from this book, it has been divided into four parts. Part One includes Chapters 1 and 2 and provides the necessary background about IPv6 and the Internet. This information is essential for understanding the subsequent parts. Readers already familiar with the Internet specifications and the details of IPv6 may choose to quickly browse through this part and start reading the next.

Part Two provides a detailed analysis of the Mobile IPv6 protocol, starting with an overview of the protocol's operation in Chapter 3. Chapter 4 introduces the topic of security on the Internet, describes the issues involved in securing communication between two machines, and presents the current state of the art in this area. The aim of this chapter is to help you understand the issues involved in securing Mobile IPv6 and the assumptions made in this part of the protocol design. Chapter 5 discusses in detail the security threats, requirements, and design of Mobile IPv6 security.

Part Three is dedicated to providing a theoretical evaluation of the performance of Mobile IPv6 handovers. First, the impact of Mobile IPv6 handovers on the transport layer protocols (and consequently applications) is presented in Chapter 6. Chapter 7 presents several optimizations aimed at improving Mobile IPv6 performance. Chapter 8 discusses some of the ongoing and future work related to mobility management for IPv6. Some of these topics are not exclusive to IPv6 or Mobile IPv6; however, we chose to include topics related to IP mobility in general, as they were found to be within the scope and spirit of this book.

Part Four focuses on deployment issues for IPv6 and Mobile IPv6. Chapter 9 presents some of the most important mechanisms designed to allow IPv6 to be deployed in an IPv4 Internet. Chapter 10 uses 3GPP networks as an example for a third-generation cellular network, provides an overview of its architecture, and illustrates how IPv6 can be deployed in that network.

## Target Readers

This book is self-contained. You need not have a strong background in the Internet or IPv6; however, basic knowledge about the Internet is needed. All of the necessary background information is provided either in the first part or in the relevant chapters. However, readers with an engineering or computer science background will have a deeper understanding of the technical issues presented in this book.

The book starts by explaining the fundamental principles of the Internet and gradually introduces more complex issues related to IPv6, mobility management, and wireless networks. Therefore, it is suited to a wide range of readers, including network practitioners, Internet developers, engineers studying the latest developments in the Internet and IPv6 mobility technologies or refreshing existing knowledge, final year undergraduate students of engineering or computer science disciplines, and postgraduate students of the same disciplines. This book is also suited to researchers in the area of mobility management, as it helps them understand the requirements and assumptions that led to the development of Mobile IPv6. Understanding these assumptions is one of the most important ingredients for a successful research program.

## Terminology and Typographic Conventions

In this section we present some basic terms that are used in this book. Other terms are explained upon introduction.

| | |
|---|---|
| Access router (AR) | A router, one hop away from a node. The access router receives traffic sent from various hosts and forwards it to the next destination, which may be another router or the ultimate receiver (host). |
| Bad Guy | Someone who is doing something that he is not authorized to do. |
| Host | A node that implements the Internet protocol suite and is not a router. |
| Link | The communication media connecting different hosts and routers. |
| Lower layers | The layers below the Internet layer. This includes the physical layer and the data link layer. |
| Neighbors | Nodes that share the same link. |
| Node | A device that implements IPv6. Nodes can be hosts or routers. |
| Router | A node that forwards traffic not addressed to itself. Sometimes we use the term intermediate node as a synonym. |
| Upper layers | Layers above the IPv6 layer (e.g., transport layers like TCP and UDP, or applications) |

The terms *datagram*, *packet*, and *IP packet* are all synonyms in this book. Likewise, the terms *octet* and *byte* are synonyms. The term *internet* refers to a group of networks joined by routers, while *Internet* refers to the global IP network.

Protocol fields or functions are presented in *italics*. The term *well-known* refers to a reserved number or a protocol field. The Internet Assigned Numbers Authority (IANA) is the authority responsible for managing reserved numbers and protocol fields. The **bold** format is used for emphasis.

In this book we use the terms he, she, him, and her in an arbitrary manner without bias to one particular gender. I prefer to use these terms arbitrarily instead of using the gender neutral, but grammatically incorrect they, their, and them when referring to a single person.

## INTERNET STANDARDS

The Internet Engineering Task Force (IETF) is the organization responsible for producing Internet standards. The IETF is an open forum consisting of individuals working with vendors, operators, researchers, and other interested individuals.

IETF participants meet three times a year; however, most of the work is usually done on mailing lists. There are no membership fees required, but participants pay a registration fee in order to attend meetings.

From the bottom up, the IETF consists of over 100 working groups. Each group addresses one particular problem. Working Groups are divided into separate Areas. Each Area is managed by two Area Directors. The group consisting of all Area Directors and the IETF chair is called the Internet Engineering Steering Group (IESG). The IAB is mostly interested in architectural issues related to Internet protocols. The IAB also handles appeals from IETF participants against the IESG. Both the IESG and IAB are chartered by the Internet Society (ISOC). ISOC also charters IANA to manage unique number assignment for Internet protocols.

Currently, there are eight Areas in the IETF: General area chaired by the IETF chair, Applications, Internet, Operations and Management, Routing, Security, Sub-IP, and Transport. The Mobile IP and IPv6 working groups are currently part of the Internet area. The v6ops working group is responsible for IPv6 deployment and operations issues and is part of the Operations and Management area.

The IETF standards progress is generally based on rough consensus and running code. That is, in order for a specification to progress, IETF participants need to agree (not unanimously) and the specification should be known to work in a running implementation. However, stricter requirements on implementations are imposed depending on the status of the specification. Each specification is documented in a Request For Comment (RFC). There are four types of RFCs: Experimental, Informational, Standards Track, and Best Current Practice (BCP). Standards Track RFCs are the most deployed RFCs on the Internet. They have different maturity levels, starting with Proposed Standard (PS) as an entry level, then progressing to Draft Standard (DS) and Internet Standard when they meet the maturity requirements for these levels—for example, when they are widely deployed and have all features of the protocol implemented and tested for interoperability.

BCP RFCs document useful current practices within the Internet community that can help the rest of the community. Informational RFCs are seen as suggestions by the author(s) without any obligations for implementations. Experimental RFCs act as an archival mechanism for the outcome of research or development efforts of a particular working group or an individual.

Each RFC is assigned a number by the RFC editor. An RFC cannot be modified and republished with the same number; a new number must be assigned in this case.

In order to suggest the publication of an RFC, authors must document their proposals in Internet drafts. An Internet draft can be based on an individual or a working group submission to the IETF. Individual submissions are usually files with names written in the following format: draft-authorsnames-workgroupname-...-version.txt, where "..." is usually the subject addressed by the draft and "version" is a number representing the draft's revision. Working group submissions' file names are written in the following format: draft-ietf-workgroupname-...-version.txt. Internet drafts have no official status and do not in any way represent IETF consensus. However, working group drafts typically reflect the group's desire to work on a particular problem or consensus on the solution proposed in the draft. Internet drafts are valid for six months, after which they expire, get updated, or progress to become an RFC.

You can read more about the IETF at *www.ietf.org*. All current RFCs and drafts can be reached from this link.

## Things Change

Many of the references used in this book are not Internet Standards. In fact, most of the RFCs are in either PS or DS status, and some references are working group or individual drafts. It is therefore very likely that some of the solutions in this book will change in a year or two. It is also likely that better solutions are developed at a later stage. In particular, all of the concepts presented in Chapter 7 are currently either working group or personal drafts. However, I tried to present the main concepts and philosophies behind the current proposals. Understanding these concepts will enable you to have a deep understanding of IP mobility, even if the bits sent on the wire are modified by the Mobile IP Working Group in future revisions.

All of the Internet drafts referenced in this book can be found at *www.awprofessional.com/titles/0201788977*. At this same address, you can send comments via the "contact us" link where the author welcomes and encourages comments on any aspect of this book.

# ACKNOWLEDGMENTS

Writing this book while keeping a demanding job, a busy travel schedule, and trying to maintain a minimum level of social life has not been easy. To complicate things further, the Mobile IPv6 standard has been a moving target for the last two years! I'm confident that I could not have finished this book without the support of many people. I just hope I can remember all of them in this section. If I miss some, they have my sincere apologies. First, I'd like to thank Michael Strang for offering me this project and for his support and patience with me and the moving target that I am writing about. Thanks to Mary Franz for her helpful tips and professional management of the production stages and to Jennifer L. Blackwell, and Kerry Reardon for their excellent editorial comments.

Karim ElMalki is an invaluable friend and colleague, and even a patient teacher sometimes. I'd like to thank him for his support in general and in particular for his efforts in reviewing the entire book and making extremely useful remarks that have significantly improved the quality of this book. I'd also like to thank Gianluca Verin very much for his support and for his very useful comments on the first three chapters. Thanks to Wolfgang Fritsche, Suresh K. Satapati, Ravi Prakash, Allen Briggs, and Matt Impett, who reviewed the entire book and provided excellent comments that helped me improve it.

I want to thank Erik Nordmark for the enlightening discussions that I had with him over the past few years. A lot of his knowledge has contributed to my ability to explain the technology presented in this book.

I'm very grateful to my colleagues at my previous employer, Ericsson Research, who have always been welcoming and generous with their knowledge and support. In particular, I'd like to thank Conny Larsson and Mattias Pettersson for reviewing parts of earlier versions of the book, Mats Naslund and Fergal Ladley for their enlightening comments on Chapter 4, Yuri Ismailov and Reiner Ludwig for their comments on the TCP sections, and especially Yuri for reviewing it twice! Thanks to Jan Holler and Olle Viktorsson for allowing me to use Ericsson's computer resources to write the first draft of this book and for understanding my need to take time off at critical stages of writing. And thanks to Andras Toth for understanding my commitment to this project and supporting me throughout the last year. Also, many thanks to Scott Corson from Flarion Technologies for giving me the time and resources needed to finish this book after I joined Flarion.

Ainkaran Krishnarajah provided substantial comments and contributions to Chapter 10 and helped me simplify the discussion on WCDMA; without him that chapter would have contained all those difficult radio topics!

On a more personal note, I want to thank my family and friends all over the planet for their encouragement and support during the last 12 months while I was writing this book. In particular, I want to thank Karim ElMalki and Gianluca Verin, who treated this project as if it were their own and relentlessly supported my work; Ainkaran Krishnarajah for the continuous encouragement and reassurance; and Angela Spink, who showed me how to take life as it comes and how to keep focused when the going gets tough.

Finally, despite having been reviewed by several experts, this book might not be error-free. I'm obviously responsible for any errors in the book and I encourage readers to send me their remarks.

*Hesham Soliman*
*November 2003*

PART ONE

# INTRODUCTION


- Chapter 1 Introduction
- Chapter 2 An IPv6 Primer

ONE

# Introduction

This chapter provides background information on the Internet's design and operation. Almost all of the concepts presented in this chapter are valid for IPv4 and IPv6.

We first present the Internet protocol suite, commonly known as the TCP/IP protocol suite, then analyze the different layers within the stack. Then we show an overview of the Internet routing architecture, which is essential for understanding some of the assumptions behind Mobile IPv6. Finally, the need for IPv6 and Mobile IPv6 is discussed.

## 1.1 The Internet Protocol Suite

Networking is a complex topic that covers several tasks and scientific disciplines. In order to simplify and solve the networking problem, it is usually divided into subproblems, represented as layers. Each layer is responsible for specific tasks. Each layer provides a service to the layer above it. As far as implementations of the protocol suite are concerned, the network and transport layers are usually implemented as a part of the operating system (kernel space), whereas applications are implemented as processes that run in the *user space*. The Internet protocol suite is divided into the five layers shown in Figure 1–1.

It is important to understand the role of each individual layer in order to understand the Internet protocol suite as a whole:

> The **physical layer** is responsible for sending and receiving bits over a wired or wireless interface. Some of the responsibilities of the Physical layer include modulation/demodulation, power measurement and synchronization. A device can contain several physical interfaces.

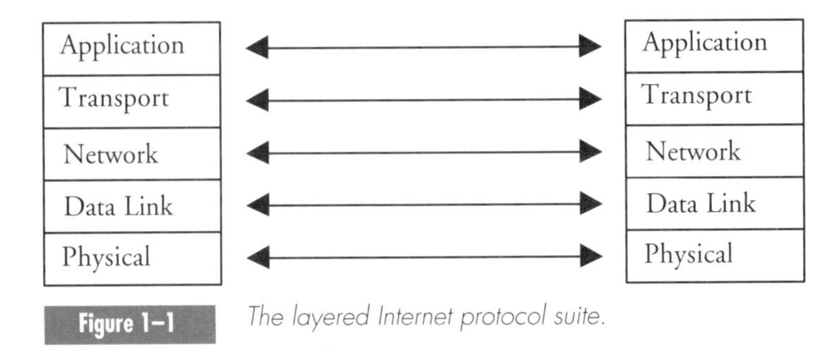


Figure 1–1 The layered Internet protocol suite.

The **data-link layer** is often called the *link layer*, or *layer 2*. This layer is responsible for link control and Media Access Control (MAC). Other responsibilities include error detection and error correction. In some sophisticated wireless link layers, security is provided in the form of encryption over the wireless link. Ethernet is one example of a data-link layer.

The **network layer** (also called the IP layer) provides connectivity to other upper layers. The network layer is responsible for forwarding packets from one node to another. Within the Internet protocol suite, the network layer consists of IP and the Internet Control Message Protocol (ICMP). IP provides an unreliable forwarding service to upper layers. That is, there is no guarantee that a packet sent will actually arrive at the intended destination, and there is no error detection or correction. On the other hand, ICMP is used to send signaling information between two devices. The term *signaling* refers to information about the connection between two devices, or errors encountered while sending data between those devices. That is, signaling contains information that is not part of the data transfer between two devices. Chapter 2, presents ICMPv6 messages in detail.

The **transport layer** is responsible for controlling the flow of data between communicating entities. Within the Internet protocol suite there are three different transport protocols: the traditional Transmission Control Protocol (TCP), User Datagram Protocol (UDP), and the recently developed Stream Control Transmission Protocol (SCTP). TCP provides a reliable transport service to applications. To do this, it includes functions that allow it to detect packet losses, delays, and errors. TCP uses retransmissions to guarantee delivery. In contrast, UDP is an unreliable transport protocol that does not guarantee delivery or error detection. UDP simply acts as an interface between the application layer and the IP layer. Why would any application want to use UDP? Some applications would rather lose information than delay them by allowing retransmissions; these applications are called real-time applications. As an analogy, consider watching a live football game on TV, broadcast via satellite. When the program is disrupted because of some transmission problems, viewers would rather continue watching the new video frames sent live than wait for another minute or two before the old scenes are retransmitted so that they can watch every second in the game. Real-time video conferencing over IP has the same requirements.

Another reason for preferring UDP to TCP is that applications may want to perform their own flow control mechanisms because they are better suited to the special needs of that particular application.

SCTP, like TCP, provides a reliable transmission service to applications. In addition, SCTP includes new features that make it more suited to traditional telephony signaling applications. SCTP also includes a multistreaming function that allows it to send the same stream more efficiently to multiple receivers.

The **application layer** is responsible for taking input from a user or a machine and formatting it before sending it via the lower layer. Some of the most common applications on the Internet are File Transfer Protocol (FTP), TELNET for remote login, Simple Mail Transfer Protocol (SMTP) for email, and Hypertext Transfer Protocol (HTTP) used for web browsing.

As illustrated in Figure 1–1, each layer communicates with its remote *peer* at the same layer, using one or more protocols. For instance, TCP allows peers to communicate on the transport layer, while IP allows peers to communicate on the network layer.

The strength of the layering model is that every layer can be modified or expanded without affecting others (with some minor exceptions), and more importantly, each layer is independent from the one below it or above it. For instance, applications know nothing about forwarding, link-layer control, or modulation, which are done in different layers. Similarly, neither the network nor the transport layer needs to know anything about the content of a particular application in order to fulfill their tasks.

### 1.1.1 Networking with the Internet Protocol Suite

When the telecommunication industry first started considering telephone networks toward the end of the 19th century, telephone operators started to connect every house to all other houses that it might wish to communicate with, ending with a fully meshed network with $N(N-1)$ unidirectional wires—or $N(N-1)/2$ bidirectional wires—where $N$ is the number of telephones that may talk to each other. It is clear that as *N* increases, this type of networking will not scale. This was also evident to telecommunication pioneers who started developing the concept of telephone exchanges (or central offices in the U.S.). On this level, the same problem exists in IP networks. If everyone needs to talk to everyone else, we will need aggregation points that can connect Internet devices to each other. These aggregation points are called *routers*. Routers forward IP packets between hosts; the forwarding is based on the information included in the network layer. Figure 1–2 shows how two hosts can communicate through an IP router.

An IP router forwards packets based on the content of the IP header. Just as we cannot directly connect every host to every other host in a very large network, we also know that there are several physical and link-layer technologies used in different deployment scenarios. Hence, forwarding on the IP layer is considered advantageous due to its independence of the underlying technology, which allows two hosts with different physical interfaces (as illustrated in Figure 1–2) to communicate. As seen in Figure 1–2, when the router's link layer receives a packet, it processes the link-layer header, then passes the packet to the IP layer. The IP layer decides where the packet should be forwarded. In this example the packet will be forwarded to a host sharing the same physical link with the router. It is pertinent to note that routers do not need to modify IP packets in order to decide where they should be forwarded; they simply forward packets on paths that will eventually lead them to their intended destinations. However, some of the fields in the IP header are modified by routers (for reasons other than forwarding), as will be shown in Chapter 2.

Routers have two or more interfaces, where an interface is the node's point of attachment to the link. Any node that has more than one interface is called a *multihomed* node. A router, by definition, is multihomed. A host can also be multihomed. The distinction between a host and a router is that a router forwards packets intended for nodes other than itself, while a host receives packets destined to it. A multihomed host need not be a router provided that it does not forward packets from one interface to another.

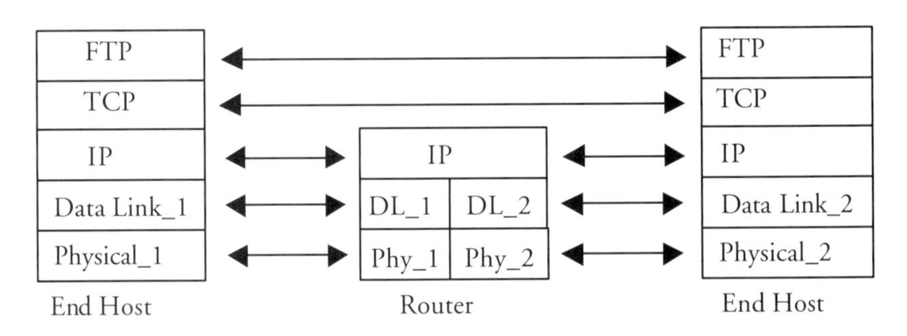


**Figure 1–2** Communication through a router.

## 1.2 IP Addresses

Each node on the Internet (host or router) has at least one address. Just like a home address is used to locate a house, an IP address is used to locate a node. Each IP header contains the address for the sender and the intended destination. An IPv4 address consists of 32 bits, while an IPv6 address consists of 128 bits, allowing for an extremely large number of addresses (millions of addresses per square meter on earth).

There are three different types of IPv6 addresses: *unicast* addresses are assigned to a single interface, while *anycast* and *multicast* addresses are assigned to multiple interfaces. IPv6 also introduces the concept of scoped addresses. An address can be suitable only on a local link (link-local address) but not beyond, or within an IPv6 *site* (site-local addresses), or globally. The validity of an address within a particular domain (link, site, or global domain) means that a node can use such address to communicate with another node located within the scope of that domain. More scopes were introduced for multicast addresses. IPv6 addresses and their textual representation are discussed in detail in Chapter 2.

## 1.3 The Domain Name System

Routers use IP addresses to locate nodes on the Internet and forward packets to those addresses. Addresses are not permanent; they might change for several reasons, such as network renumbering due to reorganizing a network or merging two networks.

Nodes on the Internet are usually assigned names (e.g., *www.example.com*) that are known to those wishing to reach them. Names must be resolved into addresses before communication starts in order to allow hosts to send IP packets to such destinations and to allow routers to forward packets between nodes; this is done through the Domain Name System (DNS). The DNS is a distributed database that is used to resolve nodes' names into IP addresses and vice versa. Applications use a standard protocol to query the DNS; the query can provide a node's name and request an address, or it can provide the node's IP address and request a name. The latter type of query is called a *reverse DNS lookup*

The domain names are hierarchical. Each node is assigned a *label*; the *domain name* of a node can be found by starting with the node's label and moving up in the tree while adding labels for each node until the *root* of the tree is reached. The root DNS does not have a label associated with it. Below the root are the *Top Level Domains* (TLDs), which are divided into two types: *Generic* domains and *Country* domains.

Labels are up to 63 characters long and are always separated with *dots*. An example of a DNS tree is shown in Figure 1–3.

Following this description, in Figure 1–3, the node Mozart would be registered in the DNS as Mozart.example.se.

The *arpa* domain is a special one reserved for name-to-address resolution. There are more generic domains that are not shown in our example. The label *com* indicates commercial, *edu* refers to educational organizations, *gov* refers to a government organization, *net* indicates networks, and *org* is assigned to other organizations. Each country is assigned a two-letter name, such as *au* for Australia, *se* for Sweden, and *uk* for the United Kingdom. It is quite common for country domains to include their own generic domain names under the country's tree (e.g., example.**com**.au).

Every node in the tree must have a globally unique domain name; however, the actual label assigned to each node could be reused within different branches of the tree (e.g., research.example.se and research.example.uk can both exist). Applying our mechanism of finding a node's domain name, we would see that the domain name of the node with a label Mozart is Mozart.example.se., which is also called the Fully Qualified Domain Name (FQDN).

The structure of the naming system infers delegation within each tree. The top-level domain names are managed by a single entity that allocates unique labels to lower levels in the hierarchy. The term *zone* refers to a subtree within the hierarchy—for instance, example.se. The authority for example.se is then delegated to the administrator of that domain. The administrator can then divide that tree further under the label "example" (e.g., public.**example**.se). It is also up to the administrators to provide a *name server* for their zones. Name servers contain information about nodes within their domain—that is, their names and IP addresses. Name servers must also know the IP address of the root of the tree. The

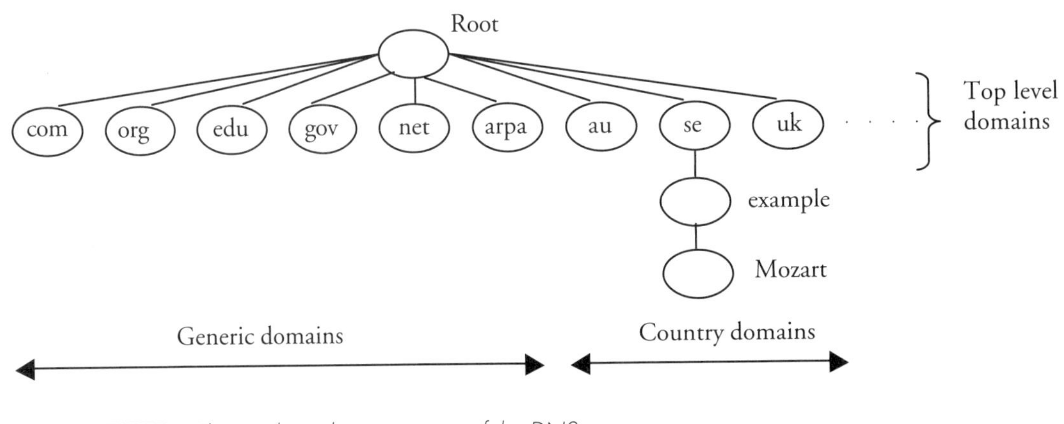


**Figure 1–3** *The tree-based organization of the DNS.*

root, on the other hand, knows all the names and IP addresses of all the authorities responsible for all of the second level domain names.

Each zone is served by a *Primary* DNS and at least one Secondary DNS to achieve redundancy (i.e., in case the Primary DNS fails). Hosts within that domain will query one of the DNSs in order to resolve the IP address of another host that they wish to communicate with. The query is initiated by the application, which calls a *resolver* (through a pseudostandard library interface) to query the DNS using a standard DNS protocol. When the query arrives at the DNS, it may or may not be able to resolve the host's name. For instance, if Mozart.example.se is trying to resolve an address for Bach.example.se, which is served by the same DNS, the local DNS will return a reply containing that address. However, if the query is for Bach.example.com, the local DNS will query the DNS responsible for the .com domain. Since the local DNS knows the IP address for the root, it can forward the request to it; the root then replies with the IP address of the DNS responsible for the .com domain. The local DNS then sends its request for Bach.example.com to the .com DNS. If that DNS cannot resolve the address, it returns the IP address of the DNS responsible for example.com, which should be able to return the address for Bach.example.com. Hence, a single DNS request can result in several recursive requests in order to find the right DNS to query. To minimize the number of messages sent and received due to DNS queries, a fundamental feature of the DNS is *caching*. Whenever a DNS receives an IP address with a corresponding name, it caches this information for a period of time (ranging from a few hours to two days depending on the information included in the DNS record for the host being queried) so that it can immediately reply with this information the next time it gets a query for the same host. This feature becomes relevant when we discuss mobility.

The DNS contains a number of *Resource Records* used for different purposes. For the purposes of this book, we only need to know about those records used for name-to-address resolution. IPv4 addresses are stored in *A records*, while IPv6 addresses are stored in *AAAA records* (i.e., four times the size of the IPv4 records, since IPv6 addresses are four times the size of IPv4 addresses). Any node can have multiple addresses and consequently multiple resource records in the DNS. For instance, a node can have several A or AAAA records or a mix of both (if the node implements both IPv4 and IPv6).

# 1.4 Host-to-Host Communication

So far we have presented the layered model of the Internet protocol suite, addresses, and how the mapping between names and addresses is done on the Internet using the DNS. Equipped with this knowledge, we can see how two hosts can communicate and learn how hosts manage such communication internally between the different layers.

## 1.4.1 Encapsulation

Before an application sends information intended for another host, it must get that host's address through the DNS or perhaps through a user interface. The application then requests a connection from one of the transport layer protocols by opening a *socket* that allows it to communicate with another application. A *socket* is a record that includes information about a connection between two applications.

The socket Application Program Interface (socket API) is a group of functions that are used by applications to communicate to the lower layers in the TCP/IP stack. Opening a socket involves storing information about the applications in both hosts, the IP addresses of both hosts, and the transport protocol that they will use to communicate (e.g., TCP). The information about the application is represented by a *port* number. Each application has a port number associated with it. Figure 1–4 shows the encapsulation process.

The port number space is 16 bits wide, allowing 64K ports to be used. Port numbers between 1 and 1023 are reserved for well-known applications.

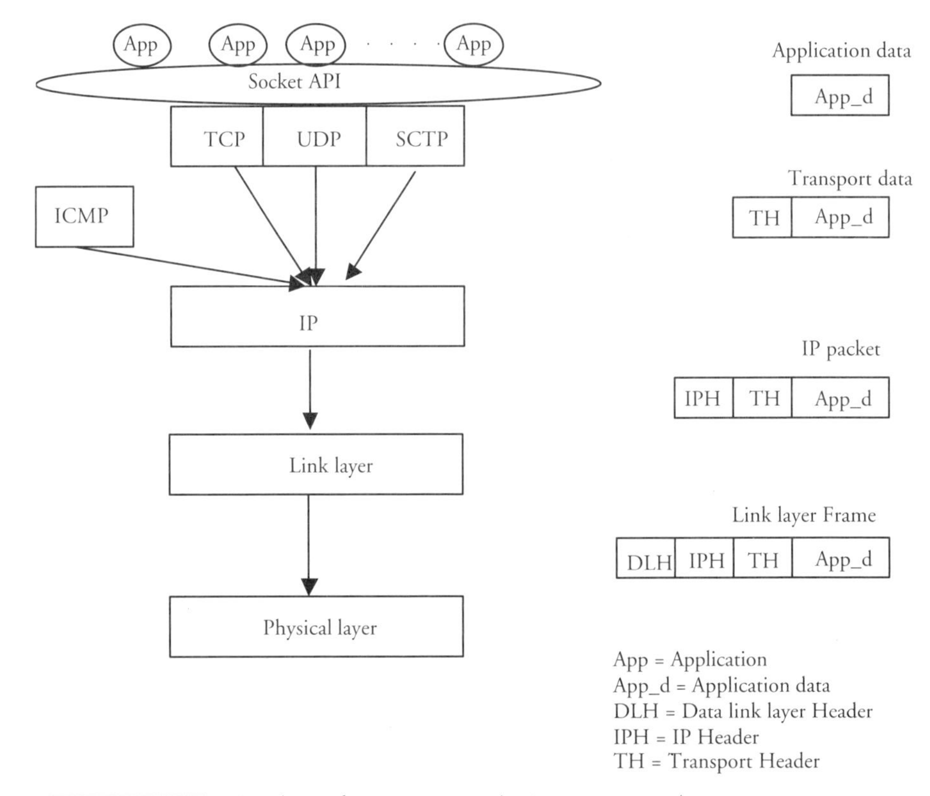


**Figure 1–4** *Sending information using the Internet protocol suite.*

For instance, an HTTP server is always allocated port 80, FTP servers use port 21, and TELNET servers use port 23. Reserved port numbers are used to allow a client's application to reach a server, an action that requires the client to know both the IP address and port number of the remote server.

Applications that do not have reserved port numbers (typically clients that do not need to be reached) are allocated *ephemeral* port numbers when opening a socket. Port numbers are used by transport layer protocols to identify applications, just like routers use addresses to forward packets to other nodes. We discuss this in more detail when we look at the demultiplexing process in the next section. It is important to note that port numbers are associated with a particular transport protocol. For instance, port number 2000 can be used for both UDP and TCP by two different applications simultaneously. The same application can also use the same port number to run over both UDP and TCP.

After opening a socket, an application sends information to the transport protocol it uses. Each layer adds a header that contains information needed by the same layer in the peer's receiving device in order to process the information up the protocol stack. The transport protocol (e.g., TCP) adds its header and then passes the information down the stack to the IP layer, which adds its header and then passes the IP packet to the link layer, which adds its header, forming a link-layer frame, and passes the frame to the physical layer. The information passed from TCP to the IP layer is called a *TCP segment*. When UDP is used, the information is called a *UDP datagram*.

ICMP also has its own header, and it also sends its messages to the IP layer. The IP layer adds its own header to ICMP messages in the same way that it does for transport layer protocols. In order for IP to know which protocol sent a particular message (e.g., ICMP or TCP), it includes an 8-bit protocol number. Protocols above the IP layer are allocated unique numbers out of this 8-bit space. For instance, UDP is allocated number 17.

### 1.4.2 Demultiplexing

When a host receives a link-layer frame from the physical layer, the link layer processes the link-layer header and then removes it to produce the IP packet. The IP packet is passed up the stack to the IP layer, which processes the IP header, then passes it to the appropriate transport protocol or ICMP. This decision is based on the protocol number included in the IP header. If TCP were used, the IP layer produces the TCP segment that was produced by the sender's TCP. When TCP receives the segment, it processes the TCP header, then removes it to produce the original application data produced by the sending application. From the TCP header, TCP can see which application should receive the packet. TCP knows this because its header includes the port numbers of both the sending and receiving applications. Figure 1–5 shows the demultiplexing process.

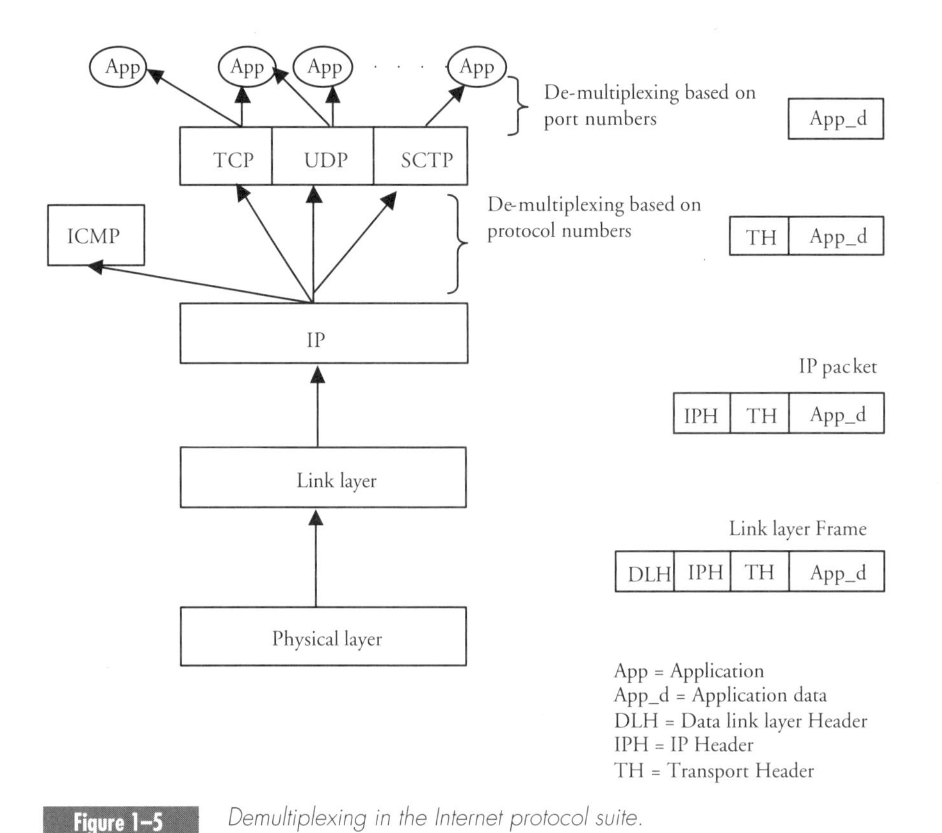


**Figure 1–5** *Demultiplexing in the Internet protocol suite.*

From this demultiplexing process, we can see that five different parameters are needed in order to pass packets from one application to another: the sender's IP address, the receiver's IP address, the transport protocol number, the sending application's port number, and the receiving application's port number. IP addresses are needed to forward the packet through the network. The protocol number is needed to allow IP to pass the received data to the appropriate transport layer protocol. Port numbers are needed to allow transport layer protocols to send the application data to the right application. These five parameters are stored in the socket that identifies the connection. In theory, the IP addresses are not required above the IP layer: port numbers would be sufficient to allow the transport layer to demultiplex the information. However, when the TCP/IP protocol suite was first designed, TCP and IP were combined in one layer. Later, when the two layers were separated, the transport layer maintained knowledge of IP addresses, and this was also included in UDP. Hence, these five parameters are always assumed to identify a connection. If any of these parameters change (possibly because the node was assigned a different IP address), the connection will break (i.e., communication will no longer be possible) and will need to be reestablished.

# 1.5 Routing in the Internet

*Routing* is the function responsible for delivering IP packets from a sender to a receiver. The routing function is divided into two components: the *routing intelligence* and the *forwarding mechanism*. All routers implement both components. The intelligence part involves communication between routers to inform each other about the best way of forwarding an IP packet to a particular destination or a set of destinations; that is, it allows routers to determine the best way of forwarding an IP packet (to the next router or the final destination if it is on the same link) given a particular destination address. This is done on the Internet using *routing protocols*, which are implemented in the user space, like applications. The routing intelligence produces a *routing table*, which lists the next-hop for a given destination address.

The forwarding mechanism is based on the information gained from the routing protocols (routing table). Forwarding refers to the mechanism of sending a packet to the *next-hop* (the next node) that a packet should be sent to in order to eventually get to its destination. Each router keeps its forwarding information stored in a routing table. The routing table determines the next-hop for a given destination address. To understand the need for routing and get an overview of its operation, let's start by considering the network in Figure 1–6.

In Figure 1–6, we have four different shared links (e.g., Ethernet); each is connected to one router, except Link_3, which is connected to two routers. Why do we have more than one link? This can be due to several reasons; perhaps the hosts on each link are generating a lot of traffic that makes it difficult

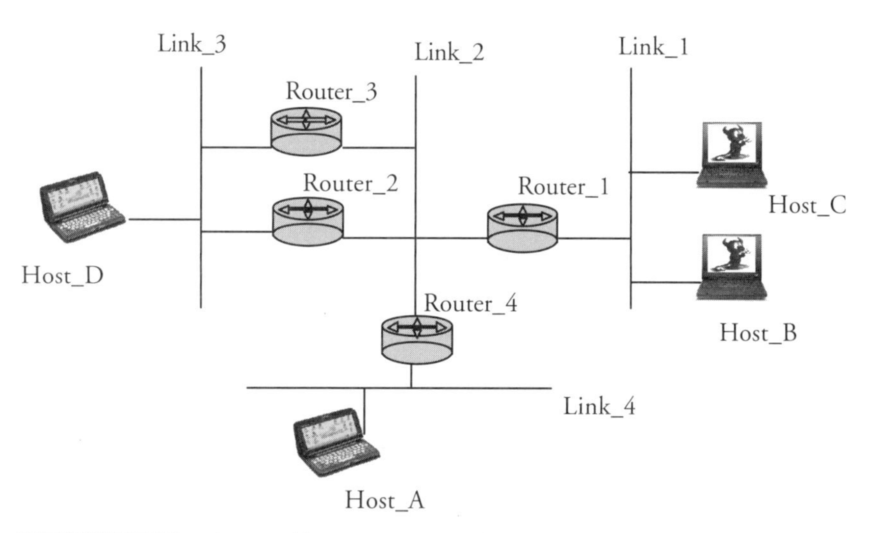


**Figure 1–6** *The need for routing protocols.*

to accommodate more hosts. Another possible reason is that these links are too far away (geographically) to replace them with one physical link. Since we have more than one link, we need routers between them to forward IP packets from one link to another.

Suppose that Host_C wants to communicate with Host_D. Host_C knows through the IPv6 Neighbor Discovery protocol (we discuss this in the next chapter) that Host_D is on a different link; therefore Host_C must send all IP packets destined to Host_D to Router_1. When Router_1 receives a packet destined to Host_D, it needs to know where this packet should be forwarded, since Host_D is not sharing a link with Router_1. Let's suppose that Router_1's routing table is manually configured (i.e., a *static route* is added in its routing table) to forward any packets sent to Host_D to Router_3. Following this assumption, packets will be forwarded from Router_1 to the next-hop, which is Router_3 in this case. This is done without changing the destination address in the IP packet but by including Router_3's link-layer address as a destination address in the link layer's header. When Router_3 receives the packet, it sends it to Host_D, which it knows, through Neighbor Discovery, to be on Link_3.

Now suppose that instead of having four links, we have a larger network consisting of 60 links and 65 routers. Trying to configure each router with the next-hop for every possible destination is clearly a tedious and unrealistic mechanism, especially in a very large network. More importantly, manual configuration cannot work around all router failures. For instance, suppose that after configuring Router_1 with the above information, Router_3's link to Host_D failed, resulting in Host_D becoming unreachable through Router_3. Host_D is still reachable through Router_2, but Router_1 does not know this. Routing protocols enable Router_1 to discover such failures and act on them by rerouting packets through another router that can reach Host_D. When routing protocols are used between routers, each router informs other routers about the links or hosts that it is able to reach and the *cost* of forwarding packets to those destinations through this router. The cost is a parameter associated with a particular destination. This parameter depends on several factors, such as the distance (in terms of hops) and the bandwidth consumption on the link. This allows routers to pick the best path for a particular destination. It also allows them to work around other router failures by finding an alternative path to the destination dynamically.

The Internet is a loose cooperative effort of a number of Internet Service Providers (ISPs) that run IP routing within their networks. An ISP's network can vary in size and functions. Some ISPs are concerned with providing Internet access to home users, enterprise networks, wireless networks, and so on. Other larger ISPs are concerned with providing Internet connectivity between smaller ISPs through leased lines or service agreements. That is, for some ISPs, the customer is the end user of the Internet; for others, the customer is another ISP that wants to connect its customers to the Internet. From this description we gather that some ISPs connect directly to each other to allow their customers to

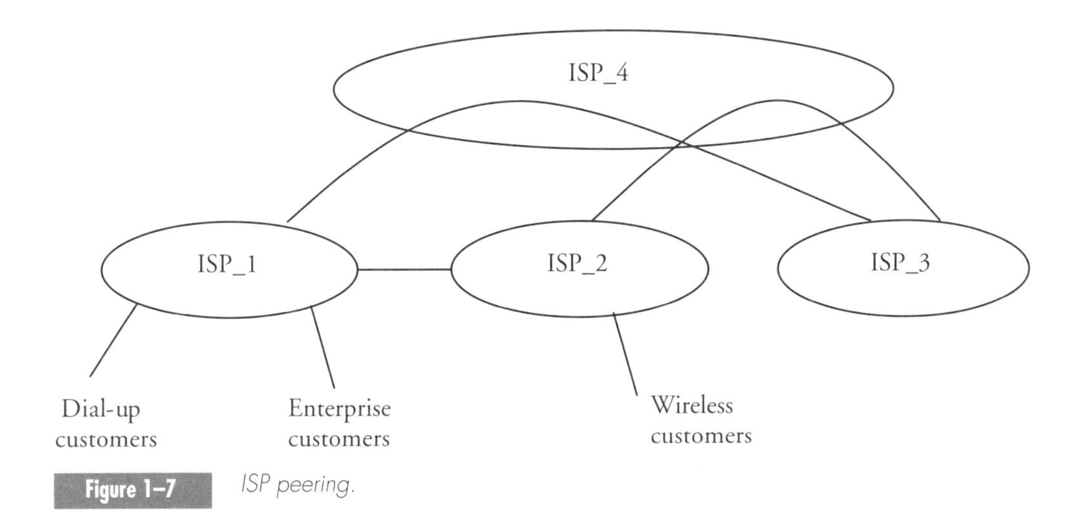


**Figure 1–7** *ISP peering.*

communicate, while others connect through a third ISP. Figure 1–7 shows an example that reflects these different scenarios.

In Figure 1–7, ISP_1 is directly *peering* with ISP_2 by running a routing protocol between the two routers on the "edge" of each domain. ISP_1 peers with ISP_3 through ISP_4. In other words, ISP_1 pays ISP_4 money to forward IP packets originating from its network to nodes in ISP_3's network. In this context, *network* means a group of links, including hosts, connected to each other by routers. Similarly, ISP_2 is not directly connected to ISP_3; instead, it gets this connectivity through ISP_4. In this configuration, ISP_4 is said to provide *transit* connectivity service to ISP_1 and ISP_2. The term *transit* implies that ISP_4 is forwarding packets that are not destined to any node in its own network. ISP_2 is providing a nontransit service to ISP_1 because it is only peering to forward packets addressed from ISP_1 to hosts inside ISP_2's network.

One way to categorize routing protocols is to divide them into two families: Interior Gateway Protocols (IGPs) and Exterior Gateway Protocols (EGPs), which are also called intradomain routing protocols and interdomain routing protocols respectively. Examples of IGPs are the Open Shortest Path First (OSPF), the Intermediate System to Intermediate System (IS-IS), and the Routing Information Protocol (RIP), with OSPF and IS-IS being the most commonly used IGPs on the Internet. RIP is generally used for very small networks. An example of an EGP is the Border Gateway Protocol (BGP), which is **the** EGP used on the Internet today for peering between ISPs.[1]

[1] To confuse matters, there is also a routing protocol called the Exterior Gateway Protocol (EGP), but clearly the use of the term EGP here is meant to refer to the family of interdomain protocols.

IGPs are typically used within domains administered by the same ISP to route packets between nodes inside that domain using the best possible path. EGPs are used between different IGP domains, or *Autonomous systems* (AS).

The distinction drawn between IGPs and EGPs is based on some assumptions about how Internet routing should work. The main assumptions are the following:

- Routing between ISPs requires special *policies*. That is, ISPs may wish to enforce special conditions on routing other ISPs' traffic. One example of such policy can be discussed if we reconsider Figure 1–6. ISP_2 is peering directly with ISP_1 in order to allow their customers to communicate. However, ISP_2 would not necessarily want to forward traffic between ISP_1 and ISP_3 through its network because this would consume too many resources (e.g., bandwidth). Hence, ISP_2 would need to configure special policies on its border router that peers with ISP_1 to make sure that it does not accept these packets.
- IGPs were not designed to allow dynamic policies such as the one described above, while BGP includes provisions for these policies in the protocol.
- It may not be possible to design IGPs to handle this kind of policy-based routing.

These assumptions do not reflect every routing expert's opinion. However, it seems that enough people believe in them; otherwise we would not have had these two different families of routing protocols.

One of the most important features that allow IP routing to scale to a large number of nodes on the Internet is *route aggregation*. To understand route aggregation, we use an example based on the configuration shown in Figure 1–7. Suppose that ISP_1 had 100 links connected to its network from different customers. Each link is allocated a pool of addresses, which we refer to in this example as *subnet numbers*. In order for these hosts to communicate with others in ISP_3's network, the border router in ISP_1's network must communicate to its peer router in ISP_4's network (e.g., through BGP) that all these subnet numbers are in its network and therefore any packets destined to any address derived from any of these subnet numbers should be forwarded to this router. Each one of these 100 subnet numbers creates an *entry* in the routing tables of routers in ISP_4's network, resulting in 100 entries for ISP_1 only. Now, 100 is not a very large number considering that the Internet contains over 300 million hosts. However, it is obvious that this approach will not scale for very large numbers of links—hence, the need for aggregating these routes in order to create fewer entries in the routing tables.

Route aggregation can be achieved on the Internet today by allocating address blocks to ISPs in a hierarchical manner. Let's consider this in the context of a 128-bit wide IPv6 address. The address is divided into two parts, one that identifies a network (in this case the network can be as large as ISP_1's domain), called the *prefix*, and another that identifies the host within the network. The prefix is a variable length of the leading most significant bits in an address. For instance, the prefix can be the most significant 28 bits in the address. IPv6 allows the prefix to be up to 64-bits wide. Using this address division concept, if we allocate ISP_1 a 32-bit prefix that it can use to derive addresses within its network, ISP_1's BGP router will only need to inform its peer in ISP_4's network that any address starting with those 32 bits belongs to ISP_1. This will create one entry in ISP_4's routing table instead of the catastrophic 100 entries created when route aggregation was not used.

As we can see, route aggregation is in fact based on two important concepts:

1. An ISP is allocated a short prefix that can be used to generate enough addresses for all hosts in its domain (this should be overestimated to allow for future expansions in the network).
2. All the hosts inside the ISP's domain must be assigned addresses that start with the prefix allocated to their ISP. Said differently, all the addresses within the ISP's network should belong to its topology, or be *topologically correct*. If one address does not belong to the ISP's topology (i.e., is derived from a different prefix), a special routing table entry needs to be created for this address. This routing entry is also known as a *specific route* because it identifies a single host. Transit ISPs usually drop specific routes because they break route aggregation and consume a scarce resource (i.e., routing table memory).

The availability of a very large address space in IPv6 allows for maintaining route aggregation in the Internet backbone. With IPv6, ISPs can get large pools of addresses (i.e., short prefixes) for their customers in order to maintain route aggregation while allowing their networks to expand in the future. This is currently not feasible with IPv4 due to the scarcity of IPv4 addresses and the consequent difficulty for ISPs to obtain addresses consistent with their original address allocation.

## 1.6 Client-Server Versus Peer-to-Peer Communication

Most of the communication on the Internet today relies on a *client-server* model. In this model the host initiating the communication is the client, and the other host is a server. The server needs to be reachable through the DNS. The host resolves the server's name through the DNS and initiates communication. The server is typically listening to a well-known port number (e.g., port 80 for HTTP web browsing) and maintains connections with many hosts on the Internet.

In order for a server to be reachable, it needs to have a stable IP address. That is, an address cannot be changing too often. With DNS caching, if the address changes too often (e.g., every 10 minutes), the clients attempting to reach a server will often get the wrong address and therefore will not be able to communicate with the server.

The rapid growth of the Internet combined with the diversity of applications developed to run on the Internet protocol suite are expected to lead to a peer-to-peer model of communication in addition to the existing client-server model. Peer-to-peer communication implies that any node can be a server for any other node. This is effectively the same model used in telephony networks today. People need to be reachable, so telephone companies provide them with stable telephone numbers to allow others to call them.[2] Enabling the peer-to-peer communication model is one of the important motivators for both IPv6 and Mobile IPv6, as will be shown later in this chapter.

## 1.7 The Need for IPv6

When IPv4 was designed, a 32-bit address was considered more than enough for the lifetime of the Internet. In theory, a 32-bit address allows approximately 4 billion nodes on the Internet. However, since IP addresses are not allocated in a consecutive manner (see earlier discussion on route aggregation), in practice, a 32-bit address will allow only around 1 billion nodes on the Internet, or perhaps more if addresses were allocated to hosts on a temporary basis. However, with the rapid growth and popularity of the Internet, new applications are emerging, which need hosts to be reachable. For instance, gaming and telecommunication applications like voice and video are all on the horizon of near-future applications. There is a clear need for a large address space in order to allow for the evolution of the Internet.

In 1992 the Internet Activities Board (IAB)—now called the Internet Architecture Board—realized the expected shortage in IP addresses and the need for a new revision of the Internet Protocol (IP). Originally, the Connectionless Network Protocol (CLNP, defined by ISO) was thought to be the most probable candidate. However, after further discussions, a new area was established at the IETF to decide on the next-generation IP. Several proposals were considered within the IPng (IP next generation) area, including the CLNP-based proposal known as TUBA (TCP and UDP over Big Addresses), Simple IP (SIP) designed by Steve Deering, and Paul's IP (PIP), Paul being Paul Francis, the author of the proposal. In 1994 the IPng Directorate decided that a merger between

[2] We do not wish to imply that a telephone number is used by telephone networks in the same way that IP networks use IP addresses. The point is that to be reachable, you need a stable address.

SIP and PIP would form the basis for the next-generation Internet protocol. The new protocol was named IP version 6. However, most of today's IPv6 features are based on SIP.

Like any upgrade of an existing technology (with the benefit of not being too restricted by backward compatibility), the developers of IPv6 designed the new protocol to avoid many of the known problems associated with IPv4. In addition, some new features were included. Although the main driver for the development of IPv6 was the lack of address space, the new technology also provides a number of useful features such as stateless address autoconfiguration, efficient implementation in silicon, new capability for flow identification, and several other features that are discussed in Chapter 2.

Having learned the difficulties of designing a mobility management protocol to fit into the existing IPv4 Internet, mobility was considered from the start when designing IPv6. From the outset of the design process, mobile devices were considered to be extremely important. In fact, mobile devices were expected to form the largest constituency of the Internet, an expectation that is rapidly developing into a reality. In Chapter 2 we present the IPv6 protocol. In Chapter 3, we see how some of the features in IPv6 have helped in the design of Mobile IPv6.

## 1.8 What Is IP Mobility?

The subject of IP mobility has been somewhat controversial or confusing for some newcomers to this field. Let's explore what IP mobility means, why it happens, and what problems it could cause. Then, we look into the main requirements that led to the design of Mobile IPv6.

We can define IP mobility as the change in a node's IP address due to the change of its attachment point within the Internet topology. This change may be caused by physical movement, such as someone moving her computer from one room to another or someone sitting in a moving vehicle that traverses different links.

IP mobility can also occur due to a change in the topology, which causes a node to change its address. In this case IP mobility occurs without any physical movement. Let's look at those scenarios in more detail by considering the network in Figure 1–8.

We already saw the need for route aggregation between ISPs and how it can be solved by allocating a short prefix to ISPs, and we saw that the maximum prefix length in the current IPv6 address is 64 bits. Hence, if an ISP is allocated a 32-bit prefix, it can break that prefix up between its links and allocate a 64-bit prefix to each link. Hosts on each link will use that link's 64-bit prefix to configure their addresses (more on this in the next chapter). Routers on the link will only need to announce to other routers (using an IGP)

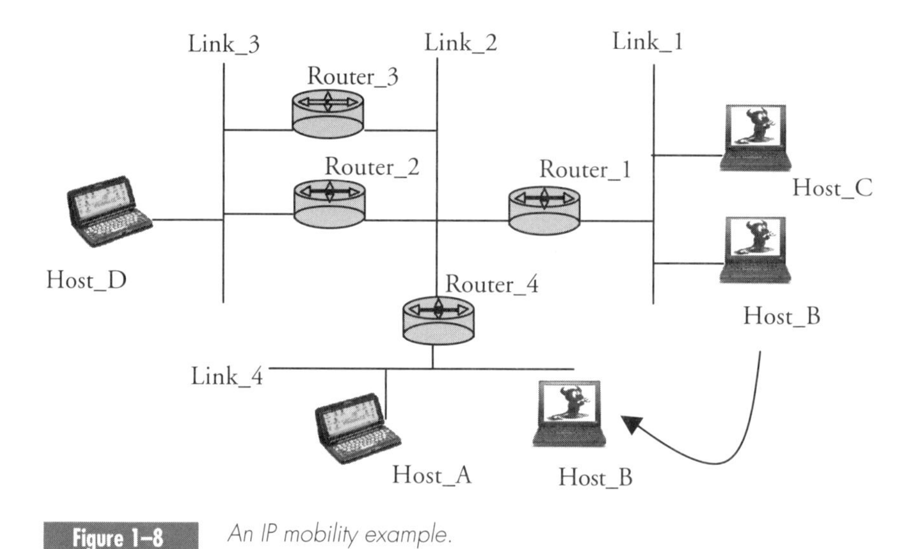


**Figure 1–8** *An IP mobility example.*

that any packets with a destination address starting with their 64-bit prefix should be forwarded to them.

Now, suppose that while Host_B is communicating with Host_D, it decides to move from link_1 to link_4. Three different problems are now obvious:

- Host_D is unaware of this movement; therefore, it will continue to send IP packets to Host_B's address on link_1. Clearly the packets will not be forwarded to anybody and will be dropped.
- Host_B's address is no longer valid because it is not derived from the prefix assigned to link_4. It is based on the prefix assigned to link_1. Hence, Host_B must form a new address. In the meantime, Host_B will not receive any packets sent to the old address.
- If Host_B derives a new address from link_4's prefix, it will have to change the socket containing information about its connection with Host_D, but this is not possible because both hosts need to maintain the same set of information about a particular connection. Therefore, the connection will be terminated, since the two peers will not be able to communicate.

All these problems have occurred because the IP address of a host has changed while it has active connections with another host or hosts. If route aggregation were not needed (i.e., a host need not derive its address from the prefix assigned to a link), or the socket was defined in a different way that didn't include IP addresses, these problems would not come about. This is the problem that Mobile IPv6 solves, with one addition: allowing mobile nodes to be reachable by offering them a stable address.

### 1.8.1 How Important Is Mobility?

You might wonder whether mobility is an important problem—yes, it is. You might also wonder if such problems are likely to occur—yes, they are. The need for an IP mobility management solution is motivated by the following:

- **Proliferation of wireless devices:** Wireless networks have been traditionally used within the context of telecommunication networks that offer voice services. However, this is no longer the case. There are various wireless technologies that range in their support from Local Area Networks (LANs; e.g., IEEE 802.11b) to Wide Area Networks (WANs; e.g., third-generation cellular systems) and all use IP. The popularity and diversity of these technologies allow hosts to move freely within large geographical areas and choose certain technologies over others. This vast geographical movement is likely to cause IP mobility. Clearly, users will not accept that all their connections are dropped every time they move. Mobile phone users do not accept this today, and there is no evidence that they will accept this in the future for IP connectivity.
- **Reachability:** As peer-to-peer applications become more popular, hosts will need to be reachable independently of their location. If hosts continuously moved within the Internet topology, their addresses would change every time they moved, making them unreachable.

These two reasons provide solid motivations for the need to solve this problem. Furthermore, they provide some basic requirements for the final solution. These requirements are discussed below in more detail.

### 1.8.2 Where Do I Need Mobility Management: Layer 2, Layer 3, or Upper Layers?

We saw how the networking problems are divided into subproblems, each represented by a layer in the protocol suite. We also know that IP mobility is a problem that needs to be solved by one of those layers. The question is, which layer?

To answer this question we need to revisit the problem itself. Our problem was that a host changed its IP address and, as a result, could not receive any packets. Why? Because the address included in those IP packets no longer belonged to that host. This problem is not peculiar to a particular application or link layer technology—it will affect **all** connections regardless of their applications, transport layer protocol, or link-layer technology. Which layer in the protocol suite is responsible for routing and is independent of any transport layer or link-layer protocol? The IP layer. Therefore, the problem is clearly an IP layer problem.

Let's check this logic by observing what would happen if other layers in the protocol suite were to solve this problem:

- **Mobility management on the Link layer:** Almost every Wireless LAN (WLAN) and WAN technology already includes mobility management. For instance, IEEE802.11b allows devices to move within its network without breaking ongoing connections. However, since it is possible for a node to move between two different links connected to different routers, IP mobility will be invoked. Furthermore, nodes may also move between different link layers (e.g., a multihomed host that moves between a cellular link and IEEE802.11b) and get a new IP address. A link-layer solution alone would not be sufficient in all cases. This is not to say that link-layer mobility management is not needed. However, we need to understand that the scope of link-layer mobility management solutions is limited to the link that uses such technology and therefore cannot be used as a general solution.
- **Mobility management on the Transport layer:** Another way of solving this problem would be to add a new protocol on the transport layer that can modify sockets and inform the peer's transport layer protocol of such modification when a host moves and changes its IP address. This solution will work for the transport layer it is designed to work on; for instance, TCP can be modified to do this, but applications that do not use TCP will still lose their connections after movement. Recall that there are three different transport layer protocols used today. Having to modify each one of them to support mobility is a significant task. New transport protocols are also being developed. They would also need to support mobility if this approach were adopted. From an architectural point of view, it is more efficient to design one protocol that will work independently of the transport layer protocol being used by applications.
- **Mobility management on the Application layer:** The same argument can be used here as the one used above for transport layer solutions. Applications come and go; there are thousands of applications today and more will come. It would be a significant burden to require each one to develop its own mobility management protocol, not to mention the inefficiency involved in running multiple protocols that solve the same problem.

We can clearly see that within the existing Internet protocol suite, the IP layer is the most suited for solving the mobility problem. Mobile IPv6 is designed to handle this problem on the IP layer (see Chapter 3).

### 1.8.3 Mobile IPv6: Main Requirements

We have seen that IP mobility is needed to allow connections between hosts to continue after a host changes its IP address. Furthermore, hosts need to be reachable in order to allow peer-to-peer communication for future applications. We can now derive the main requirements that should be satisfied by Mobile IPv6. Understanding these requirements is important for future developments in IP mobility. The main requirements are shown below.

- **Session continuity:** It is important to ensure that connections will not break when a host changes its IP address—just as phone calls are not immediately dropped when people move around with their mobile phones.
- **Reachability:** Reachability is likely to be as important as any other aspect of mobility. Most users would likely find reachability crucial. After all, there is generally little point in allowing a device to be mobile if no one can reach it! Based on our understanding of the DNS operation, stable addresses are key in order to allow hosts to be reachable.
- **Application independence:** To ensure smooth integration with current and future applications, Mobile IPv6 must not require any changes or add requirements to applications. This requirement is crucial for the success of the protocol and its ubiquity. Mobile IPv6 addresses this requirement by "hiding" mobility from upper layers (transport and application layers).
- **Lower layers independence:** The IPv6 protocol can be used over most link layers. Since Mobile IPv6 is expected to be part of several IPv6 hosts, it is essential that it is designed in a way that does not introduce any link-layer dependencies or further requirements other than those imposed by IPv6.
- **End-to-end signaling:** All nodes on the Internet can be divided into two categories: hosts and routers. Routers forward IP packets based on the content of the destination address in the IP header. Hosts are responsible for maintaining connections, flow control, and so on. Routers know nothing about the connections that are open between any two different hosts. Separating the roles and putting the intelligence in the end hosts ("smart hosts and dumb network" approach) allowed the Internet to scale to support hundreds of millions of hosts. Mobile IPv6 should not deviate from this approach. In fact, keeping the end-to-end model is crucial for Mobile IPv6 deployment. If Mobile IPv6 requires changes in Internet routing, or many special nodes all over the Internet, instead of relying on end

hosts to control mobility, its deployment is guaranteed to be delayed significantly. In addition, deviating from the end-to-end model of communication will add scalability limitations on the Internet, as it is likely to concentrate the load on special nodes that will become bottlenecks when the number of hosts increases.

The requirements presented here form the essential basis for the design of Mobile IPv6. The details of the design are discussed in Part 2.

## 1.9 Summary

This chapter introduced the basic principles of the Internet protocol suite using the layered model. We learned how two nodes can communicate using the layered Internet protocol suite and the DNS. We also saw that routing is needed in order to scale the Internet to hundreds of millions of nodes. Moreover, we discussed the importance of route aggregation, which allows the routing infrastructure to scale to a very large number of machines supported by many ISPs. Using this knowledge, we introduced mobility and the problems that it introduces. Finally, we presented the fundamental requirements that must be met by an IP-layer mobility management protocol like Mobile IPv6.

This chapter served as an introductory step to those who are not familiar with the basic operations of the Internet protocol suite. The Internet fundamentals presented in this chapter are used to explain the mobility problem in the current context and justify the requirements that were placed on Mobile IPv6.

## Further Reading

[1] Huitema, C. *Routing in the Internet*. Prentice Hall, 1995.

[2] Perlman, R. *Interconnections: Bridges, Routers, Switches and Internetworking Protocols*, 2nd ed. Addison-Wesley, 2000.

[3] Stevens, W. *TCP/IP Illustrated*, Vol. I. Addison-Wesley, 1994.

T W O

# An IPv6 Primer

IPv6 provides a number of improvements over the currently deployed IPv4-based Internet. The main motive for the introduction of IPv6 was the forecasted IPv4 address space depletion. This is addressed in IPv6 by extending the address space to 128 bits, a significant increase from the existing 32-bit IPv4 address. Apart from increasing the lifetime of IPv6, a 128-bit address simplifies address autoconfiguration, potentially reduces the number of routing entries in backbone routers, and allows for new features like scoped and Cryptographically Generated Addresses (CGAs), as described in later chapters. New types of addresses are defined in IPv6 (e.g., scoped addresses); other existing addresses are enhanced with new fields (e.g., multicast addresses).

In addition to enhancements in addressing, IPv6, being a new Internet technology, allowed the developers to include recently developed IP security (IPsec) protocols to the IP layer in every IPv6 implementation. IPsec allows nodes to protect connections by encrypting and authenticating data transferred between two communicating nodes.

To simplify the processing of IP packets in intermediate and end nodes, IPv6 headers and options are always aligned on their 64-bit boundaries. This allows fast and efficient processing of both headers and options. Furthermore, additional simplification was made to the IPv6 header and options encoding to allow for simpler processing and ease of introduction of new options. The latter feature was one of the main obstacles that prevented the addition of new header options in IPv4.

This chapter introduces IPv6 and presents a number of related features like extension headers, security, ICMPv6, and Neighbor Discovery. Finally, a communication example shows how those protocols are used by hosts to send and receive IPv6 packets.

## 2.1 The IPv6 Protocol

The IPv6 specification is defined in RFC 2460. The specification includes the header format, extension headers, and their processing rules. The IPv6 header (see Figure 2–1) contains the minimum amount of information necessary for two nodes to communicate. Extension headers contain additional options for hosts or routers receiving the IPv6 packets. An IPv6 packet may contain one or more extension headers when necessary for the processing of such packet.

The *version* field is 4 bits wide; it includes a number corresponding to the IP version for this header. In this header, it must always be set to 6.

The *traffic class* is an 8-bit field corresponding to the *Type of Service* (TOS) field in the IPv4 header. The use of this field allows for Differentiated Services (DiffServ). That is, hosts or routers can set this field to indicate that certain packets require priority forwarding over others (e.g., real-time versus best-effort forwarding).

The *flow label* is a 20-bit field. This is discussed in more detail in section 2.1.3.

The *payload length* is a 16-bit field indicating the number of octets following the IPv6 header. The payload length includes any extension headers following the IPv6 main header. Hence, a maximum payload length of 65,536 octets is possible for any IPv6 datagram. A datagram larger than the maximum payload length is called a *jumbogram*. Jumbograms are discussed later in this chapter.

The *next header* is an 8-bit field used to describe the protocol following the IPv6 main header. The following header may be a transport layer header

| 8 bits | 8 bits | 8 bits | 8 bits |
|---|---|---|---|
| version / traffic class | | flow label (20 bits) | |
| payload length | | next header | hop limit |
| source address (128 bits) | | | |
| destination address (128 bits) | | | |

Figure 2–1 IPv6 header format.

(e.g., TCP) or one of the extension headers shown later in this chapter. A unique value is reserved for each header type. For example, the value 17 is reserved for the UDP header, and 6 is reserved for TCP. Hence, the IPv6 layer can multiplex 255 different transport layers or extension headers. The decimal value 59 is used to indicate that there is no next header.

The *hop limit* is an 8-bit field used to limit the maximum number of hops allowed for a single IPv6 packet in a similar manner to the Time To Live (TTL) field in IPv4. The field is set to a value between 0 and 255 by the host originating the packet. It is then decremented by one by each router forwarding the packet. If the hop limit value is decremented to zero, the packet is discarded. This field is needed to avoid routing loops—that is, cases where packets are forwarded indefinitely in a loop of routers.

The *source address* field contains the 128-bit address of the host originating the packet.

The *destination address* field contains the 128-bit address of the ultimate receiver of the packet or an intermediate node in case the *routing header* is included in the packet. The details of the routing header processing are described later in this chapter.

The IPv6 header removed some of the fields that were previously included in the IPv4 header (e.g., the Checksum field) and added new fields. The following sections discuss the rationale for some of these design choices.

### 2.1.1 Why Doesn't the IPv6 Header Contain a Checksum Field?

This question is frequently asked by people when they are introduced to the IPv6 header, especially since the IPv4 header contains a checksum field. The IPv4 checksum field was designed to make sure that routers verify whether the IP header was modified along the path to the ultimate destination. The checksum value covered the entire IPv4 header. Packets may be modified due to transmission errors on links or by a malicious node along the path. However, it is observed that link layers typically perform checksums on the entire link layer frame, which includes the IP header. Hence, the benefit of having a checksum on the IP layer to detect transmission errors is not clear. It is also worth noting that the verification of the checksum is a time-consuming process that increases the forwarding latency of IP packets. When considering the pros and cons above, it was seen that checksums were not necessary on the IP layer for detecting transmission errors.

The other cause for changes in IP packets is the existence of malicious nodes along the path. Such malicious nodes may modify packets to disrupt the communication between two nodes. However, IP layer security mechanisms have been developed and are required to be supported by all IPv6 nodes. These security mechanisms allow communicating nodes to detect any changes in any part of the IP packet (including the IPv6 header). It is important to note that a checksum field would not be helpful in detecting changes in the IP header by malicious nodes. A malicious node could easily recalculate the checksum after modifying the packet, and the change would go unnoticed.

### 2.1.2 Do We Need a Larger Payload Length Field?

The IPv6 header allows for a maximum payload size of 65,535 octets (a 16-bit field). If a link allows nodes to send larger packets, it would be desirable to allow the IPv6 header to carry more than the maximum payload size. Jumbograms are packets whose size is larger than 65,535 octets. When these packets are sent, the payload length in the IPv6 header is set to zero. In addition, an option is included in the Hop-by-Hop extension header (see section 2.2.1) that includes the real payload size, which is represented in a 32-bit field. This allows for a maximum packet size of 4,294,967,295 octets.

Jumbograms are expected to be rarely used on the Internet simply because most links on the Internet do not allow for these packet sizes.

### 2.1.3 The Flow Label

The evolution of the Internet is expected to allow new real-time services to be deployed on top of the existing best-effort applications (e.g., email). Real-time applications (e.g., voice) are sensitive to delays and end-to-end delays. The delay sensitivity of real-time services requires special handling in intermediate nodes (routers) forwarding data between the communicating hosts. To be able to provide the necessary treatment, routers need to be able to identify flows. IPv6 introduces the *flow label*, which allows intermediate routers to identify flows in an efficient and fast manner, eliminating the need for inspecting upper layers' headers.

The flow label is used to identify a flow. A flow is essentially one or more connections between two nodes. Based on this definition, we can deduce that a flow can be uniquely identified based on the source address, destination address, and the flow label. That is, if two nodes communicate using globally unique addresses, a flow label value (unique for the two nodes) would provide a globally unique identification for such flow when combined with the source and destination addresses of the two nodes.

Flow identification becomes necessary when end nodes need to communicate information about such flow to intermediate routers. For instance, on bandwidth-limited links (e.g., cellular links or low bandwidth fixed links), it is important for nodes to request a reservation of bandwidth appropriate for the services that the end node wishes to use. In addition, end nodes can request different treatment for different flows. For example, voice traffic is expected to be forwarded immediately, without delay, while email can be delayed for short periods without having any observable negative implications on the service when seen by the communicating nodes. Therefore, it is important for end nodes to be able to communicate their requests to routers. The communication is done via a signaling protocol; the Resource reSerVation Protocol (RSVP) can be used to communicate these requests. In addition to communicating the requests, it is important that nodes identify the flows that the requests apply to. This is where the flow label becomes important. The flow label can be used to enable flow identification by end nodes and intermediate routers to be able to provide the Quality of Service (QoS) needed by the different applications.

In IPv4, a connection is identified by the source and destination addresses, the protocol number (e.g., TCP or UDP), and the source and destination port numbers. However, in order for an intermediate router to identify a connection, it would need to scan the IP packet to read the protocol number and port numbers. This process adds more delay to forwarding of IP packets. The delays are even worse for IPv6 due to the possibility of having several extension headers, as is shown in the following section. Furthermore, IP packets may be encrypted for confidentiality (more on security in 2.2.4.1 and 2.2.4.2). In this case, routers will not be able to read the encrypted port numbers, which renders flow identification impossible. Therefore, having a field in the IPv6 header can be very useful in providing an efficient mechanism for nodes to identify flows. One important application for identifying flows is flow identification for QoS purposes. There may be other applications that require flow identification and might end up using the flow label for this purpose.

The flow label is set to an arbitrary number by the sending node (remember that the actual number is meaningless—what matters is the way it is used) and should reach the ultimate destination with the same value.

## 2.2 IPv6 Extension Headers

Extension headers are defined to encode certain options that are needed for processing of the IPv6 packet, or in some cases, subsequent packets. Encoding options in extension headers allows for efficient forwarding by routers. Routers must minimize the amount of time needed in order to parse the header and forward the packet on the correct route. The increase in forwarding speed will certainly enhance a router's performance and help promote certain vendors over others. To enhance the forwarding speed, routers typically have a *fast path* used for handling normal IP packets that do not require any exception processing. The fast path is usually implemented efficiently in hardware. A *slow path*, on the other hand, is used for the cases where the IP packet requires some exception handling. By exception we mean additional processing to the simple forwarding based on the destination address. Exceptions are typically handled in software due to their complexity and for additional flexibility (e.g., new features will require software upgrades, which is simpler than modifying and upgrading the router's hardware). Processing IP options is considered an exception to the normal destination address–based forwarding.

The benefits of encoding options in extension headers can be best illustrated when comparing it to option encoding in IPv4 in light of the above discussion about router forwarding efficiency. In IPv4, the header size depends on the number of options inside it, unlike IPv6, which has a fixed header size (40 bytes), shown in Figure 2–1.

Consider a router receiving an IPv4 packet including one or more options: the router would first need to be aware that the packet is carrying IP options; this can be easily seen from the IPv4 header size field. Following this step, the router must parse the IP header to find out which options require processing by the router itself as opposed to processing by the ultimate receiver of the packet. Note that the router cannot simply place a packet on a slow path because it contains options; these options may be intended for the ultimate receiver only, so such a premature decision may result in unfair treatment of certain packets simply because they contain options.

The process of parsing the header and its options takes some time and can impact the overall performance of the router if it receives these packets frequently. This performance penalty has forced many router designers to ignore IPv4 options. Consequently, it became difficult to add any new features in IPv4 options, as they would most likely be ignored. In effect, IPv4 options have become virtually useless.

IPv6 avoids these problems by defining a fixed-size 40-byte header and extension headers for additional options. The extension headers' definition and ordering are intended to avoid the problems associated with IPv4 options. Each extension header may contain several options. While RFC 2460 does not define all options for each extension header, it does define the behavior and processing rules for each one—for instance, the entity processing the header and the security requirements for a particular header. Hence, new options must be carefully included in the appropriate extension header. Each extension header starts with an 8-bit field indicating the type of the following one (the type of the first header is included in the main IPv6 header). This is done to allow nodes to skip the processing of headers when such processing is not required or when the header type is unknown. The next field is an 8-bit field indicating the length of the header in octets (clearly this field is only needed for variable length extensions headers). The extension headers and their formats are described in the next section.

### 2.2.1 The Hop-by-Hop Options Header

Figure 2–2 shows the Hop-by-Hop header. It includes options that need to be processed by every node along the packet's delivery path—that is, by every router receiving and forwarding the packet along the path to the packet's ultimate destination as well as by the final receiver.

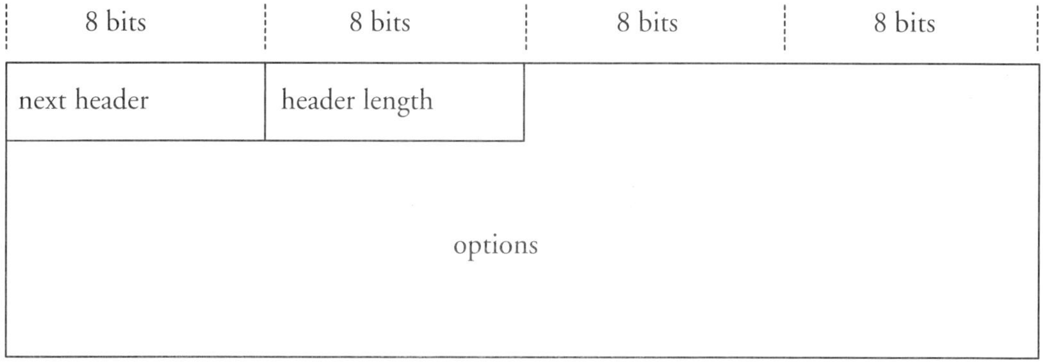

| 8 bits | 8 bits | 8 bits | 8 bits |
|---|---|---|---|
| next header | header length | | |
| options | | | |

Figure 2–2 *Hop-by-Hop header.*

Like all variable-length IPv6 extension headers, this header starts with two 8-bit fields indicating the type of the following header (next header) and the actual header's length in octets (header length).

The Hop-by-Hop header can be used by any application that requires installing state in every node along the path to a particular destination as well as to the final destination of an IP packet. An example of such state is information about special treatments of certain IP connections/flows. For instance, a node may request that certain flows get faster forwarding services or more bandwidth than other flows. This would require that all routers along the path that the packet is expected to travel through become aware of such request.

The *router alert* option was designed to alert all routers on-path about important information contained in a packet; therefore it is included in this header. Protocols like RSVP and other QoS signaling protocols can utilize the Hop-by-Hop extension header and the router alert option to install states in routers along certain paths.

### 2.2.2 The Routing Header

Figure 2–3 shows the Routing Header. The Routing Header provides a functionality similar to the *loose source routing* option in IPv4, but with some enhancements. When using this header, the sender of the IPv6 packet places a number of IPv6 addresses through which the packet should be relayed. When using the routing header, the sender includes a number of IPv6 addresses in this header (including the final destination of the packet). The destination address in the IPv6 header is the address of the first node that will receive the packet; the rest of the relays are included in the routing header. When the node whose

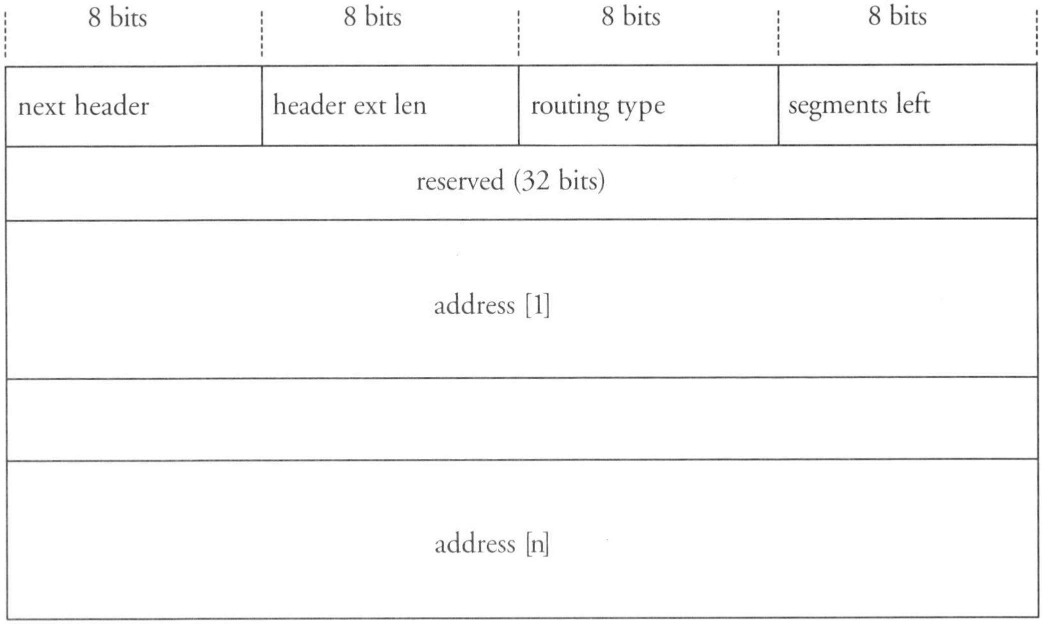

| 8 bits | 8 bits | 8 bits | 8 bits |
|---|---|---|---|
| next header | header ext len | routing type | segments left |
| reserved (32 bits) | | | |
| address [1] | | | |
| | | | |
| address [n] | | | |

Figure 2–3 *The routing header.*

address is in the destination address field of the IP header receives the packet, it removes the destination address (its own), replaces it with the next address in the routing header, and reduces the *segments left* field by one. Its own address is placed in the bottom of the list. Hence, the total number of addresses visited by a packet containing the routing header will be included in the routing header received by the final address (i.e., address [n] above). When the packet arrives at the final receiver (the last address in the routing header), the segments left field will be reduced to zero. The receiver then proceeds with the processing of the following header(s).

The *next header* field in this header contains the type of the header that follows. The *header ext len* is the length of the header in 8-octet units, not including the first 8 octets. The *routing type* field is used to define the type of routing header. For instance, a Type 0 routing header would contain 0 in this field.

The *segments left* field contains the number of segments (addresses) that were not visited by the packet but are listed in the routing header. This field is decremented by one every time the packet is delivered to one of the addresses in the list. That is, when the packet is finally received at Address[n], the number of segments left should be set to zero.

*address[1]...address[n]* are IPv6 addresses that the packet is intended to be relayed to. The *reserved* field may be used in the future for additional protocol flags, but it is currently set to zero and ignored by the receiver. The next section describes how the Routing Header can be used by applications.

#### 2.2.2.1 USING THE ROUTING HEADER

The *ping* program provides a very simple and powerful tool that can be used by network administrators to find out if a node is reachable and to measure the Round Trip Time (RTT) between two nodes. The node initiating the ping sends an ICMP *echo request* message (ICMP is discussed later in this chapter), which causes the receiver to send an *echo reply*. The time difference between sending the request and receiving the reply is approximately the RTT between the two nodes in question. Let's see how the Routing Header can extend this feature by considering the example in Figure 2–4.

Consider this scenario: A network administrator decides that he wants to measure the RTT between Host1 and Host2 when packets are routed through links A, C, H, H, C, A. Host1 can do so by adding a Routing Header to the ICMP packet (the ping program can request this through the sockets API). When being sent by Host1, the headers in the IP packet would look like this:

```
src: Host1
dst: Router3
Routing header: Host2, Router3, Router1
Segments left: 3
```

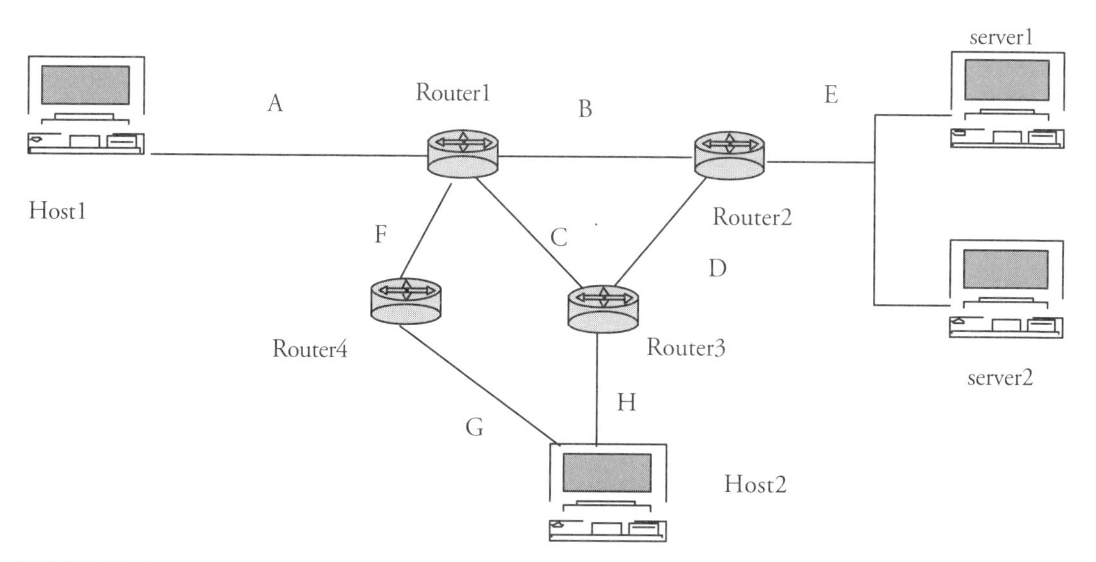


**Figure 2–4** *The use of the Routing Header.*

Note that we set the destination field to Router3 in this case to avoid having it forwarded on link F, since we're not interested in that path; we also assume that since the packet has to be sent to Router1 first, there is no need for Host1 to include Router1 in the destination address.

Router3 will receive the packet, replace the destination address by the Host2's address obtained from the routing header, and send it to Host2 through standard routing. The packet leaving Router3 would look like:

```
src: Host1
dst: Host2
Routing header: Router3, Router1, Router3
Segments left: 2
```

When received at Host2, routing header processing will cause the destination address to be replaced by Router3's address, and the packet will be forwarded to Router3. The packet now looks like this:

```
src: Host1
dst: Router3
Routing header: Router1, Router3, Host2
Segments left: 1
```

A similar processing at Router3 will result in the following packet:

```
src: Host1
dst: Router1
Routing header: Host2, Router3, Router3
Segments left: 0
```

Now the packet received at Router1 will require no further processing of the routing header, as the *segments left* is 0. Router1 will then process the *next header* (ICMP echo request) and send the reply back to Host1. Host1 will now have the reply and the RTT for the entire path.

#### What Are the Security Hazards of the Routing Header?

A more recent analysis of the use of the routing header [20] has shown that it may not always be as good as in the example above. Let us consider the same network in Figure 2–4. Now let's assume that the network administrator wants to prohibit Host1 from reaching server2 (server2 might have confidential information). To achieve this, the network administrator has configured a forwarding rule *(filter)* on Router2 that would cause it to drop any packet with the source address of Host1 and the destination address of server2. Let's also assume that the administrator allows Host1 to reach server1. Now, if Host1 is aware of server2's IP address, it can send the following packet:

```
src: Host1
dst: server1
```

```
Routing header: server2
Segments left: 1
```

This would result in routing the packet to server1, which after processing the Routing Header will send the packet to server2. This would happen despite the network administrator's filter on Router2.

#### *How Can We Avoid This Problem?*

There are different ways to stop this problem from happening. The network administrator could add a policy that forbids Router2 from forwarding any packet with a Routing Header. However, this would also eliminate the usefulness of the Routing Header. A less radical approach would be to add another rule to Router2 and require it to parse every packet to see if it contains a Routing Header, and if so, to make sure that packets sourced from Host1 and containing server2's address in the routing header are dropped.

A third approach is to require all hosts to have the routing header processing as a configurable feature. The default behavior would be to not forward a packet with a Routing Header to any other devices (forwarding to itself would be allowed). This would stop server1 from forwarding any packet to any other node (apart from itself) based on routing header processing.

### 2.2.3 The Fragmentation Header

Link layers typically have a maximum frame size that limits the amount of data that can be placed in a link-layer frame and consequently in an IP datagram. For example, the maximum size for an IP datagram on an Ethernet link is 1500 octets. This figure is known as the link's Maximum Transmission Unit (MTU). When a host sends an IP packet to another host on the same link, the link MTU must be known. If the packet size exceeds the MTU, fragmentation must be used. Fragmentation in this case would take the entire IP packet and produce two or more packets whose size is equal to or less than the link MTU.

However, if a host is communicating with another host from which it is separated by several links, Path MTU (PMTU) becomes the important factor. PMTU is a parameter describing the minimum MTU of all links between the two hosts. For example, in Figure 2–5, the PMTU between Huntsman and Redback is 1500. Therefore, the maximum packet size that can be sent between Huntsman and Redback is 1500 octets. If Huntsman needs to send a 4000-octets packet to Redback, this packet would have to be fragmented to three different packets. Except for the third packet, the length of each fragmented packet must be a multiple of 8 octets. The fragmented packets received at Redback will then be reassembled into one 4000-octet packet and passed to the transport layer.

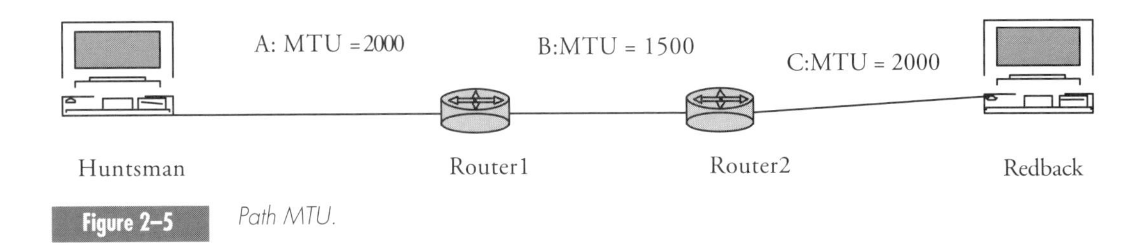


**Figure 2–5** *Path MTU.*

Each fragment containing part of the original message will also contain a Fragmentation header. The Fragmentation header does not include a *length* octet, as it has a fixed length of eight octets. The fields in this header are described below. Figure 2–6 shows the format of the Fragmentation header.

> The *fragment offset* field shows the order of the fragmented packet within the *original*, unfragmented packet. In the example shown above, this field would be set to 0, 1, and 2 for the first, second, and third packets respectively.
>
> The *res* field is a 2-bit reserved field for future use.
>
> The *m* flag indicates whether the fragment received is the last fragment (if set to 0) or whether there are more fragments to come (if set to 1). This flag is used to inform the receiving host whether reassembly of the fragmented packet can start.
>
> The *identification* field should be set to a unique value for each large packet between two communicating nodes. Hence, the same identification value should be used for each of the fragments to allow the receiving host to reassemble the received fragments appropriately. For example, consider the case where Redback and Huntsman have two ongoing connections, which are causing

| 8 bits | 8 bits | 13 bits | 2 + 1 bits | |
|---|---|---|---|---|
| next header | reserved | fragment offset | Res | m |
| identification | | | | |

**Figure 2–6** *The Fragmentation header.*

Huntsman to send large packets to Redback. If both applications happened to send a large packet, the IP layer would fragment each packet. For Redback to be able to reassemble the received packets belonging to each connection, it needs to be able to identify fragments belonging to each packet. Hence, an identification field is needed.

Note that the identification field should be unique only for a pair of communicating IP addresses. That is, Huntsman may use the same identification values used with Redback, with another host receiving fragmented packets from Huntsman. A simple wraparound counter at the sending end for the identification field (for each destination address) would allow for $2^{32}$ unique values. This should avoid any conflicts between the identities of the different fragments.

#### 2.2.3.1 HOW IS FRAGMENTATION DONE IN IPV6?

In IPv6 only the sending host can perform fragmentation. Intermediate routers cannot fragment large packets (unlike IPv4, which allows fragmentation by intermediate nodes). This feature of IPv6 was designed to enhance the performance of routers by reducing forwarding delays. The IPv6 specifications require that all link layers support a minimum MTU of 1280. Any link layer with a lower MTU value must be able to fragment IP packets on the link layer. That is, they must hide the true MTU value (if less than 1280) from the IP layer. On the other hand, hosts should be able to benefit from a PMTU value larger than the minimum, as it would allow them to fully utilize the network bandwidth. The PMTU discovery mechanism was designed to allow IPv6 hosts to discover PMTU and decide on packet sizes accordingly. Hence, an IPv6 host has two choices:

1. Assume a PMTU of 1280 and not fragment any packets. If a link on the path has an MTU less than 1280, it must fragment the packet on the link layer. Or,
2. Perform PMTU discovery and decide whether fragmentation is needed on a per-destination basis.

If a host chooses option 2 and then decides to send a large packet, the packet should be fragmented as shown in Figure 2–7.

In order to fragment a large packet, the host needs to identify the *unfragmentable part* of the IP packet; this includes the IPv6 header and all existing extension headers that may be processed by intermediate nodes (e.g., Routing Header). Following this step, the host fragments the packet into two or more fragments, each (except for the last fragment) containing an integer multiple of 8 octets of data. The unfragmentable part and the Fragmentation Header are then appended to each fragment. Each fragment is sent to the receiver,

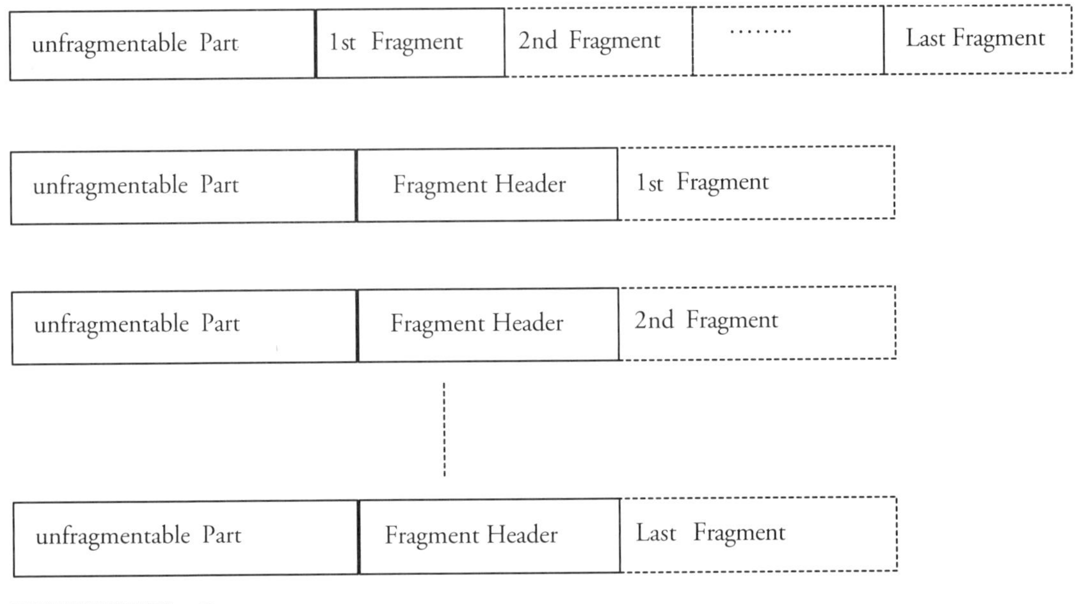


**Figure 2–7** *Fragmentation process.*

where reassembly is performed based on the contents of the Fragmentation Header.

When the packet is reassembled at the receiving end, it is passed to upper layers in order to recover the data sent by the peer's application.

#### 2.2.3.2 HOW IS PMTU DISCOVERY PERFORMED?

We saw earlier that a node can assume PMTU of 1280 or discover the real PMTU value to allow for the maximum throughput on a particular path. Most links support an MTU that is larger than the IPv6 minimum MTU; therefore, it is useful for nodes to discover the PMTU, especially when a large amount of information is being sent (e.g., a large file sent using FTP). This allows nodes to send larger packets and consequently utilize the network bandwidth more efficiently.

PMTU discovery is done on the transport layer based on knowledge gained from ICMP and IP. The link MTU is discovered through Neighbor Discovery messages (discussed in section 2.6). Transport layers are informed of the initial maximum message size they can send by subtracting the IP header's size from the link's MTU. Hence, a transport layer can start by sending messages based on the link MTU. The message size is increased gradually until a router along the path to the destination sends an ICMP error message to the sending host informing it that the packet sent was too large and including its link's MTU in that message.

This information is received by the ICMP layer and communicated to upper layers. Upon receiving this message, the PMTU is set to the new size included in the ICMP error message. Note that this message can be received more than once if another link, further along the path, has a lower MTU than a previous link. Hence, the PMTU is continuously modified until a packet the size of the minimum MTU of all links reaches the destination. The PMTU can then be stored in the IPv6 node and associated with that particular destination.

A path to a destination may change with time due to changes in the network or router failures, resulting in an increase or a decrease in the PMTU. A reduction of PMTU will be detected immediately when an ICMP error message is received. However, an increase in PMTU is difficult to detect. The only way of detecting that would be to increase the size of a message and see if an ICMP error is received. However, this mechanism is not recommended, as it would disturb ongoing connections.

The PMTU specification recommends that a node never increase its PMTU based on a received ICMP error message containing a large MTU. This is done to avoid the case where Bad Guy sends this message deliberately to a host, causing it to increase its PMTU beyond the real value to force a router to drop its packets.

It is worth noting that in some cases upper layers are unable to modify the sizes of their messages due to an application's requirements. If the message sizes are larger than the PMTU, the IP layer will fragment the packets as shown earlier.

### 2.2.4 IP Layer Security

The widespread use of the Internet and its ease of access is one of the reasons for the increasing need for security protocols. In fact, security is very rapidly becoming an integral function of any protocol under development. IPv6 was designed with security in mind from day one, mandating the support of authentication and encryption in all IPv6 implementations. This chapter considers only the basic information required to understand the purpose and use of the Authentication Header (AH) and the Encryption Security Payload (ESP). More details on security are covered in Chapter 4.

Before we consider the details of the Authentication and ESP headers, we need to first understand what authentication and encryption mean.

Authentication is the process by which an entity (a machine or human being) verifies the claimed identity of another entity. This might not be needed in all scenarios of communication; however, certain cases, which are associated with some sort of authorization, require the authenticator to authenticate an entity so that it can authorize the entity to do something. For example, to be authorized to withdraw money from my bank account, I must be able to produce some evidence (acceptable by the bank) to show that I am the person that I claim to be (e.g., passport or identity card).

Encryption can be defined as the process by which an encryptor receives a readable string (readable by a human or a machine) and shuffles its contents in a way that makes this string unreadable. Obviously, the shuffling needs to be reversible; otherwise, the original information could not be recovered. Hence, the need for a documented algorithm that can be used to shuffle the data, and when performed at the receiver, it can recover the original message. Such encryption algorithm would be negotiated between the two nodes during the Security Association (SA) establishment process.

Authentication and encryption require some knowledge between the communicating parties. We refer to the process by which this knowledge is established as the *secure handshake*. A secure handshake allows two parties to establish their identities and trust in order to verify the origin of the information for a specified amount of time. Secure handshakes are described in Chapter 4, as are the details of how the sender can protect the information being sent. We refer to the knowledge of identities, trust establishment between two corresponding nodes, and the duration of such trust as the security association. Once a security association is established by a secure handshake, two nodes can communicate in a secure manner by protecting the information exchange.

#### 2.2.4.1 THE AUTHENTICATION HEADER

The Authentication Header is a mechanism defined to protect a packet's integrity following the establishment of a security association. The integrity of a packet's content is verified by checking that the packet was received from the claimed sender (i.e., the source address in the IPv6 header). The Authentication Header protects most of the packet (some exceptions are shown below), so any modification to any part of protected information in the packet along the path it traverses will be detected by the receiver. In addition, the Authentication Header protects against replay attacks by including a sequence number field, which is incremented every time a packet is sent. Replay attacks take place when an attacker stores one of the packets and resends it later (e.g., a packet containing a password to gain network access). Since the sequence number will not be higher than the last received number, the packet will be dropped by the receiver without checking whether it was modified en route.

When using an Authentication Header, the sending node calculates a checksum over the entire packet, referred to as the Integrity Check Value (ICV). The ICV is calculated on certain parts of the packet that are not expected to change until the packet reaches the final destination. These fields are called the *immutable* fields. Other fields that may change en route to the final destination (e.g., Traffic Class field, Hop limit, and some options) are called *mutable* fields and are not included in the ICV calculation. Between those two cases are fields that may change en route, but their final value is predictable (e.g., the Routing Header). For those fields, the ICV is calculated over the predictable final content—

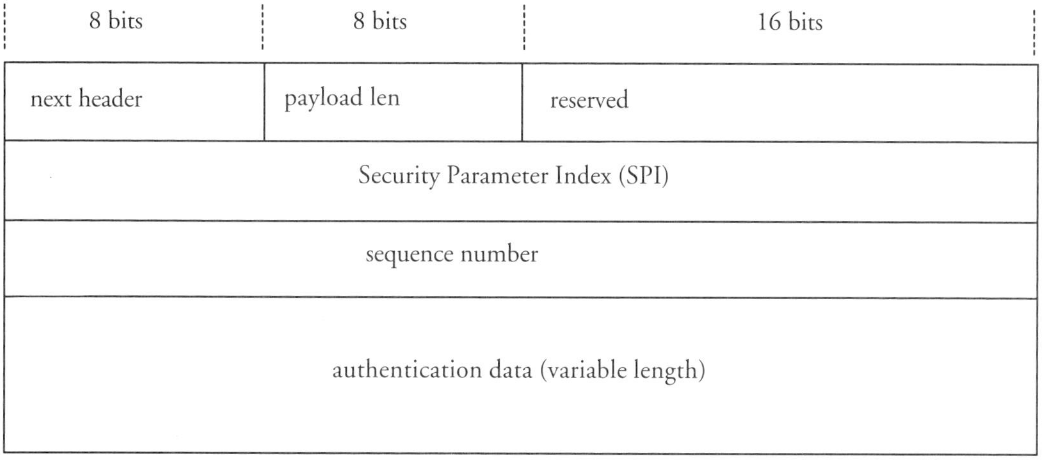

| 8 bits | 8 bits | 16 bits |
|---|---|---|
| next header | payload len | reserved |
| Security Parameter Index (SPI) | | |
| sequence number | | |
| authentication data (variable length) | | |

**Figure 2–8** *The Authentication Header.*

that is, the fields' content when received by the ultimate receiver of the packet. Figure 2–8 shows the format of the Authentication Header.

The *payload length* indicates the number of 32-bit words included in this header starting after the first 32-bit word. The *reserved* field is not used.

The *Security Parameter Index (SPI)* is an arbitrary 32-bit number negotiated between the nodes involved in the SA establishment to be able to identify a security association (when combined with the IP addresses of the communicating nodes). The combination of the SPI and two IP addresses sharing a security association is unique. When a security association is established, the details of this association (algorithm, keys, and lifetime) are stored in a Security Association Database (SAD). Since several security associations can exist between two nodes (each SA may be associated with a different level of authorization), communicating nodes must identify the right security association when sending and receiving packets.

The *sequence number* field is a 32-bit counter that is incremented for each packet. The sequence number cannot be cycled. Therefore, for two nodes to continue communication, after the highest value of the sequence number has been reached, they need to establish a new security association. In other words, each security association will allow approximately 4 billion packets to be exchanged between the two nodes before it needs to be renewed.

The *authentication data* field is a variable-length field that contains the ICV calculated by the sending end. The receiving node is expected to recalculate the ICV based on the algorithm negotiated during the SA establishment. If the ICV matches that calculated by the sending node (included in the received header), the packet is authentic; otherwise, the packet is dropped. ICVs and Message Authentication Codes (MACs) are discussed in Chapter 4.

The decision as to which packets should be protected by an Authentication Header depends on a defined policy negotiated when the security association is established or configured in both nodes. The security policy for a certain node is defined in the Security Policy Database (SPD). The SPD must be located at both, the sender and the receiver. The SPD is a database containing policies detailing whether an IP packet needs to be protected. The policies are based on several parameters referred to as *selectors*. Selectors include IP addresses and protocol numbers. For instance, an SPD entry may contain a policy stating that all TCP traffic between addresses A and B must be protected by AH. The policy would also point to an SA in the SAD to indicate what keys, algorithms and SPI should be used when protecting the packet. Hence, all packets sent or received by a node must go through the SPD to check whether a packet should be protected and, if so, how. Such policy is set during the SA establishment.

### 2.2.4.2 THE ENCRYPTION SECURITY PAYLOAD HEADER

Unlike the Authentication Header, the aim of ESP is to provide confidentiality. That is, its purpose is to ensure that no one en route can understand the content (located after the ESP header) of an IP packet protected by ESP. This is achieved by encrypting the packet at the sending end. The ESP header is a result of an encryption process performed by the sending node. The header contains the information required in order to retrieve the original packet at the receiving end and verify its content. This includes the SPI and sequence

| 8 bits | 8 bits | 8 bits | 8 bits |
|---|---|---|---|
| Security Parameter Index (SPI) | | | |
| Sequence Number Field | | | |
| Payload Data (variable length) | | | |
| | Padding (0 - 255 bytes) | | |
| | | Pad Length | Next Header |
| Authentication Data (variable length) | | | |

**Figure 2–9** *The ESP header format.*

number fields, which have the same meaning discussed earlier for the Authentication Header. Figure 2–9 shows the format of the ESP header.

The *payload data* contains the payload that needs to be encrypted by ESP, followed by some padding, if required, and the next header field, which indicates the protocol type for the encrypted payload. The *authentication data* field has the same meaning as in an Authentication Header. However, authentication in ESP is optional. Another difference between ESP and Authentication headers is that when ESP is used, authentication and encryption do not cover the entire IP packet but only those headers following the ESP header (e.g., other extension headers in addition to the payload). Hence, the IPv6 header is not authenticated when using ESP. Clearly, the IP header could not be encrypted either; otherwise, routers would not be able to route a packet to its destination address!

## 2.2.5 The Destination Options Header

The Destination options header contains options that should be processed by the node whose address is the destination address in the IPv6 header. The destination address can be an address of an intermediate node (recall the routing header processing) or the address of the ultimate receiver of the IP packet. Figure 2–10 shows the format of the destination options header.

The format of the Destination options header is the same as the Hop-by-Hop options header. The difference between the two headers is that the former is processed by nodes that the packets are destined to, while the latter is processed by every node along the path to the ultimate receiver.

## 2.2.6 Ordering of the Extension Headers

The ordering of processing of extension headers is very important for communicating nodes. RFC2460 recommends that the extension headers be arranged in the following order:

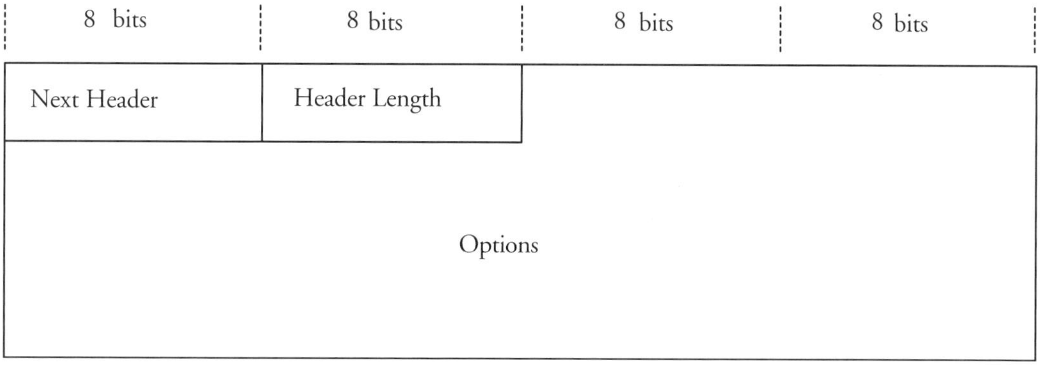

| 8 bits | 8 bits | 8 bits | 8 bits |
|---|---|---|---|
| Next Header | Header Length | | |
| Options | | | |

**Figure 2–10** *The destination options header.*

- Hop-by-Hop header
- Routing Header
- Destination options
- Fragmentation Header
- Authentication Header
- Encryption Security Payload
- Destination options
- Upper-layer headers (e.g., TCP, UDP, SCTP, and Mobility header)

This ordering of the extension headers is clearly intended to simplify the processing of packets by intermediate nodes as well as by the ultimate receiver of the packet. Routers forwarding IPv6 packets will only need to read the *next header* field in the IPv6 header. If it indicates that the next header is the Hop-by-Hop header, the routers should process the packet (most likely on a slow path); otherwise, the packet is forwarded based on the contents of the destination address. If a packet contains the router's address in the destination field (probably because a Routing Header is also included if the router is not actually the ultimate receiver), again the router will process this packet in the slow path.

The Destination options header is the only header that can be found in two different places within the daisy chain of extension headers. This is because the Destination options header is designed to contain options intended for the receiver of the packet. Hence, if more than one node is intended to receive the packet (a Routing Header is included), the destination options could be placed after the Routing Header. In this scenario, all the nodes included in the Routing Header must process the destination options field . The sender may also wish to include options intended for the ultimate receiver of the packet (or the last address in the Routing Header if one exists). In this case, the sender should include these options in the final header, following the ESP header (if it exists) and before the upper layers' headers. This placement will also allow these destination options to be encrypted if they contain confidential information.

Based on the ordering of extension headers, it can be noted that starting from the Fragmentation Header downwards, all headers are processed by the final receiver of the IPv6 packet only. Designers of future extension headers must consider the placement of such headers within this protocol architecture.

The *Mmobility Header* is defined by Mobile IPv6; its use and content are described in Chapter 3, and Chapter 5.

## 2.3 ICMPv6

The Internet Control Message Protocol for IPv6 (ICMPv6) is used to carry IP control messages for various purposes. ICMPv6 is defined to carry information between hosts, between routers, or between hosts and routers. It is also used to carry Neighbor Discovery protocols, as shown in section 2.6.

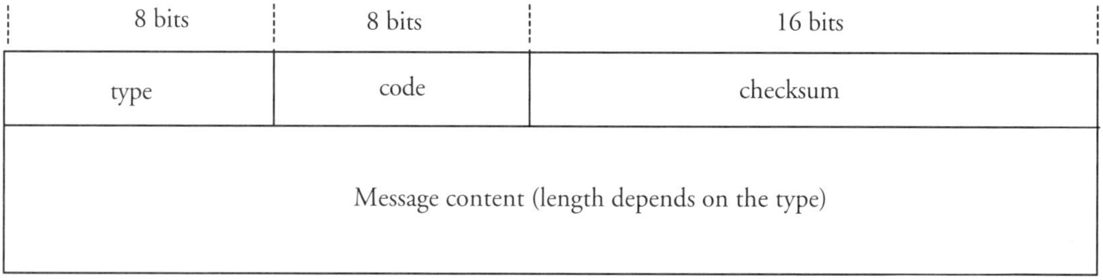

| 8 bits | 8 bits | 16 bits |
|---|---|---|
| type | code | checksum |
| Message content (length depends on the type) | | |

**Figure 2–11** *ICMPv6 generic message format.*

The protocol defines a generic message format that includes a message *type, code,* and *checksum*. The *type* field indicates the purpose of the message; values from 0 to 127 are reserved for error messages, while larger values are reserved for other messages. The *code* field is used to provide a finer granularity for information conveyed by the message. That is, a particular message type can include more details about the message, which are encoded in the *code* field. The *checksum* field is calculated over the source and destination addresses in the IPv6 header, the next header field (which should be set to 58 for ICMPv6), and the entire ICMPv6 message. When calculating the checksum field, the actual field is set to zero. The generic ICMPv6 message format is shown in Figure 2–11.

Note that it is possible to add options to such messages, each including its type and length fields. This concept is used by the Neighbor Discovery specifications, as discussed later in this chapter.

The basic ICMPv6 message format and general processing rules are detailed in RFC 2463. Other specifications, such as Neighbor Discovery, make use of such message definition by defining new message types. In this section, we discuss the base ICMPv6 specification and the types of messages defined in it. In later sections, other messages are discussed as deemed relevant.

## 2.3.1 ICMPv6 Error Messages

The ICMPv6 specification defines four error messages; each message is assigned a separate type. The code field is also used by some of these messages to convey more information about the error. The error types are as follows:

1  Destination unreachable.
2  Packet too big.
3  Time exceeded.
4  Parameter problem.

The *Destination unreachable* message is used to inform the sender of an IP packet that the final destination could not be reached. This message can be sent by a router along the path to the intended destination, for instance, due to a failure in the routing system that renders the final receiver unreachable or because the destination address does not exist (e.g., host failed or never existed). Alternatively, the message can be sent by the final receiver itself. For example, it is possible that the sender was attempting to reach a certain application (e.g., FTP) that is not supported by the final destination. Hence, the term *destination* may refer to either the end node as a whole or to a particular port number (identifying an application) on that node. This is a typical case that illustrates the use of the *code* field. To allow for different granularities of destination, the code field is used to indicate, more accurately, what went wrong. The following values for the code field are used to further elaborate on why the destination was unreachable:

0 No route to destination.

1 Communication with destination is administratively prohibited.

2 This code value is not assigned.

3 Address unreachable.

4 Port unreachable.

The *Packet too big* error is used by routers to indicate to the sender that the packet size has exceeded the router's link MTU. This feature is used by PMTU discovery, as discussed earlier in section 2.2.3.2.

The *Time exceeded* error is used by routers to indicate to hosts that the hop limit field in the IP packet has reached a value of zero and the packet will not proceed further. Upon receiving this error message, the IP layer notifies upper layers.

The *Parameter problem* error is sent by the ultimate receiver of a packet to indicate to the sender that an error was encountered when processing one of the fields in the IP header, extension headers, or options within one of the extension headers. This error message includes a pointer field to point to the part of the packet where the error has occurred. The *code* field is used to indicate whether the error encountered was in the IPv6 header (when set to 0), an extension header (when set to 1), or in one of the options inside an extension header (when set to 2). When this error message is received, the IP layer implementation is required to notify upper layers.

Note that ICMPv6 error messages must not be sent in response to other ICMPv6 error messages. Furthermore, sending ICMPv6 errors is subject to rate limiting. Rate limiting is the process by which a node limits the frequency of sending ICMP error messages to a certain node or to limit the rate of sending ICMP in general. Rate limiting is done to avoid congesting links with such messages as well as to reduce the chance of overuse of a node's processing power, as it would keep busy sending ICMP errors and become unable to process other legitimate packets. This is important to avoid Denial of Service (DoS) attacks in which an attacker purposely sends erroneous IP packets to trigger ICMP errors, consuming processor power and link bandwidth resources, which reduces the ability of a node to process legitimate packets. To perform rate limiting, a node may either set a certain value to indicate the allowable rate of ICMP errors that can be sent to a certain destination (e.g., one packet/second for destination A) or, alternatively, set a value that indicates the allowable percentage of link bandwidth that can be used by ICMP errors (e.g., 1 % of link bandwidth). The method used by an implementation depends on the critical factors for a node—that is, processing power and bandwidth.

## 2.3.2 ICMPv6 Informational Messages

One of the most powerful diagnostic programs used on the Internet today is the *ping* program. The program is used to confirm, in a very simple manner, whether a destination (anywhere on the Internet) is reachable. It relies on sending a message *(echo request)* to a particular destination (specified by the user), which triggers a reply by the receiver of that message. The program also measures the RTT for the original message and reply, hence providing reachability confirmation as well as the time it took for the message to get to the other end and back.

ICMPv6 acts as a carrier for such information by defining two messages: *echo request* (type 128) and *echo reply* (type 129). The messages are identical except for the *type* field value in each one. The message format is shown in Figure 2–12.

The *identification* and the *sequence number* fields are set by the sender of the echo request and reflected back, with any data included in the message, in the echo reply message.

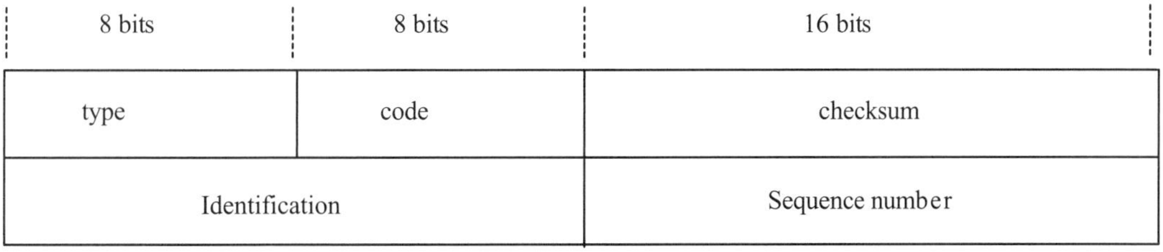

| 8 bits | 8 bits | 16 bits |
|---|---|---|
| type | code | checksum |
| Identification | | Sequence number |

Figure 2–12 *ICMPv6 echo request and echo reply message format.*

## 2.4 Tunneling

Tunneling is the process by which a node (host or router) encapsulates an IPv6 packet in another IPv6 header, resulting in a packet containing two or more (if the encapsulation is done more than once) IPv6 headers. The node originating the tunnel is called the *tunnel entry point*. The node terminating the tunnel is called the *tunnel exit point*. When seen by the original IPv6 header, the tunnel resembles a virtual point-to-point link, starting at the tunnel entry point and ending at the tunnel exit point. This is similar (when seen by the original packet) to encapsulating IPv6 into a link-layer protocol.

Several scenarios require a host or a router to send IP packets encapsulated in another IP header. Consider Figure 2–13, where Huntsman is sending IP packets to Redback. Assume that Router1 is configured with a certain policy that requires it to enforce that certain types of traffic are delivered to Redback through Router4 (e.g., this path is more secure). One way to guarantee that Router1 meets this policy is to overwrite the destination address in the original packet (from Huntsman, with the destination address set to Redback) and write Router4's address. However, this would lead to several problems: What does Router4 do when it receives this packet? How does it discover that it was originally intended for Redback? There would need to be additional signaling between routers to allow Router4 to recover the original destination address. In another scenario, Router1 may not trust Router2 and may prefer to protect all traffic addressed to Router4 using IPsec (AH or ESP). However, this would require changing the original packet sent by Huntsman, which would break any IPsec header that might be included in the original packet. Hence, this problem must be solved, which is what tunneling does.

Tunneling can be performed by the original sender of the packet or by an intermediate node along the path that the packet traverses to the final receiver. Tunneling can be a very useful tool in several scenarios. Mobile IPv6 uses

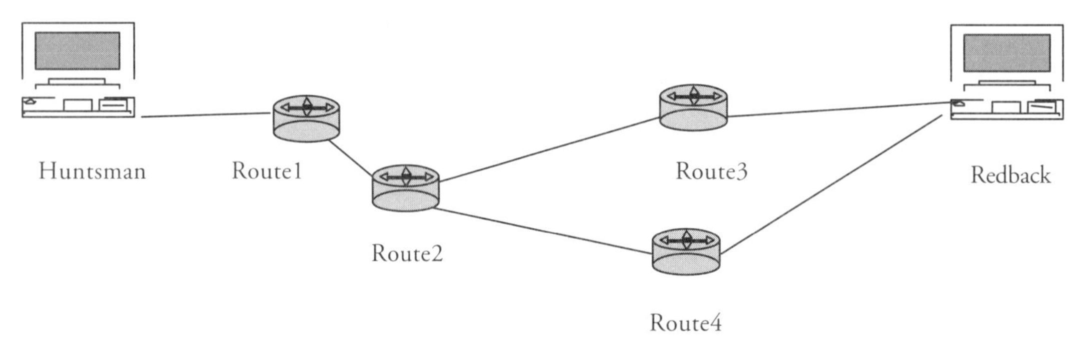


**Figure 2–13** A scenario where tunneling can be used.

| IPv6 Header | TCP Header | Data |
|---|---|---|

Original IPv6 packet sent by the source

| New IPv6 Header | Original Header | TCP Header | Data |
|---|---|---|---|

Tunnelled IPv6 packet

**Figure 2–14** *IPv6 tunneling.*

tunneling, as will be seen in Chapter 3. Figure 2–14 illustrates the process by which tunneling is performed.

Tunneling is an extremely useful tool for Virtual Private Networks (VPNs). For instance, let's consider a case where the networks behind Router1, Router3, and Router4 belong to the same corporation and happen to be separated by several routers on the Internet; for instance, two different sites belonging to the same corporation would want to ensure that traffic between Router1 and Router3 or Router4 is secured from any malicious nodes along the path. Therefore, the corporate network administrator would want to be sure that all IP packets between the two sites are protected. One way of meeting this requirement is to use IPsec (e.g., ESP) between the edge routers. The only way to protect these packets without modifying the original content is to encapsulate them in secure tunnels. In this case, an edge router (e.g., Router1) would encapsulate the original packet (e.g., sent from Huntsman) in another IPv6 header, using its own address as a source address and Router4's address as a destination address. IPsec can cover the entire packet, as shown in Figure 2–15.

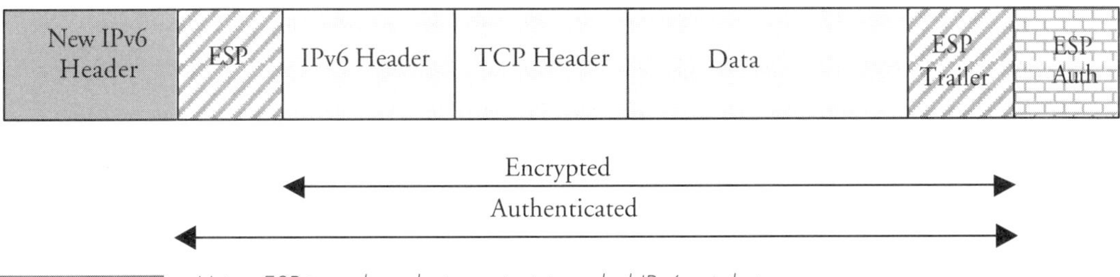


**Figure 2–15** *Using ESP tunnel mode to protect tunneled IPv6 packets.*

In this scenario, Router1 is the *tunnel entry point*; Router4 is the *tunnel end-point* or *tunnel exit point*. The IPv6 header sent by Huntsman is called the *original header,* and the IPv6 header resulting from the tunneling process is called the *tunnel header*. Note that tunnel headers may also contain some extension headers. Figure 2–15 shows the contents of the tunneled packet when using ESP. In this example, ESP is used in *tunnel mode* because it protects the content of the tunneled packet.

The receiver of the tunneled packet shown in Figure 2–15 decapsulates the packet after verifying the ESP header. This step recovers the original packet received at the *tunnel entry point*. The original packet can then be forwarded to the intended destination.

### 2.4.1 What Happens to other Fields in theTunnel Header?

Apart from the source and destination addresses, which are set to the tunnel entry and exit points respectively, the tunneling node needs to set the rest of the fields in the tunnel header. The fields in question are the *type of service*, the *flow label*, the *next header*, the *payload length*, and the *hop limit*. All fields except the next header field can be copied from the original header. Alternatively, the tunneling node may be configured to set these fields to certain values when tunneling the original packets. However, a tunneling node must be configured in a way that informs it whether it should copy the original values or set new values given by the node's administrator. In the absence of tunnel extension headers, the next header field must be set to a value of 41, indicating *IP in IP encapsulation*. However, if extension headers are used in the tunnel header, the next header field is set to the appropriate value depending on the type of the following header (e.g., 0 for hop-by-hop headers, 60 for destination options headers). The final extension header (preceding the original IPv6 header) should contain a next header field of 41.

### 2.4.2 How Many Times Can an IPv6 Packet Be Tunneled?

*Nested encapsulation* refers to tunneling of an already tunneled packet. Nested encapsulation may occur when a packet traverses a path that includes more than one node acting as tunnel entry points for the same packet. This may have negative consequences in some cases. For instance, if node A is configured to tunnel packets to node B, which was misconfigured to tunnel the same packets to node A, these nodes would continue tunneling until the packet size became larger than the PMTU between the two nodes. On links with large MTU, this could take a long time and cause bandwidth inefficiencies. For this reason, a *tunnel encapsulation limit* option was defined. This option includes a value indicating the number of times that a packet can

| Option Type | Option Length | Tunnel encapsulation limit | Padding |
|---|---|---|---|

Figure 2–16 *Tunnel encapsulation limit destination option.*

be tunneled. The option is included by the tunnel entry point and processed by the tunnel exit point. The option is included in the Destination options extension header. Figure 2–16 shows the format of the encapsulation limit option.

A value of zero in the tunnel encapsulation limit field indicates that the packet should not be encapsulated before being decapsulated at the receiving node. That is, nested encapsulation is not allowed.

## 2.5 IPv6 Addresses

IPv6 addresses are assigned to interfaces and not to nodes. A node may contain one or more physical interfaces (e.g., an Ethernet and a dial-up interface). A physical interface is the hardware responsible for attaching a node to a link. It is also possible for a node to have a nonphysical (virtual) interface. For instance, a tunnel entry point can be presented to the IP layer as an interface despite that it uses a physical interface to send IP packets and that such physical interface is presented to the IP layer as another interface. The use of such virtual interfaces becomes clearer when we discuss Mobile IPv6 in Chapter 3.

Each interface may be assigned an IPv6 address. An IPv6 implementation may assign the same address to more than one interface (e.g., for load-sharing traffic between two interfaces). However, such implementation would present only one interface (logical, as opposed to multiple physical interfaces) to the IP layer to maintain the concept of one IPv6 interface for each address.

IPv6 defines a number of addresses of different types and scopes. Each address type can be used on different scopes (e.g., link-local or site-local). The scope of an address essentially defines the domain in which such address is valid for allowing nodes to communicate with each other.

The content of all IPv6 addresses can be broken into two main parts: a *prefix* and an *interface identifier*. The prefix defines the number of bits (starting from the most significant bit) that are assigned to a particular link. The prefix identifies a link (or subnet) and is therefore used by routers to deliver IPv6 packets to the right link. The remaining part of the address (when subtracting the number of bits in the prefix) is the interface identifier. As the name suggests, it identifies the interface to which the IPv6 address was assigned. Now let's discuss the different IPv6 address types and scopes.

### 2.5.1 Textual Representation of IPv6 Addresses

Due to the length of IPv6 addresses, they are represented in hexadecimal format. IPv6 addresses are divided into eight parts, each containing two octets *(words)* and separated by a colon (:). The following is an example of representing IPv6 addresses in text:

```
3ffe:0200:0008:72AB:1434:0:0:1    (1)
```

It can also be written as follows, without the leading zeros in each field:

```
3ffe:200:8:72AB:1434:0:0:1
```

When several words consist of zeros, a (::) can be used to compress those fields. However, :: must appear only once in an IPv6 address. For instance, the following address,

```
3ffe:0:0:5:104A:2A61:0:0
```

can be written as

```
3ffe::5:104A:2A61:0:0
```

or

```
3ffe:0:0:5:104A:2A61::
```

However, the same address **cannot** be represented as follows:

```
3ffe::5:104A:2A61::
```

The reason is obviously because the last representation does not tell us how many words are set to zero, and therefore it is open to misinterpretation.

The following addresses are discussed in the next sections:

| | |
|---|---|
| `::` | The *unspecified* address |
| `::1` | The *loopback* address |

IPv6 addresses are written in a way that represents the address and its prefix as follows:

```
Address/prefix length in bits
```

Based on this format, we can rewrite the address in (1), which has a 64-bit prefix, as follows:

```
3ffe:0200:0008:72AB:1434:0:0:1/64 or,
3ffe:200:8:72AB:1434::1/64
```

IPv6 addresses can be categorized into three types, each consisting of a number of scopes: unicast, anycast, and multicast.

## 2.5.2 Unicast Addresses

Unicast addresses are assigned to a single interface. Typically, every device on the Internet today is assigned a unicast address. Unicast addresses have four scopes: node-local (loopback address), link-local, site-local, and global.

- The *loopback* address is valid only within one node and is built by setting all of the most significant 127 bits to 0 and setting the least significant bit to 1. As the name suggests, the *loopback* address allows different applications running on the same node to communicate with each other and allows any other node-local communication. Why would applications on the same host use IP instead of internal function calls to communicate? Because this feature makes the software design independent of the node in which the application is running. When applications communicate using the TCP/IP suite, they can be later moved to different nodes without changing the application's code. Only the destination address would need to be changed.
- A *link-local* address can be used to communicate only with nodes sharing the same link. Routers on-link cannot forward IPv6 packets with link-local source or destination addresses beyond the scope of the link. The format of the link-local address is shown in Figure 2–17.

  When generating link-local addresses, the interface identifier is generated by the interface that owns the address.

| 10 bits | 54 bits | 64 bits |
|---|---|---|
| 1111111010 | 0 | Interface identifier |

Figure 2–17 Link-local address format.

| 10 bits | 38 bits | 16 bits | 64 bits |
|---|---|---|---|
| 1111111011 | 0 | Subnet ID | Interface identifier |

Figure 2–18 Site-local address format.

- *Site-local* addresses are unique and useful only within an IPv6 *site*. IPv6 packets with site-local source or destination addresses cannot be forwarded beyond the site border. The size of a site is defined by the network administrator. It can be as small as a home network or as large as a multinational corporation. However, a wise administrator would probably limit the size of a site to a building or a campus area. The format of a site-local address is shown in Figure 2–18.

  Since sites can contain several links, a 16-bit field is reserved to allow site administrators to *subnet* their sites. For instance, the administrator of a school of engineering may wish to break up the school's site in several subnets, such as electronics lab, optics lab, and so on. This can be done by assigning a unique subnet identity to each one of those subnets. The uniqueness of subnet identities is limited to the site. That is, the same subnet identities can be reused within other sites.

- *Global* addresses, unlike local addresses (link and site scopes), are unique within the scope of the Internet. Every IPv6 node is likely to be assigned one or more global addresses. Global addresses are constructed by concatenating a globally unique prefix (assigned to a link) to an interface identifier unique within that link's scope. The format for a global address is shown in Figure 2–19.

  The *global routing prefix* is assigned to a site consisting of a number of links and is used by routers in the Internet to route packets to the site. The *subnet id* field is assigned to a link within the site. The *interface identifier* uniquely identifies an interface on a link. The interface identifier is not required to be unique beyond the link. The sum of $N + M$ is always equal to 64, except for addresses

| N bits | M bits | 128 - N - M bits | | | |
|---|---|---|---|---|---|
| Global routing prefix | Subnet ID | | g | u | Interface identifier |

Figure 2–19 Global address format.

starting with 000 in the three most significant bits. Those addresses are discussed in section 2.5.6.

The *u*-bit and *g*-bit are the least significant bits in the most significant byte of the interface identifier. The *u*-bit indicates whether the interface identifier is globally unique (when set) or locally (within the link) unique. Getting a globally unique interface identifier requires some form of an authority that allocates interface identifiers. For instance, the IEEE EUI64 format can be used to derive a unique interface identifier. RFC 2462 describes a method that allows a node to use its 48-bit Ethernet address (being globally unique) to compute a globally unique 64-bit interface identifier. The reason for having the *u*-bit was that it might be useful for people to get globally unique interface identifiers. However, there are currently no applications that make use of a globally unique interface identifier. Therefore, the *u*-bit does not seem to add any value, but it has generated enough controversy within the IETF, which made it worth mentioning in this book! The *g*-bit indicates a group or individual interface identifier. Note that the same format for the interface identifier applies to all unicast addresses of various scopes.

#### 2.5.2.1 CHALLENGES IMPOSED BY SCOPED ADDRESSES

The idea of scoped addresses implies that there is a domain in which these addresses are valid and that packets containing these addresses cannot be forwarded beyond the border of such domain. For instance, packets containing link-local addresses (in the source or destination fields) cannot be forwarded beyond the link that they originated from. The same concept applies to *site* borders for packets sent containing site-local addresses. The border control functions must be clearly included in routers. Therefore, a router on a particular link must ensure that it does not forward packets containing link-local addresses. To do that, the router checks the source and destination addresses in the packet. This is a deviation from standard IP routing, which should inspect the destination address only for forwarding purposes. However, in this case the source address is not checked for forwarding per se; it is checked to enforce a rule. If the source address were not checked and packets were forwarded containing a link-local source address, the receiver would assume that the sender is on the same link and could end up replying to a different node located on its link because it has the same link-local address. This problem is fairly simple for link-local addresses: a router's default behavior would be to assume that each interface is located on a separate link. Therefore, no packets containing link-local addresses should be forwarded from one interface to another. However, for site-local addresses, this is not as simple. A *Site Border Router (SBR)* does not know automatically which interface belongs to which site because a site contains a number of links. For instance, if a router has four different interfaces, there are various configuration permutations, some of which are the following:

1. Each interface belongs to a separate site.
2. Any combination of two interfaces belongs to the same site.
3. Three interfaces belong to the same site and the fourth does not.

Therefore, it is essential to configure an SBR with the mapping between interfaces and site names. Clearly the site names need only be unique within the router; two different nodes can have different names for the same site without having any impact on their communication with site-local addresses.

A more significant impact of scoped addresses can be illustrated by examining site-local addresses and their impact on certain applications in more detail. Let's consider the network in Figure 2–20 as an example. In this case a *site* is configured by a network administrator, and an *SBR*, on the edge of the site, is configured to not forward any packet containing site-local addresses. Hosts within the site are allocated both site-local addresses (to communicate with other nodes in the site) and global addresses for communication outside the site. Let's assume that the site administrator has included both site-local and global addresses in the local DNS.

Now assume that a host outside the site is attempting to communicate with Host1. It will query the DNS. However, Host1 has two AAAA records in the DNS, one for its global address and another for its site-local address. Since

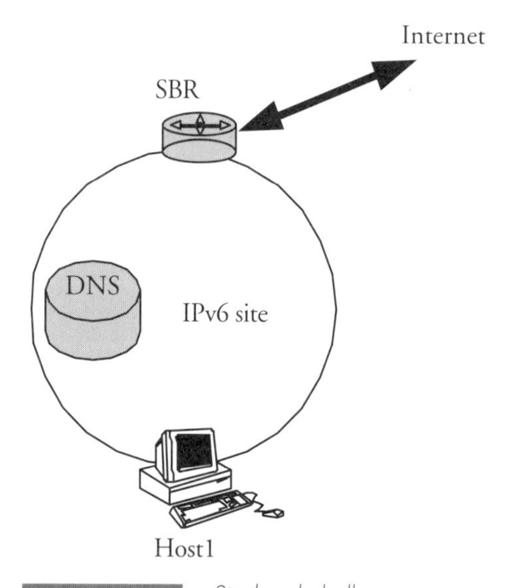


**Figure 2–20** *Site-local challenges.*

the other host is outside the site, there is no point in providing the site-local AAAA record in the reply; only the global address should be given because it is the only address that will allow the two hosts to communicate. This is another restriction for name servers; they need to treat DNS queries coming from outside the site differently from those originating within the site. A DNS implementing this feature is called a *two-faced DNS* because it gives a different response depending on the location of the node querying the DNS. This problem can be generalized to any application (the DNS is one example) that refers IP addresses between nodes.

With these complications in mind (and more devils in the details), are site-local addresses worth the trouble? Said differently, when are site-local addresses useful? Every year or two there are probably a few thousand emails on this topic on the IPv6 mailing list in IETF. It is also worth noting that the IPv6 working group is considering deprecating site-local addresses from the architecture. We do not cover the entire discussion here; instead, we present three main scenarios in which site-local addresses can be useful:

- Disconnected sites: Some sites—for example, a number of boards connected together using the TCP/IP suite to form a large system with multiple links—may never connect to the Internet. These boards might be part of a larger device, such as a robot. In this case there would be no need for these boards to get globally unique IP addresses.
- Intermittent connection to the Internet: Home networks are a good example of a network that is not always connected to the Internet. An ISP might charge a user based on connection time (e.g., dial-up service), so the user does not connect to the ISP unless she needs to use the Internet. However, if this home user has a small network (containing more than one link) inside the house, she will not be able to use it (e.g., transfer a document from one device to another using IP) unless the ISP offers it IP addresses. However, if site-local addresses are used, the home network can be an IPv6 site and devices inside this network can communicate independently of the connection to the ISP. This example applies even if the ISP does not charge based on the connection time; the user might like to use her home network even when the link to the ISP fails.
- Renumbering: One of the reasons for renumbering a network is a change of ISPs (recall the discussion on route aggregation in Chapter 1). When the network is renumbered, nodes are forced to derive their addresses from the new prefix. Some connections are expected to survive for a very long time (e.g., long-lived TCP connections). Therefore, ongoing connections will be broken. This problem is similar to mobility except that renumbering should rarely occur. If the site administrator wants to keep some important connections up while renumbering, she can do so by configuring devices that need long-lived connections to use site-local addresses for such connections.

Whenever site-local addresses are used together with global addresses within the same site, and both types of addresses are configured in the site's DNS, a two-faced DNS is needed. Therefore, only disconnected sites will not require a two-faced DNS.

## 2.5.3 Multicast Addresses

Multicast addresses are assigned to more than one interface. A packet addressed to a multicast address is copied and delivered to each interface assigned a multicast address. Multicast addresses are useful in several scenarios. For instance, a server might announce a broadcast video session (e.g., a speech by a corporate executive) at a certain time. Such server would announce a multicast group (a multicast address of an appropriate scope) that others can join to receive the video broadcast. Each node joining the multicast group would receive a copy of every packet sent by the server. The Multicast Listener Discovery (MLD) protocol was defined to allow nodes to join IPv6 multicast groups. The format for the multicast address is shown in Figure 2–21.

The *flags* field is used to indicate whether a multicast address is permanent (i.e., well known) or transient (i.e., assigned for a period of time). A permanent multicast address can be assigned to a particular server (e.g., DNS) for server discovery purposes. For instance, a reserved site-local multicast group may be assigned to all DNS servers in a site to allow clients to discover them dynamically without the need for manual configuration. On the other hand, if a video server were to broadcast a session for a period of time, after which the group members will no longer need to connect to the server, a temporary multicast group address may be more suitable. The format for the *flags* field is as follows:

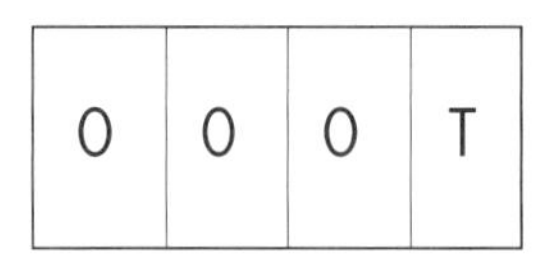


T = 0 indicates that the address is permanently assigned. T = 1 indicates transient assignment.

| 8 | 4 | 4 | 112 bits |
|---|---|---|---|
| 11111111 | flags | scope | group ID |

Figure 2–21 The multicast address format.

The *scope* field identifies the scope of the multicast group. The following values for the scope and their meanings are currently defined:

| | |
|---|---|
| 1 | interface-local scope |
| 2 | link-local scope |
| 3 | subnet-local scope |
| 4 | administration-local scope |
| 5 | site-local scope |
| 8 | organization-local scope |
| 14 | global scope |

Link-local and site-local scopes have the same meaning as for unicast addresses. Interface-local scopes are used for communication between functions on the same interface. An organization scope is larger than a site scope but smaller than a global scope. That is, an organization can contain several sites. Administration-local scope is limited to a network with a single administrative entity, and its scope must be configured by a network administrator. Administration-local scope is larger than subnet-local but smaller than site-local scope.

It is important to note that multicast addresses must not be included in the source address field of an IP packet; this would cause the receiver of such packet to reply to a number of nodes (depending on the size of the group). Bad Guy could use this mechanism to flood many nodes with unwanted packets by sending a packet including their multicast address in the source address field.

#### 2.5.3.1 LINK-LOCAL MULTICAST ADDRESSES

Link-local multicast addresses are useful for multiple access links like Ethernet. These link layers have their own addresses that are valid within the scope of the link layer (i.e., within a single link). Each interface on those links is configured to extract packets sent to any of the link-layer addresses that it is configured to listen to, including multicast frames.

Link-local multicast addresses are useful for autoconfiguration of IPv6 addresses. When a node boots, it can configure an address, unique within the link, then use this address to discover other neighbors (including routers) to be able to configure itself with globally unique addresses. To execute these steps, the following types of messages must be supported:

- A message to all nodes on the link: These messages are needed to broadcast important information on the link. IPv6 reserves the *all-nodes multicast address* for this purpose. Any message intended for all nodes on the link is sent to this address. Hence, every IPv6 node is required to be a member of this multicast group. That is, every Ethernet driver must be configured to receive messages to this multicast group. The *all-nodes multicast address* is represented as `FF02:0:0:0:0:0:0:1`.

- A message to all routers on the link: These messages are necessary in order to discover routers and global prefixes that are valid on a link. All IPv6 routers are required to join the *all-routers multicast address* in order to receive these messages. The link-local *all-routers multicast address* is `FF02:0:0:0:0:0:0:2`, and the site-local *all-routers multicast address* is `FF05:0:0:0:0:0:0:2`.
- A message to test address uniqueness: When a node configures a link-local address, it needs to ensure its uniqueness within the link. This can be done by sending messages to all nodes on the link to check whether they were assigned the same address. However, this consumes unnecessary bandwidth; ideally we only want the node that has the same address to receive the message. The *solicited node multicast address* is used for this purpose. This address is computed by taking the least significant 24 bits from a node's unicast address and appending them to the reserved 104-bit prefix: `FF02:0:0:0:0:1:FF::/104`. For instance, the solicited node's multicast address corresponding to the IPv6 unicast address `3ffe:200:1:5:1234:ABEF: 1256:F140` is `FF02:0:0:0:0:1:FF56:F140`.

The use of these addresses becomes clear when we discuss Neighbor Discovery messages in section 2.6.

### 2.5.4 Anycast Addresses

Anycast addresses are assigned to more than one interface. However, unlike multicast addresses, a packet addressed to an anycast address will be delivered to only one of the interfaces assigned such address. When an IP packet arrives at a router connected to a link with more than one node that has an anycast address, the router will deliver to only one of those nodes. The choice of which node the packet is sent to depends on the router's implementation. For instance, a router may choose to select the node based on a round-robin algorithm, delivering a packet to one node, the next packet to another node, and so on.

Anycast addresses can be useful for providing redundancy in cases where a packet can be received and processed by more than one node. Anycast addresses are also used for discovering some services like the home agent function in Mobile IPv6, as is shown in Chapter 3. Figure 2–22 shows the format of anycast addresses.

Anycast addresses have the same scopes as unicast addresses and are allocated from the same block of addresses. In addition, RFC 2526 describes a mechanism for reserving anycast addresses that can be used independently of their topological location. This is done by setting the most significant 57 bits in the interface identifier to 1 (except for the *u*-bit) and leaving the least significant 7 bits to be used for anycast address allocation. This mechanism allows for allocating 128 (7 bits) different anycast addresses. The 64-bit prefix remains the

| N bits | 128 - N bits |
|---|---|
| Subnet prefix | Interface identifier |

Figure 2–22 Format for anycast addresses.

same as with unicast addresses. Hence, a node configured with an anycast address can append its global prefix to the anycast interface identifier and become reachable by any other node on the Internet. The hexadecimal value 7E (in the least significant 7 bits) is assigned to Mobile IPv6 home agents. Hence, all home agents (discussed in Chapter 3) should be configured with this address. Clearly, a node configured with an anycast address can also be configured with other types and scopes of addresses.

### 2.5.5 The Unspecified Address

The *unspecified* address is an address that does not indicate any specific part of the topology. All 128 bits in the unspecified address are set to zero. Since routers would not be able to forward packets to the right link, this address cannot be used as a destination address. The unspecified address is used by nodes only while attempting to configure an address in a stateless or stateful manner (see section 2.7 for more address configuration).

### 2.5.6 IPv6 Addresses Containing IPv4 Addresses

To allow IPv6 to be gradually deployed within the current IPv4 Internet, there is a need for mechanisms that will ensure a smooth migration toward IPv6. Some of these transition mechanisms require the ability to embed IPv4 addresses in IPv6 addresses. IPv4 addresses can be embedded in the least significant (rightmost) 32 bits or in the 32 bits following the first two octets. *IPv4-mapped IPv6 addresses* and *IPv4-translated IPv6 addresses* embed IPv4 addresses in the rightmost 32 bits. Both addresses start with 000 in the most significant bits. On the other hand, *6-to-4 addresses* embed IPv4 addresses in the 32 bits following the first two octets. A special prefix (hexadecimal 2002) is reserved for the first two octets of all 6-to-4 addresses. The use of these addresses and the related transition mechanisms are presented in Chapter 9.

## 2.6 Neighbor Discovery

Neighbor Discovery is one of the cornerstones of IPv6. The protocol includes several functions that cover address resolution in a similar way to the Address Resolution Protocol (ARP) in IPv4 as well as new functions such as reachability confirmation and tools to enable address autoconfiguration and collision detection. The following sections highlight the main functions in the Neighbor Discovery protocol. These functions play an important role in enabling stateless address autoconfiguration and Mobile IPv6.

When viewed by an IPv6 node, a neighbor is another IPv6 node (host or router) that resides on the same link.

### 2.6.1 Why Does a Node Need to Discover a Neighbor?

In many links, an IPv6 host typically shares a link with other hosts and one or more routers. Such links (e.g., IEEE802.3 links) may allow hosts to communicate directly with each other. These links are called *multiple access links*. In such links, devices may be aware of being neighbors on the link layer. However, for these devices to communicate on the IP layer, IP implementations must know the right link-layer addresses for the nodes it is communicating with. Therefore, it is necessary to allow the link-layer interface to place the correct destination MAC address within the link-layer frame when sending IP traffic to another node. The Neighbor Discovery protocol allows nodes to discover link-layer addresses for nodes they communicate with. This functionality is called *address resolution*.

When communicating with other off-link nodes (i.e., non-neighbors), an IPv6 node forwards its traffic to a router, which in turn forwards IP packets to the final destination or to another router along the path to such destination. The first router that receives the IP packet to forward it to the final destination is called the *default router*. To be able to communicate with other off-link hosts, every host must discover one or more on-link default routers. Neighbor Discovery allows hosts to discover their default routers in a dynamic manner.

In addition to address resolution and router discovery, the protocol enables address autoconfiguration, Neighbor Unreachability Detection (NUD), and Duplicate Address Detection (DAD). The details of those functions are described in the following sections.

All Neighbor Discovery messages use ICMPv6 as a carrier for the protocol. The integration of Neighbor Discovery with the IP layer allows for significant future developments to the protocol, such as IP layer security (without manual configuration). This is a significant improvement over the ARP protocol in IPv4.

The protocol consists of five main messages and several options for each message. Each message is identified by the ICMP *type* field. Note that Neighbor Discovery messages must not be forwarded beyond the local link. RFC 2461 recommends that nodes use link-local addresses in the source and destination fields of the IP header (some exceptions below) and requires all messages to be sent with a *hop limit* value of 255 in the IP header. This hop limit value is needed to allow the receiver to ensure that the received packet was not routed beyond the link.

### 2.6.1.1 ROUTER SOLICITATIONS

Router solicitations are used by hosts to discover one or more default routers neighboring the soliciting host. Router Solicitations are sent to the *all-routers multicast address*. This address is a link-local multicast address that must be hardcoded in all router implementations. When sending the solicitation, all routers on-link will receive it. The reception of a router solicitation will result in sending a router advertisement from each router on-link. The format for the ICMP message for router solicitation is shown in Figure 2–23.

The *type* field indicates that the ICMP message is a router solicitation (set to 133 for this message). The *code* field is set to zero, and so is the *reserved* field.

Hosts can send router solicitations frequently to discover routers or when they first attach to a link. To avoid collision cases on a shared link where many hosts are starting up at the same time and therefore all send solicitations simultaneously, Neighbor Discovery recommends that hosts delay sending solicitations by a random value that ranges between zero and one second.

### 2.6.1.2 ROUTER ADVERTISEMENTS

Router advertisements can be sent as a response to Router solicitations or regularly in an unsolicited manner. When a router is solicited, the advertisement is sent to the unicast address of the soliciting node. By contrast, an unsolicited Router advertisement is sent to the *all-nodes multicast address.*

Router advertisements allow neighbors to discover the default routers' IP and link-layer addresses. In addition to default-router discovery, router advertisements contain useful information about the local link, including the link

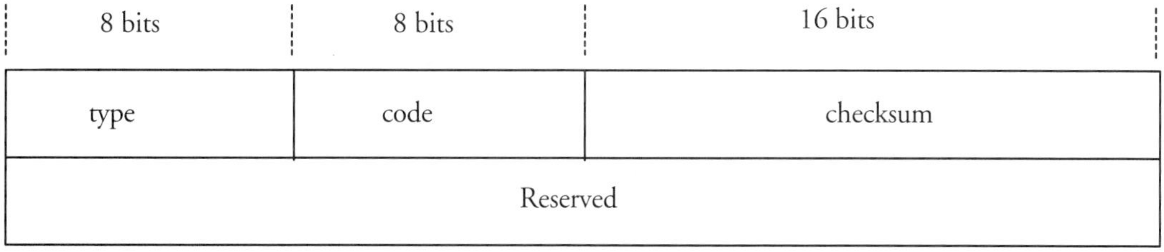

| 8 bits | 8 bits | 16 bits |
|---|---|---|
| type | code | checksum |
| Reserved | | |

**Figure 2–23** *ICMPv6 message format for router solicitations.*

| 8 bits | 8 bits | | | | 16 bits |
|---|---|---|---|---|---|
| type | code | | | | checksum |
| cur hop limit | M | O | H | reserved | router lifetime |
| reachable time | | | | | |
| retransmission timer | | | | | |

Figure 2–24 Router Advertisement message format.

prefix and MTU. This information is carried as ICMPv6 options. The router advertisement message format is shown in Figure 2–24.

The value of the *type* field for a router advertisement message is 134. The *cur hop limit* field contains the default value that on-link hosts should use in the *hop limit* field when sending IP packets other than Neighbor Discovery messages. The *M* flag indicates whether hosts should form addresses using stateless (when cleared) or stateful (when set) mechanisms. The *O* flag indicates (when set) that hosts should use stateful mechanisms to configure information other than addresses (e.g., a DNS server's address). The *h* flag is used to indicate that the router will act as a Mobile IPv6 home agent on this link. The use of this flag is discussed in Chapter 3.

The *router lifetime* field indicates the time (in seconds) during which the router can be considered a default router. The expiration of this timer does not imply ignoring any information sent by this router; it simply means that this router cannot be used as a default router.

The *reachable lifetime* indicates the time (in milliseconds) that a node can assume a neighbor is reachable. In a node's implementation, this timer can be started after receiving a reachability confirmation from a node. For instance, if this field is set to 10,000, it implies that a node can assume that a neighbor will continue to be reachable for 10 seconds after confirming that neighbor's reachability.

The *retransmission timer* indicates the frequency (in milliseconds) that address resolution of other nodes should be done.

Router advertisements can include several options; the most important one is the *prefix information* option. This option provides hosts with information about the link prefix and its lifetime, which allows nodes to configure addresses. The format for the prefix information option is shown in Figure 2–25.

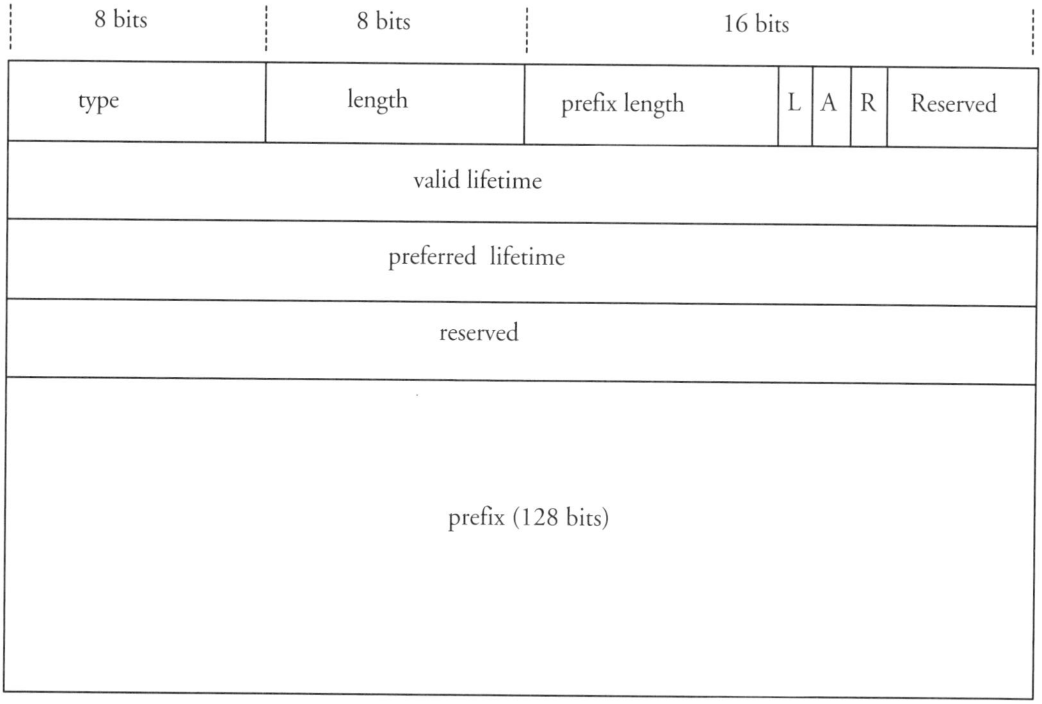

<table>
<tr><td>8 bits</td><td>8 bits</td><td colspan="5">16 bits</td></tr>
<tr><td>type</td><td>length</td><td>prefix length</td><td>L</td><td>A</td><td>R</td><td>Reserved</td></tr>
<tr><td colspan="7">valid lifetime</td></tr>
<tr><td colspan="7">preferred lifetime</td></tr>
<tr><td colspan="7">reserved</td></tr>
<tr><td colspan="7">prefix (128 bits)</td></tr>
</table>

**Figure 2–25** *The prefix option format.*

The *prefix length* field informs nodes receiving the option about the length of a prefix and consequently the length of the interface identifier that a node can use to form the entire address. The length can be any value between 0 and 128. This part of the Neighbor Discovery specification is confusing when considering that an IPv6 prefix can only be 64-bits long, which makes this field redundant. The reason for this discrepancy is that the IPv6 address architecture specification was updated after the Neighbor Discovery specification was finalized to state that an IPv6 prefix must always be 64-bits long. This discrepancy is currently being addressed by the IPv6 working group in IETF.

The *L* flag is used to inform nodes whether the prefix can be used for on-link destinations (when set). That is, it informs nodes whether they can communicate directly with other neighbors that are assigned addresses based on this prefix without having to send packets to their default router. If the *L* flag is cleared, that does not necessarily mean that the prefix is off-link; it simply means that the prefix option is intentionally vague about whether the prefix is on-link or off-link. In this case, a host should always send packets to its default router when the destination address is derived from this prefix. Why is the prefix option vague about whether the prefix is on-link? In some cases, it is useful to advertise the same prefix on more than one link; however, when this happens, hosts should be forced to send everything to their default router, which should send the information to the right destination (which might be off-link). Assigning the same prefix to more than one link can be useful in cases where the network contains more than one link and only assigned a 64-bit prefix. Recall that the maximum prefix length in IPv6 is 64 bits. Hence, if a network (e.g., a home network) is assigned such prefix by an ISP, it might need to assign it to more than one link within the house. This configuration requires routers to relay messages that are usually confined to one link (e.g., all Neighbor Discovery messages). This is especially important for messages sent to a link-local multicast address. Therefore, the *L* flag is used to force hosts to send all messages to their default router. The default router relays those messages to the right destination (if on-link with the router) or to another router on the path to the destination address.

The *valid lifetime* field is used by nodes to determine the duration of time (in seconds) for which the prefix can be considered on-link. In other words, this is the lifetime during which the information provided by the *L* flag is valid. The only way to inform hosts that a prefix is no longer on-link is to send the prefix option with the *L* flag cleared and the valid lifetime set to zero.

The *A* flag is used to indicate whether a node can use stateless address autoconfiguration (when set) to configure an address other than the link-local address. When this flag is cleared, hosts should use stateful address autoconfiguration to configure an address other than the link-local address.

The *preferred lifetime* field is the length of time (in seconds) during which addresses formed from a prefix can be preferred for use as source addresses. When this timer expires, hosts should prefer addresses derived from other prefixes.

The *prefix* field includes the valid leading bits of the prefix (the number of valid bits is indicated by the prefix length). The rest of the bits (i.e., those that can be used as an interface identifier) are set to zero **if** the *R* flag were not set. If the *R* flag were set, the prefix field will include the prefix and the interface identifier of the router sending the advertisement. That is, if the prefix length were 64, the first 64 bits in the prefix field would include the prefix for the link, and the last 64 bits would be set to zero if the *R* bit is not set and would include the router's interface identifier if the *R* flag were set. The *R* flag is a recent addition to router advertisements by the Mobile IPv6 specification. and the reason for having it is shown in Chapter 3.

#### 2.6.1.3 NEIGHBOR SOLICITATIONS

Neighbor solicitation messages are used for several reasons: address resolution, NUD, and DAD. First, consider the format of the Neighbor Solicitation message in Figure 2–26.

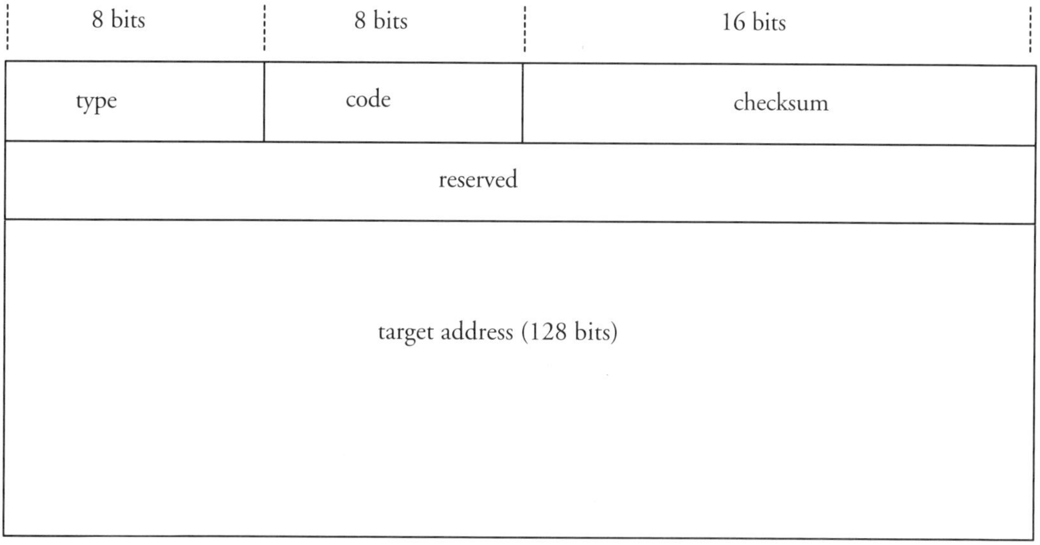

| 8 bits | 8 bits | 16 bits |
| --- | --- | --- |
| type | code | checksum |
| reserved | | |
| target address (128 bits) | | |

Figure 2–26 Message format for Neighbor Solicitations.

The *target address* field contains the IP address of the target of the solicitation. The target address is not necessarily the same as the destination address in the IP header. The reason for this mismatch can be understood by examining the use of this message for address resolution. Consider Figure 2–27, where node_A is sharing a link with node_B, and assume that this link is an Ethernet link (i.e., with multicast capability on the link layer). Node_A has a unicast IP

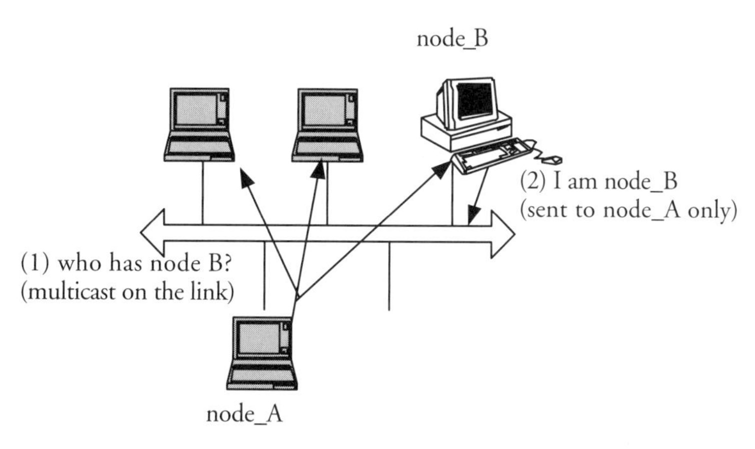


Figure 2–27 Address resolution on multicast-capable links.

address for node_B. However, for node_A to be able to communicate directly with node_B, it needs to obtain its link-layer address. To achieve this, node_A sends a Neighbor Solicitation message to node_B. The source address is node_A's address, and the destination address is node_B's *solicited node multicast address*. The target address is node_B's unicast address. When this packet is received by node_A's link-layer driver, it is sent to a link-layer multicast address. Only node_B's IP layer (the owner of the solicited node multicast address) will receive this message and reply to node_A with a *Neighbor Advertisement* (discussed in the following section). Its reply will include its link-layer address as an option in the neighbor advertisement message. The use of the solicited node multicast address allows this message to be mulitcast on the link layer (because the address looks like a multicast address to the link layer), while being processed by one node: the one that has the corresponding unicast address.

To avoid sending more messages over the link (in order to perform the same procedure in the opposite direction), the soliciting node (node_A in this example) could include its own link-layer address as an option within the neighbor solicitation message. This option is called the *link-layer address* option.

Another useful application for neighbor solicitation is the detection of duplicate addresses. A typical scenario for DAD is when a node starts up. Every IPv6 node is able to form a link-local address for use within the local link; this is done by appending the node's interface identifier to the well-known link-local prefix. However, a node cannot start using an address without ensuring that such address is unique within its scope. Until address uniqueness is tested, the address is marked as a *tentative address* and cannot be assigned to a node's interface. To test for address uniqueness, a node sends a neighbor solicitation to the solicited node multicast address computed from its own tentative unicast address. The source address of this packet contains the unspecified address. The node's tentative address is included in the *target address* field. If the address were already assigned to another node, the solicitation results in a neighbor advertisement from that node, which causes the soliciting node to try another address and perform DAD again. Note that the only way of knowing the difference between a solicitation message sent for address resolution and one sent for DAD is the content of the source address of such message. In the former scenario, the source is the soliciting node's unicast address, while in the latter, the source address is the unspecified address (::).

DAD does not guarantee that address duplication has not occurred. DAD provides an unreliable mechanism to test for address duplication—that is, "good enough" considering the size of the interface identifier (64 bits) and the low probability of collision. DAD is unreliable because it is a message that cannot get a reply to guarantee success. The sender **assumes** that there is no address duplication because it did not receive a neighbor advertisement. Consider a case where a link is disconnected due to a physical failure. If a host forms an address and attempts to test while the link is disconnected, the DAD message may not reach all nodes on the link (the link may be fragmented). Hence, when reconnected, there could be a chance of address duplication. This is particularly relevant for unreliable wireless links, which are prone to errors over the radio link. A DAD message may be lost, causing the sender to assume that the address is unique on the link.

#### 2.6.1.4 NEIGHBOR ADVERTISEMENTS

Neighbor advertisements (Figure 2–28) are sent in response to solicitations or are sent unsolicited. As explained earlier, this can be part of the address resolution process or DAD. Neighbor advertisements are also used for NUD. This process is explained below in detail.

The *R* flag is used to indicate whether the advertising node is a host or a router. Although a router is discovered through router advertisements, this flag is still useful in case a node's routing capability has been changed (e.g., an administrator has decided to change a node from a router to a host, or vice versa). When this flag is set, it indicates that the advertising node is a router.

The *S* flag is used to indicate whether the advertisement was solicited. If the flag is set, then the advertisement was solicited. The use of this flag is discussed below.

The *O* flag indicates (when set) that this advertisement should override the existing information about the advertising node. For instance, if the node's link-layer address has changed, the new address in this advertisement should replace the old one.

If this neighbor advertisement were sent in response to a solicitation, the *target address* field would contain the unicast IP address in the source address field of the neighbor solicitation that triggered this message. Otherwise, if this

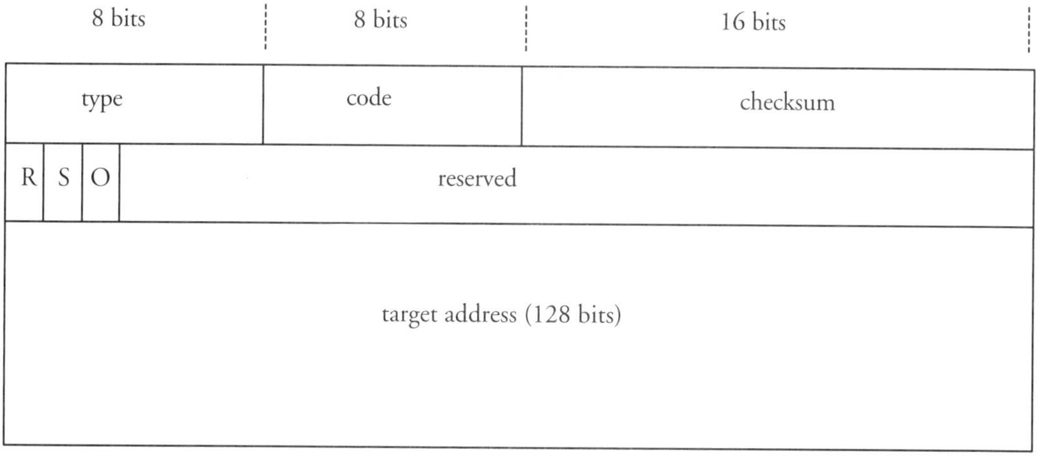

| 8 bits | 8 bits | 16 bits |
|---|---|---|
| type | code | checksum |
| R \| S \| O | reserved | |
| target address (128 bits) | | |

**Figure 2–28** *Neighbor Advertisement format.*

message were unsolicited, the *target address* field would contain the IP address corresponding to the link-layer address included in this message.

We discussed the use of neighbor solicitations and advertisements for address resolution and DAD in the last section. Another important function that utilizes these messages is the reachability detection of a neighbor. Symmetric (i.e., bidirectional) reachability of a neighbor can be determined if a node can send a message to such neighbor and receive a confirmation that the message sent was actually received. In some cases, this confirmation of reachability can be achieved when the IP layer receives "hints" from upper layers (e.g., TCP) to indicate that a particular connection with a neighbor is making *forward progress*. That is, TCP segments were recently sent to this neighbor, and the neighbor has confirmed reception of these packets. These hints can be passed from upper layers to the IP layer. However, in some other communication scenarios, such hints may not be available. For instance, unreliable transport layers like UDP do not receive acknowledgments from their peers. In addition, reachability confirmation is needed for default routers with which hosts do not typically have connections. One could argue that sending a router solicitation to a router and the subsequent reception of a router advertisement could confirm reachability of such router. While this is likely to be true in many cases, a router may have a router advertisement scheduled for sending to all nodes at approximately the same time that it receives the router solicitation. Hence, instead of replying to the soliciting hosts, it would send the advertisement to the *all-nodes multicast address*, assuming that the soliciting node would receive the advertisement. However, this advertisement can only confirm one-way communication (i.e., from the router to the host) and would not confirm to the soliciting host that the router actually received the solicitation message.

Reachability confirmation is achieved by sending a neighbor solicitation unicast to a neighbor. Since the advertisement is solicited, the neighbor will respond with a neighbor advertisement with the *S* flag set. The reception of this advertisement confirms the reachability of the solicited node. It should be noted that a reception of an unsolicited neighbor advertisement from a neighbor does not confirm symmetric reachability, as it does not prove that a neighbor is able to receive packets from the node.

Neighbor advertisements can also be sent by a *proxy* node, either solicited or not. In this case, a proxy node would send a neighbor advertisement informing the receiver(s) that the link-layer address associated with a particular IP address should be modified to a new one (that belongs to the advertising node), included in the link-layer address option. This action causes the original node's IP traffic to be forwarded to the advertising node. The format of a proxy neighbor advertisement message is exactly the same as the message shown in Figure 2–28; the only difference is between the two uses of the message; that is, proxy advertisements are sent by a node other than the one assigned the IP address. Proxy Neighbor Advertisements are crucial for the operation of Mobile IPv6 home agents, as discussed in Chapter 3.

### 2.6.1.5 REDIRECT MESSAGES

The Redirect message is used by routers to inform hosts of a better next-hop than the one the packet was initially forwarded to. This can be useful in several cases. For instance, consider a case in which a node is located on a link where more than one router is available. If the node is communicating with an off-link node, it needs to choose one of the default routers as a next-hop router. However, one of the default routers might be topologically closer to the final destination than the other routers and hence better utilized as a default router for that destination. If a host chooses a less optimal router as a next-hop, that default router can forward the packet but also send a Redirect message to the originating host, informing it of another address that should be used as a default router for that destination. In fact, the final destination may also be an on-link node, in which case the default router will inform the sending node that it should communicate directly to the destination without the need for a default router (provided that the link layer allows for direct communication). The format for the Redirect message is shown in Figure 2–29.

The *target address* field contains the IP address recommended as a next-hop for the destination address included in the *destination address* field. It should be noted that hosts might ignore Redirect messages for several reasons. For example, a host may ignore any unauthenticated Redirect message (not including the Authentication Header). For this reason, it is recommended that routers limit the rate of sending Redirect messages to avoid unnecessary waste

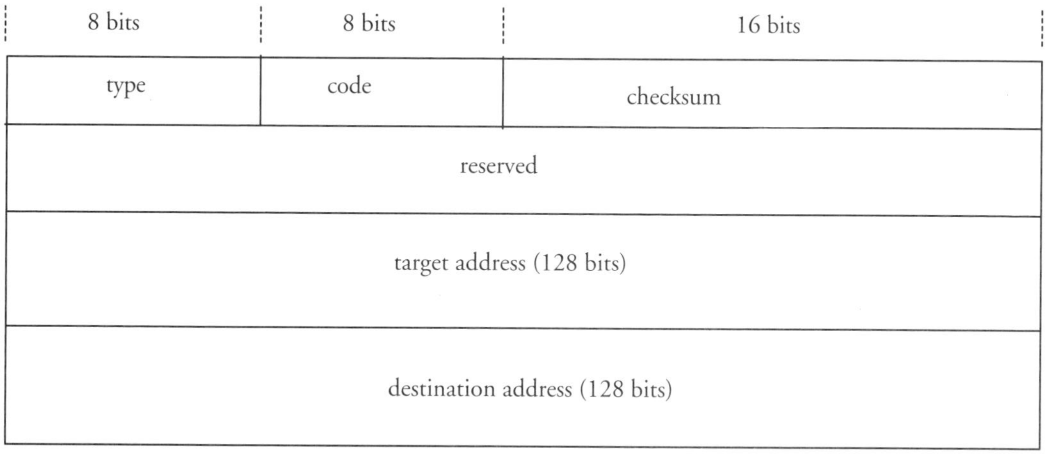

| 8 bits | 8 bits | 16 bits |
| --- | --- | --- |
| type | code | checksum |
| reserved | | |
| target address (128 bits) | | |
| destination address (128 bits) | | |

**Figure 2–29** *Redirect message format.*

of bandwidth or processing power. Rate limiting can be done by setting a value (low enough) for the amount of time that a router can send Redirect messages to a certain node or, for better granularity, by using a certain value for Redirect messages to a certain node for one destination address or an entire prefix.

#### 2.6.1.6 CONCEPTUAL DATA STRUCTURES IN HOSTS

After receiving router advertisements from one or more routers on-link and possibly neighbor advertisements, hosts will store important information extracted from the received advertisements. The main data structures discussed in relation to Neighbor Discovery are the following:

- *Neighbor cache*: The Neighbor cache contains an entry for each known neighbor on link. An entry is based on the neighbor's IP address. Each entry includes the reachability state of a neighbor, indicates whether NUD is being executed, and specifies the next time NUD should be executed. In addition, each entry includes the link-layer address of the neighbor and whether it is a router or a host.
- *Destination cache*: The Destination cache contains an entry, based on the IP address, for each destination to which packets have been sent recently. Each entry contains a mapping between the destination IP address and the next-hop that packets should be sent to (e.g., default router for off-link destinations or link-layer address for direct on-link communication) and the host's own interface that should be used to send packets; these entries include both on-link and off-link destinations. In addition, each entry may contain the PMTU to a destination and RTT measured by upper layers. This cache is updated by information received from router advertisements and Redirect messages.
- *Prefix list*: This list includes all prefixes that are valid on-link and their associated lifetimes. The well-known link-local prefix can also be stored in this list; however, its lifetime should never expire.
- *Default router list*: This list contains the IP addresses of all default routers. These addresses are extracted from router advertisements. Each default router is associated with a lifetime taken from router advertisements.

The above data is used to allow hosts to configure addresses and select next-hops for different destination addresses. In addition, this information allows hosts to determine whether a destination is located on-link, by matching the prefix of the destination address with the prefixes stored in the prefix list. If the destination is not on-link a host will choose a default router from its default router list to use it as a next-hop for off-link destination.

## 2.7 Stateless Address Autoconfiguration

The Neighbor Discovery protocol provides powerful mechanisms that allow nodes to learn all the necessary information about their links. Recall that a node needs to use an address of a larger scope than the link-local one when communicating with other off-link nodes. The prefix information option allows a node to configure an address (e.g., global) that allows it to communicate with other off-link nodes. Note that the prefix information option is not required to allow nodes to configure link-local addresses, since they already have a reserved prefix.

Using the prefix information, nodes can append an *interface identifier* to the advertised prefix to form a unique (within the scope of the prefix) IPv6 address and assign it to a certain interface. The interface identifier needs to be unique within the link. That is, since the prefix is unique within its scope (a global prefix is globally unique and a site-local prefix is unique within the site scope), appending an interface identifier to a prefix will give the node a unique IPv6 address. The uniqueness of the address is determined by performing the DAD procedure.

This address configuration process is called stateless address autoconfiguration. The term *stateless* refers to the fact that an address is configured without the need for keeping a record of such address allocation in any node (i.e., no state) except for the node that assigned the address to one of its interfaces. The entire process relies on distributed processing of the information related to the address assignment. This eliminates the need for a node that avoids address duplication by keeping records of all addresses allocated to each interface on a link.

RFC 2462 defines stateless address autoconfiguration for IPv6 hosts. It provides a method by which a node can use its 48-bit Ethernet address to generate an EUI-64 interface identifier. However, this RFC recommends that a node disable an interface when DAD fails the first time an address is tested. We will see in the next chapter that this action is too drastic and undesirable for mobile nodes. Fortunately, another RFC (RFC 3041) allows the node to generate a random number that can be used as an interface identifier. RFC 3041 recommends that nodes attempt to generate new interface identifiers upon DAD failure. This process can be repeated for up to five times until a unique interface identifier is found. This approach is mobility-friendly and preferable to the one specified in RFC 2462.

IPv6 allows for stateful address configuration using the Dynamic Host Configuration Protocol for IPv6 (DHCPv6). This protocol relies on having a DHCP server in the network that allocates addresses to hosts. DHCPv6 is also being currently extended to allocate prefixes to routers.

### 2.7.1 Ingress Filtering

Recall the discussion in Chapter 1 about route aggregation within the Internet backbone. To achieve route aggregation, nodes need to configure addresses that belong to the topology in which they are located. Hence, for an IPv6 node to have a topologically correct address, it needs to use the link's prefix to form this address. Therefore, the prefix information contained in the router advertisement is required by nodes to form an address.

Another reason for using the on-link prefix is *ingress filtering*. Ingress filtering is the process used by routers to allow traffic coming from outside the network to pass through the network—that is, to allow traffic sent by on-link hosts (arriving at a router's *ingress interface*) to be routed beyond the link. Routers typically set filters on their ingress interfaces to ensure that only nodes with topologically correct addresses can send packets beyond their links. This is necessary to avoid IP address spoofing attacks in which nodes may form a topologically incorrect address and use it to launch attacks on other nodes on the Internet. While ingress filtering would not stop nodes from stealing other topologically correct addresses, it would ensure that the location (link) from which an attack was launched is traceable.

## 2.8 A Communication Example

We can now use the knowledge gained from this chapter to illustrate the fundamental steps that need to be taken by an IPv6 host in order to send a packet to another host. In this section we take an example of an application that attempts to communicate with another application located on another host and look at the most important steps taken in the node's IPv6 layer in order to make this communication possible.

After the application resolves the peer's IP address (using the DNS or other means, like user's command), it requests to open a socket; this request includes the peer's IP address, the transport layer that the application wants to use, the peer's port number, and the IP version. The IP version is 6. Let's assume that the transport layer is TCP (the same steps are done for UDP). The application may also request the addition of certain extension headers to IP packets sent for this connection. The first step that the IPv6 implementation needs to do is select an appropriate source address that can be used for this connection. The source address selection will depend on the destination address. For instance, if the destination address is a global address, the IPv6 implementation cannot select a link-local or site-local address. The source address must be of the same scope or larger than the destination. When we discuss transition mechanisms in Chapter 9, we will see that other addresses, like 6-to-4, will need to be considered when a source address is being selected. The same point is made in Chapter 3 for Mobile IPv6 home addresses.

A more sophisticated source address selection algorithm may consider other factors, like interface preference. Every unicast IPv6 address is associated with an interface; therefore, source address selection is essentially an interface selection as well. On a multihomed host, some interfaces may be preferred to others based on the link-layer characteristics. Finally, it is important to note that source addresses must always be unicast addresses, regardless of their scope.

Figure 2–30 shows the main steps done by the IPv6 layer in order to send an application's data. When the IPv6 layer receives the data from the transport layer, an IP header is constructed based on the source and destination addresses in the socket. The *flow label* and *traffic class* fields may be set based on the application's preference. The *next header* field is set based on the next extension header included; if none are needed it is set to the protocol number of the transport layer protocol.

The need for extension headers is checked for each header. The need for extension headers depends on whether certain options need to be included in the packet; if so, the corresponding extension header is added.

Before the packet is sent, the SPD is checked for the need to include IPsec headers. If IPsec is needed for that particular destination address or for a particular protocol number included in the packet (e.g., all TCP connections to a particular destination address need to be protected by ESP), the SPD will point to the right entry in the SAD. The entry will indicate the type of protection needed (i.e., ESP, AH, or both), and the necessary protection will be executed.

Now the packet is ready to be sent; the final step is *next-hop determination*. In this example we have assumed that a multihomed host would implement a separate *destination cache* per interface. That is, since router advertisements are received on a per-interface basis, a multihomed host would have a default router per interface.

Hence, an implementation would first need to determine which *destination cache* should be searched to determine the next-hop for a packet. This can be determined based on the source address in the packet (recall that a unicast address is assigned to one interface). Following this, the on-link determination is done based on the destination's prefix; the sender will detect whether the destination is on-link (by matching its prefix to those prefixes in the prefix list), since the prefix option includes the $L$ flag to inform nodes whether a prefix is on-link. If the destination is on-link ($L = 1$), the host will check in the *neighbor cache* whether its link-layer address is known; if not, address resolution will take place to determine the link-layer address. If the link-layer address is known, the packet is passed to the link layer with the destination link-layer address.

If a destination address is not on-link or its on-link status is unknown (because $L = 0$), the host must forward the packet to its default router. Each

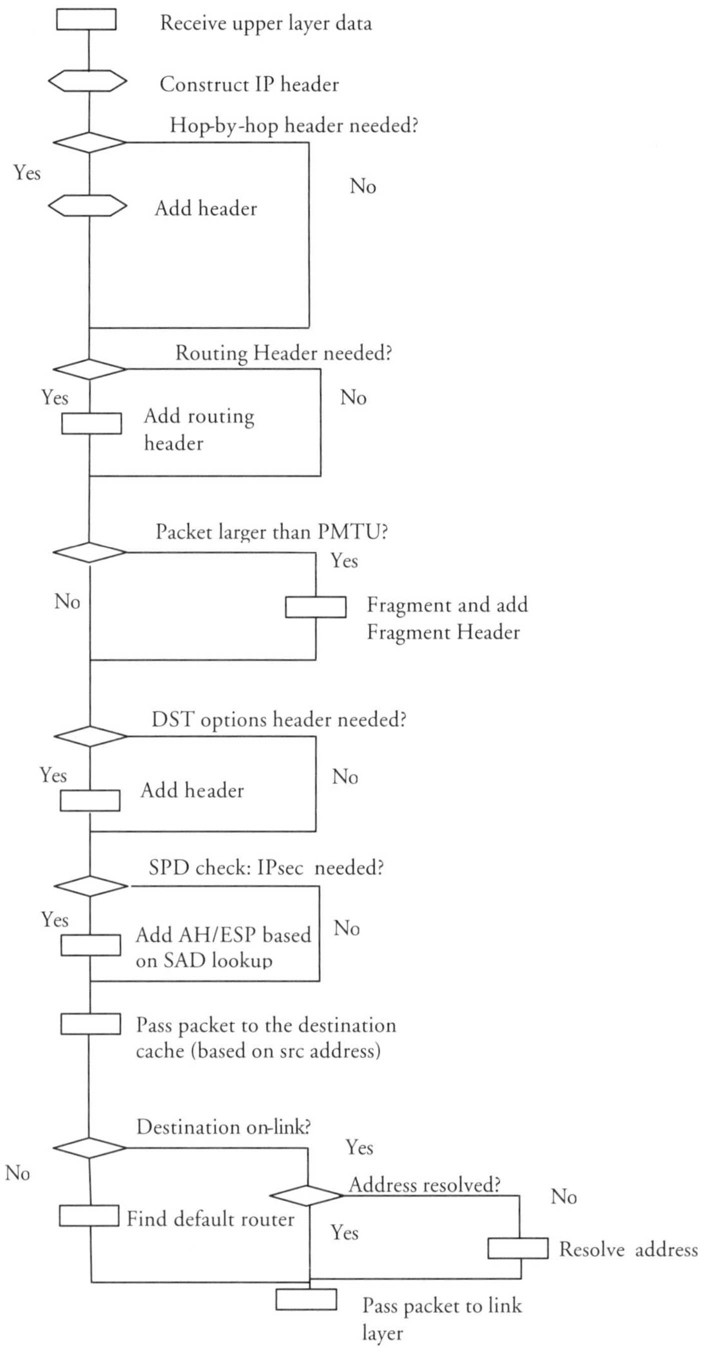


**Figure 2–30** *Main steps involved in sending packets from IPv6 hosts.*

interface has a list of its default routers and their link-layer addresses received in router advertisements. The packet is passed to the link layer with the default router's link-layer address.

## 2.9 Summary

This chapter presented a detailed introduction to IPv6, including the protocol headers, addresses, ICMPv6, tunneling, and Neighbor Discovery. Some of the features presented here (e.g., extension headers) are unique to IPv6, and some (e.g., tunneling and ICMP) are essentially upgrades of equivalent features in IPv4. The main reason behind this chapter is to familiarize you with IPv6 on a detailed enough level that you can more easily follow the rest of the book and understand the assumptions that were made during the development of Mobile IPv6.

Almost all of the knowledge gained in this chapter is needed in order to understand the operation of Mobile IPv6, which is described in Chapter 3.

## Further Reading


[1] Borman, D., S. Deering, and R. Hinden, "IPv6 Jumbograms," RFC 2675, August 1999.

[2] Conta, A., and S. Deering, "Internet Control Message Protocol (ICMPv6) for the Internet Protocol Version 6 (IPv6) Specification," RFC 2463, December 1998.

[3] Conta, A., and S. Deering, "Generic Packet Tunneling in IPv6 Specification," RFC 2473, December 1998

[4] Deering, S., and R. Hinden, "Internet Protocol Version 6 (IPv6) Specification," RFC 2460, December 1998.

[5] Draves, R., "Default Address Selection for Internet Protocol version 6 (IPv6)," RFC 3484, February 2003.

[6] Deering, S., W. Fenner, and B. Haberman, "Multicast Listener Discovery (MLD) for IPv6," RFC 2710, October 1999.

[7] Harkins, D., and D. Carrel, "The Internet Key Exchange (IKE)," RFC 2409, November 1998.

[8] Hinden, R., S. Deering, and E. Nordmark, "IPv6 Global Unicast Address Format," draft-ietf-ipv6-unicast-aggrv2-02, work in progress.

[9] Hinden, R., and S. Deering, "Internet Protocol Version 6 (IPv6) Addressing Architecture," RFC 3513, April 2003.

[10] Huitema, C., *IPv6 The New Internet Protocol*, 2nd ed. Prentice Hall, 1998.*

[11] IEEE, "Guidelines for 64-bit Global Identifier (EUI64) Registration Authority," http://standards.ieee.org/db/oui/tutorials/EUI64.html.

[12] Johnson, D., and S. Deering, "Reserved IPv6 Subnet Anycast Addresses," RFC 2526, March 1999.

[13] Kent, S., and R. Atkinson, "IP Authentication Header," RFC 2402, November 1998.

[14] Kent, S., and R. Atkinson, "IP Encapsulating Security Payload," RFC 2406, November 1998.

[15] Kent, S., and R. Atkinson, "Security Architecture for the Internet," RFC 2401, November 1998.

[16] McCan, J., S. Deering, and J. Mogul, "Path MTU discovery for IP version 6," RFC 1981, August 1996.

[17] Narten, T., and R. Draves, "Privacy Extensions for Stateless Address Autoconfiguration in IPv6," RFC 3041, January 2001.

[18] Narten, T., E. Nordmark, and W. Simpson, "Neighbor Discovery for IP Version 6 (IPv6)," RFC 2461, December 1998.

[19] Rajahalme, J., A. Conta, C. Carpenter, and S. Deering, "IPv6 Flow Label Specification," draft-ietf-ipv6-flow-label-07, work in progress.

[20] Savola, P., "Security for IPv6 Routing Header and Home Address Option," draft-savola-ipv6-rh-ha-security-03, work in progress.

[21] Thomson, S., and C. Huitema, "DNS Extensions to Support IP Version 6," RFC 1886, December 1995.

[22] Thomson, S., and T. Narten, "IPv6 Stateless Address Autoconfiguration," RFC 2462, December 1998.

* Some of the drafts and RFCs referenced in this book are now obsolete. When in doubt, always check the IETF's RFC index Web page (http://www.ietf.org/iesg/1rfc_index.txt).

PART TWO

# MOBILE IPv6

THREE

# Mobile IPv6

Consider a scenario where you had to change your place of residence on a semi-permanent basis, for instance, due to relocation of your company. One problem (of many) that you would need to solve is how you would continue to be reachable by your family, friends, bank, among other organizations that need to send you mail. In most countries, the post office can redirect your mail to your new residence for a fee. After the redirection request is processed by the post office, all mail addressed to your original residence will be forwarded to your new residence. On a very high-level, this is essentially how Mobile IPv6 provides reachability. Mobile IPv6 allows devices to be reachable and maintain ongoing connections (e.g., FTP, video, or audio) while moving within the Internet topology. To be reachable, mobile nodes are provided with a permanent IP address, called the *home address*. To maintain ongoing connections while moving, Mobile IPv6 uses the redirection function provided by the post office in our example; the Mobile IPv6 equivalent of the post office is called the *home agent*. The home agent redirects packets addressed to a mobile node's home address to its current location. Mobile nodes always update their home agent with their current location (i.e., every time they move).

In addition to reachability and maintaining ongoing connections, the protocol allows for optimal routing between mobile nodes and other nodes they are communicating with. That is, Mobile IPv6 allows a mobile device to communicate with another peer using the shortest possible path provided by IP routing.

In this chapter, we discuss the design details for Mobile IPv6 and the reasons behind the choices that were made during the protocol design. The following chapters discuss the security aspects of the protocol and the reasons for the current design. We start by discussing the basic operations required to allow mobile nodes to change their point of attachment within the Internet without dropping ongoing connections and while being reachable by potential correspondents. Following this, we show how mobile nodes can maintain knowledge about their home links (where the home address is located) while located anywhere on the Internet. Such features are intended to reduce the amount of manual configuration required in mobile nodes. Finally, we discuss how mobile nodes can optimize their communication with correspondent nodes by ensuring that packets take the shortest possible path between a mobile and a correspondent node. Before we get into the various details, we first discuss the terminology used in Mobile IPv6.

## 3.1 Mobile IPv6 Terminology

**Mobile Node (MN):** A mobile node is a node that changes its location within the Internet topology. A node's mobility could be a result of physical movement or of changes within the topology. That is, movement can be due to a device moving from one link to another because of its physical movement (e.g., a device in a car or a train) or because of changes in the topology that cause the device to be attached to a different router (e.g., router failure) while being in the same place physically. Mobile IPv6 is dedicated to host mobility; therefore, the term "mobile host" would have been more appropriate in this context. However, the Mobile IPv6 specification talks about "mobile nodes" (deceptively referring to hosts only), so we use the same term in this book to avoid further confusion.

**Correspondent Node (CN):** Any node that communicates with the mobile node. Note that the terms "mobile" and "correspondent" nodes refer to certain functions within an IPv6 node. Therefore, a mobile node can also be a correspondent node, and vice versa, depending on the context. For instance, a host can be seen as a correspondent node by the mobile (moving) node it communicates with. At the same time, the correspondent node might also move, which makes it a mobile node (because it is moving) while being seen as a correspondent node by its mobile peer.

**Home address:** A stable address that belongs to the mobile node and is used by correspondent nodes to reach mobile nodes. Like all IPv6 addresses, the home address is based on the 64-bit prefix assigned to the *home link* combined with the mobile node's interface identifier. A mobile node can have more than one home address. IP packets addressed to the home address are routed to the home link using standard routing protocols.

**Home link:** A link to which the home address prefix is assigned.

**Home Agent (HA):** A router located on the home link that acts on behalf of the mobile node while away from the home link. The *home agent* redirects packets addressed to a mobile node's home address to its current location (care-of address) using IP in IP tunneling.

**Foreign link:** Any link (other than the home link) visited by a mobile node.

**Care-of address:** An address that is assigned to the mobile node when located in a foreign link. This address is based on the prefix of the foreign link combined with the mobile node's interface identifier. There is no special format for a care-of address; it is a normal unicast IPv6 address that is configured through the stateless or stateful means described in Chapter 2. This address identifies the current location of the mobile node.

**Binding:** The association of the mobile node's home address with a care-of address for a certain period of time. That is, between the stable home address and the mobile node's current location. This allows the home agent (post office) to forward packets to the mobile node's current location. The binding is refreshed (if the timer expires) or updated when the mobile node gets a new care-of address (because it moved to a new link).

**Binding cache:** A cache stored in volatile memory containing a number of bindings for one or more mobile nodes. A binding cache is maintained by both the correspondent node and the home agent. Each entry in the binding cache contains the mobile node's home address, care-of address, and the lifetime that indicates the validity of the entry. When the binding cache is maintained by correspondent nodes, it also contains some security parameters, discussed in Chapter 5.

**Binding Update List (BUL):** A list maintained by the mobile node in volatile memory. This list contains all bindings that were sent to the mobile node's home agent and correspondent nodes. This list is maintained in order for the mobile node to know when a binding needs to be refreshed and is also used for selecting the right care-of address when communicating directly with a correspondent node. The details of the use of the binding update list are presented in this chapter.

## 3.2 Overview of Mobile IPv6

Mobile IPv6 was designed to allow nodes to be reachable and maintain ongoing connections while changing their location within the topology. To meet the requirements discussed in Chapter 1, mainly transparency to upper layers,

Mobile IPv6 uses a stable IP address, assigned to mobile nodes (home address). The home address is used for two reasons: first, to allow a mobile node to be reachable by having a stable entry in the DNS, and second, to hide the IP layer mobility from upper layers. We already saw in Chapter 1 that DNS caching implies that a node that frequently changes its IP address will not have the same IP address in all DNS servers because some of them will cache an old address until the caching period for such address expires. Therefore, in order for nodes to be reachable at the right address, the address should be stable and should not change every time they move. Hence, the need for the home address provided by Mobile IPv6. A consequence of keeping a stable address independently of the mobile node's location is that all correspondent nodes try to reach the mobile node at that address, without knowing the actual location of the mobile node. That is, whether the mobile node is physically located on the home link or not, packets are forwarded to the home link. If the mobile node is not at its home link, its home agent is responsible for tunneling packets to the mobile node's care-of address (i.e., its real location).

Since correspondent nodes attempt to connect to the mobile node's home address, sockets (in the mobile and correspondent nodes) use the home address to record such connections, so it is necessary that applications (on both nodes) see only the home address for the mobile node. Therefore, the IP layer is responsible for presenting the home address to applications running on the mobile node as a source address regardless of the mobile node's location. That is, the IP layer hides the mobility (address change) from upper layers to maintain ongoing connections while the mobile node changes its address. If the address actually changed, the information in the sockets would become invalid and connections would automatically terminate.

The home address is formed by appending an interface identifier to the prefix advertised on the home link. Like any IPv6 node, a mobile node can be assigned several addresses of different scopes. This applies to both the home and care-of addresses.

While at home, a mobile node operates like any other IPv6 node. It receives packets addressed to any of its home addresses and delivered via normal routing. Mobile IPv6 is only invoked when a mobile node is not "at home"—that is, when the mobile node is located on a foreign link.

When a mobile node moves from its home link to a foreign link, it first forms a care-of address based on the prefix of the foreign link. The care-of address can be formed based on stateless or stateful mechanisms. However, this book focuses on stateless address autoconfiguration, as it is expected to be the most widely used mechanism for mobile nodes. Following address configuration, the mobile node informs its home agent of such movement by sending a *Binding Update* (BU) message. The binding update message is one of several Mobile IPv6 messages that are encoded as options in a new header called the *mobility header*. The mobility header is the last header in the chain of extension headers and appears as an upper layer protocol (see section 3.2.1).

The binding update message contains the mobile node's home address and its care-of address. The home address is included in a new option called the *home address option*, and the care-of address is included either in the source address in the IP header or in a new option called the *alternate-care-of address option*. The home address option is part of the *destination options* extension header, while the alternate care-of address option is included in the mobility header. Chapter 5 and Chapter 7 discuss some of the typical uses of the alternate-care-of address option. For now, we only need to know that the mobile node's care-of address can be included in the source address in the IPv6 header or the alternate-care-of address option.

The purpose of the binding update is to inform the home agent of the mobile node's current address (i.e., care-of address). Therefore, the home agent needs to store this information in order to forward packets addressed to the mobile node's home address. The home agent contains a *binding cache*, which contains all bindings for the mobile nodes it serves. Each entry in the binding cache stores a binding for one home address. When the home agent receives the binding update, it performs a number of actions to validate the message; if the binding update is accepted, the home agent searches its binding cache to see if an entry already exists for the mobile node's home address. If an entry is found, the home agent updates that entry with the new information received in the binding update. Otherwise, if no entry is found, a new one is created. From this point onward, the home agent acts as a proxy for the mobile node on the home link. That is, it represents the mobile node on the home link. The home agent essentially becomes the post office offering a forwarding service to the mobile mode. To ensure this representation is understood by all nodes on the home link, the home agent sends a *proxy neighbor advertisement* addressed to the all-nodes multicast address on the link. The advertisement includes the mobile node's home address in the *target address* field and the home agent's link-layer address. Hence, the home agent ensures that any IP packet addressed to the mobile node is forwarded to the home agent's link-layer address. Since the recipients of this message cannot confirm reception to the home agent, it may resend the neighbor advertisement more than once to reduce the probability of the message being lost. The home agent also defends the mobile node's addresses in case another node configures an address that collides with a mobile node's home address (or addresses). Hence, if another node tentatively configures one of the mobile node's addresses and attempts to test it using DAD, the home agent replies to the message, indicating that the address is already assigned to another node.

At this stage, the home agent starts receiving all packets addressed to the mobile node. Upon receiving a packet addressed to the mobile node's home address, the home agent checks its binding cache to see whether an entry for the mobile node exists. The home agent searches the cache by using the mobile node's home address (the destination address in the received packet) as a key to identify each entry. When an entry is found, the packet is tunneled to the mobile node's care-of address, which is included in the mobile node's binding cache entry. The *tunnel entry point* is the home agent (source address in the outer header), and the *tunnel exit point* is the mobile node's care-of address. The tunnel is bidirectional. That is, when the mobile node sends any IP packets, it tunnels them first to the home agent, which decapsulates the packet and forwards the original one to its destination. Figure 3–1 illustrates the binding update messages and data forwarding (for ongoing connections) between the mobile node and its home agent.

The Mobile IPv6 specification prohibits the home agent from tunneling packets addressed to the mobile node's link-local address. Furthermore, the specification recommends that the default behavior of the home agent should be set to not tunnel packets addressed to any of the mobile node's site-local addresses. The only exception to these rules are packets sent by or to the mobile node that are needed for multicast group membership. The Multicast Listener Discovery (MLD) protocol requires nodes to use their link-local addresses when attempting to join multicast groups. Therefore, MLD packets need to be tunneled by the home agent.

Tunneling is required to ensure the transparency of the service provided by the home agent. This is needed to preserve the end-to-end nature of IP packets exchanged between the mobile node and correspondent nodes. Recall that routers must not modify the content of the source or destination addresses in the IP header, thereby preserving the integrity of the packet and allowing for end-to-end integrity checks (e.g., Authentication Header). Furthermore, tunneling is essential to maintain transparency for upper layers. Consider the case where, instead of tunneling, the home agent rewrites the destination address in the packet (i.e., replace the home address with the care-of address). A direct result of this action is that the packet's integrity is compromised, causing the Authentication Header (if present) to fail. In addition, the mobile node receives the packet and passes it to upper layers (e.g., TCP), which process the packet assuming that the care-of address is used to identify the connection. However, the correspondent node identifies the connection by its address, source port number, the mobile node's **home** address, the mobile node's port number, and the protocol number. Hence, if the mobile node replies to the correspondent node directly, using its care-of address, the correspondent node is unable to find the connection and consequently drops the packets.

The home agent may choose to secure the tunnel to the mobile node using an Authentication Header or ESP, depending on local policies within the home network. For instance, within a corporate network containing a home agent, other nodes within the corporate will be unaware of the mobile node's movement away from the home link, as they continue to send packets to a mobile node's home address. Nodes would typically assume that within a wired corporate network, IPsec is not required (door lock security!) due to the difficulty involved in penetrating corporate security (physically) and the typical use

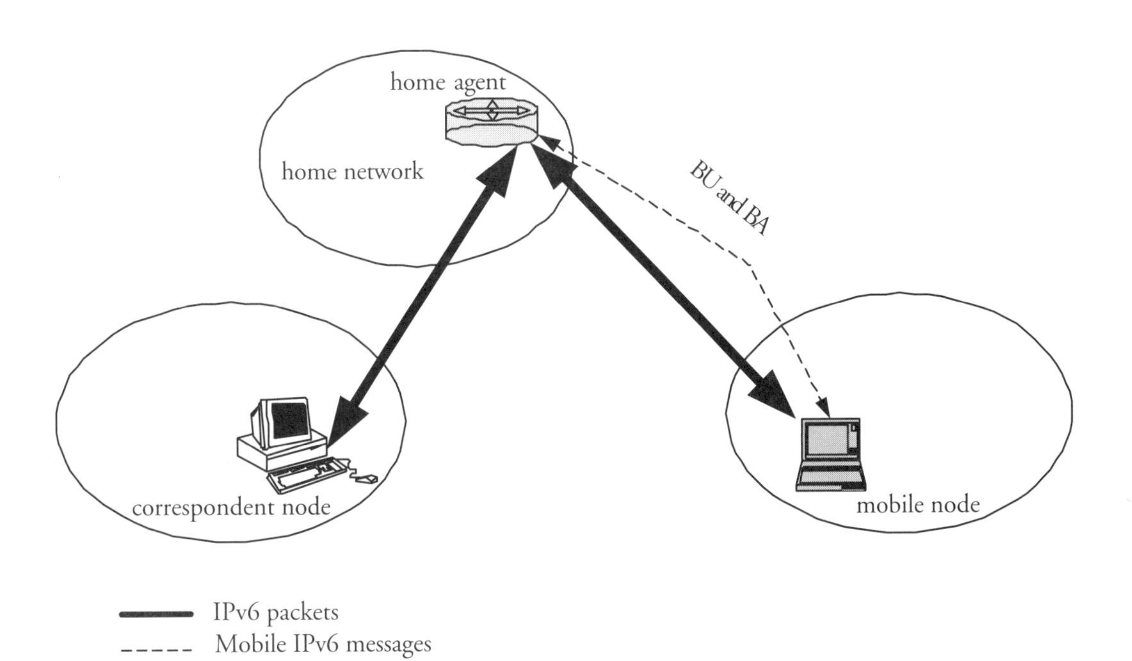


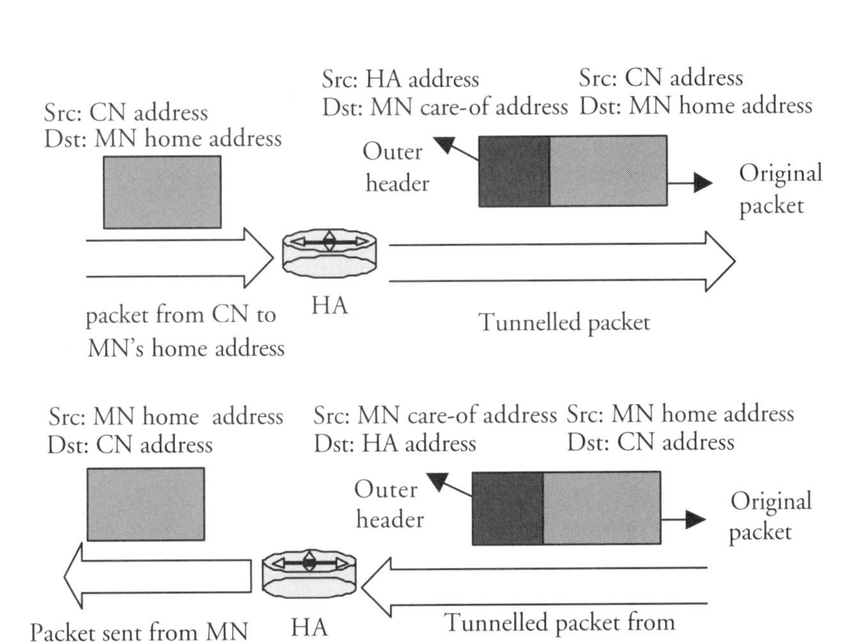


**Figure 3–1** *Mobile IPv6 operation and data forwarding.*

of Firewalls on the edge of the network (more on Firewalls in Chapter 4). In this case, a home agent can be configured to use ESP to protect all traffic tunneled to a mobile node that moved away from the home link located within the corporate network.

### 3.2.1 Binding Updates and Acknowledgments

It is crucial for the mobile node to ensure that it has a binding with its home agent. If the binding is lost or the binding update message was not received, the mobile node will be unreachable while away from home. Hence, a reliable protocol is required to install a binding in the home agent's binding cache. The home agent is required to acknowledge the binding update message sent by the mobile node. Figure 3–2 shows the binding update message format.

Mobile IPv6 defines a new IPv6 extension header, the *mobility header*. The mobility header is used to carry all Mobile IPv6 messages. It contains four permanent fields (included in all messages): *payload proto, header length, mobility header type,* and *checksum.*

The *payload proto* field indicates the type of the following header, which makes you wonder what the difference is between this field and the next header field in all IPv6 extension headers: there is no difference, it was just named differently! According to the current specification, this field is always set to the decimal value 59, indicating that the mobility header is the last header.

The header length field includes the length of this extension header in 8-octet units, excluding the first 8 octets.

The *mobility header type* field is used as a switch to indicate which message is included in the mobility header. When a binding update is included in the mobility header, this field is set to 5. It should be noted that like all extension headers, the mobility header may carry other options after the message indicated by the mobility header type field. For instance, message type 5 (the binding update) can also include the alternate-care-of address option after the binding update message.

| 8 bits | 8 bits | 8 bits | 8 bits |
|---|---|---|---|
| Payload proto | header len | MH Type | reserved |
| checksum | | sequence number | |
| A \| H \| L \| K reserved | | lifetime | |

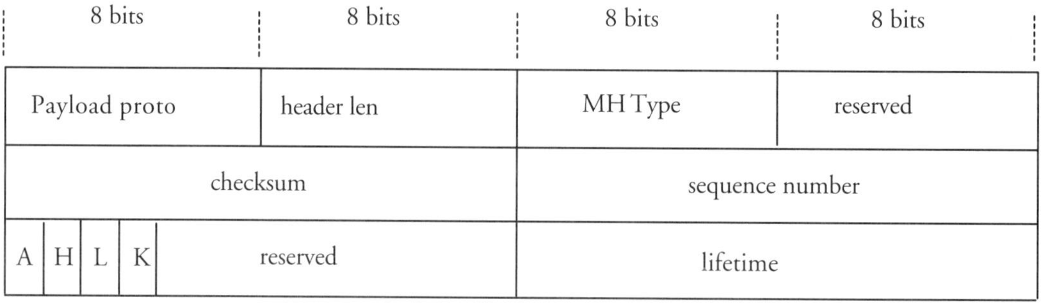

**Figure 3–2** *The binding update message format.*

The *checksum* field includes a checksum value for the entire mobility header in addition to the IPv6 header. The calculation of the checksum is slightly different when the *home address option* is included in the destination options header. We discuss this when we discuss the home address option in section 3.3.

The *sequence number* field contains a 16-bit integer used to ensure that binding updates are received and acknowledged in order. The mobile node stores the value of the last used sequence number per destination. Whenever a binding update is sent to a particular destination, the sequence number is incremented. Hence, this number represents a cyclic counter used to allow a window for the expected sequence number. That is, since the sequence number space is finite, it is inevitable that it will be repeated. However, this field ensures that consecutive binding updates are received in the correct order.

The *A* flag indicates whether an acknowledgment is expected for this binding update. When sending the binding update to the home agent, the *H* and *A* flags are set to indicate that the binding update is sent to a home agent (the *H* flag) and that an acknowledgment is required (the *A* flag).

The *L* flag is used by the mobile node to indicate (when set) to the home agent that its link-local address' interface identifier is the same as that included in its home address. Hence, the home agent can defend the link-local address by appending the interface identifier to the well-known link-local prefix.

Binding updates need to be integrity-protected to avoid Bad Guy from stealing a mobile node's traffic by sending a binding update on its behalf. We discuss the security issues in detail in chapters 4 and 5. The *K* flag indicates whether the protocol used to establish a security association between the mobile node and its home agent must be rerun every time the mobile node moves. If the *K* flag is cleared, the protocol will not survive movement; that is, the security association will need to be reestablished when the mobile node gets a new care-of address. Otherwise, if the flag is set, the protocol will survive movement.

The *lifetime* is a 16-bit field that indicates the requested lifetime (in 4-second units) for the binding cache entry created by the receiver of the binding update. When this value is set to zero, it indicates a request to delete the binding cache entry for a home address. A binding cache entry remains valid for the duration of the lifetime field; if the mobile node does not refresh the binding cache entry, the entry is deleted after the lifetime expiry.

When the home agent receives the first binding update from a mobile node, it performs DAD for the mobile node's home addresses included in the binding update. If DAD fails, the mobile node is informed (when receiving the binding acknowledgment) and is unable to use the home address. Duplicate address detection is a useful tool for avoiding address collisions while the mobile node is away. For instance, while the mobile node is moving away from its home link, another node might be arriving at that home link and forming an address that collides with the mobile node's home address. Although this scenario is highly unlikely (considering the probability of collision for a 64-bit interface identifier), DAD would detect this case. A more likely case of address collision could take place if the mobile node's binding has expired and not been updated for some time (for instance, the mobile node is turned off). During this period, another node could visit the mobile node's home link and generate an address that collides with the mobile node's home address.

As a side note, it is worth mentioning that DAD is not performed when the mobile node is removing a binding cache entry (i.e., lifetime is set to zero). In this case the mobile node is sharing the link with the home agent; therefore, DAD can cause an infinite loop because both the mobile node and the home agent will defend the same addresses. We revisit this scenario in section 3.2.5.

After a successful DAD operation, and provided that there is nothing stopping the home agent from accepting the binding (e.g., wrong message format or lack of memory resources), the home agent copies the contents of the message into an existing binding cache entry or creates a new one if this was the first binding update. If a binding cache entry already exists, the home agent updates it with the contents of the new message. Hence, a binding update message overwrites an existing binding, resulting in a single entry per home address. Table 3–1 shows an example of a binding cache.

**Table 3–1** *Binding Cache Entries in the Home Agent*

| Home address | Care-of address | Sequence no. | Lifetime | Flags |
|---|---|---|---|---|
| `3ffe:200:8:1:A:B:C:D` | `3ffe:200:1:5:A:B:C:D` | 11 | 250 | A/H/K/L |
| `3ffe:200:8:1:D:E:F:9` | `3ffe:100:3:1:D:E:F:9` | 2000 | 400 | A/H/L |

For the home agent to be able to tunnel packets to the right care-of address, the address must be contained in the binding cache entry associated with the mobile node's home address. The care-of address can be retrieved through two different fields: the source address included in the IP header or an alternate-care-of address option included in the mobility header containing the binding update. Retrieving the address from the source address field in the header is clear: the home agent can simply copy the source address in the IPv6 header, to the binding cache entry. However, in some cases (discussed in the chapters 5 and 7), a mobile node may wish to associate the binding with a care-of address other than the one included in the source address field in the IP header. For those cases, Mobile IPv6 allows the mobile node to include the care-of address in a separate option, the alternate-care-of address option. The option format is shown in Figure 3–3.

The alternate-care-of address option simply consists of an IPv6 address that the mobile node wishes to use as a care-of address. The *option type* is a number that informs the receiver about the content and format of the option. The *option length* indicates the number of bytes following this field.

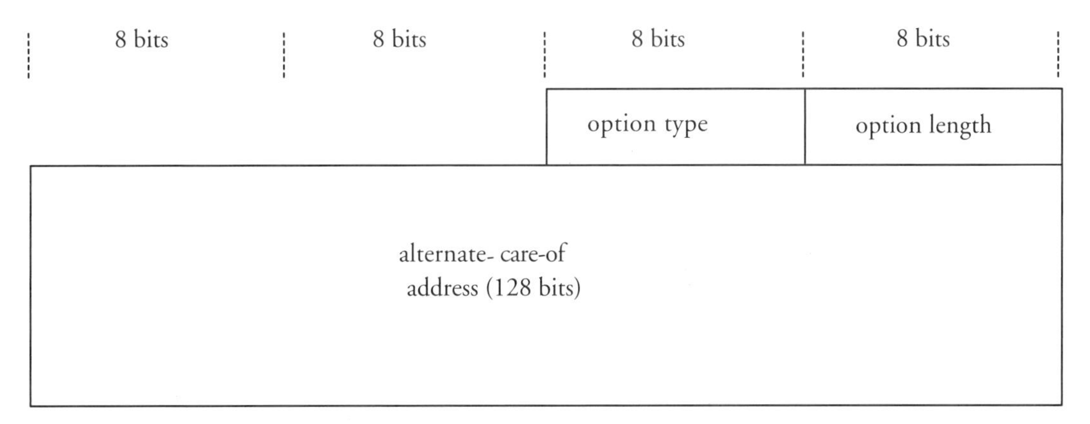


Figure 3–3 *The alternate-care-of address option format.*

Binding updates sent to the home agent must request an acknowledgment to ensure that a mobile node receives confirmation for the bindings it has established with the home agent. The *binding acknowledgment* message provides this information to the mobile node by including a *status* field that indicates the success or failure of the binding. Values below 128 indicate success; the rest of the values indicate the reasons for failure. The message also contains a *K* flag that indicates (when cleared) that the protocol run between the home agent and mobile node to secure binding updates does not survive movement.

The binding acknowledgment (Figure 3–4) synchronizes the lifetime of the binding between the mobile node and its home agent. A home agent might reduce the lifetime proposed in the binding update, subject to loading or other policies. Hence, the lifetime fields in the binding update and acknowledgment messages may not have the same values. If the lifetime of the binding is changed in the binding acknowledgment, the mobile node must store the new value returned by the home agent and ignore its own suggestion sent in the binding update message.

Finally, the binding acknowledgment contains the sequence number of the binding update being acknowledged in order to allow the mobile node to track which binding update is being acknowledged. For instance, the mobile node may have sent more than one binding update to its home agent and therefore would need to know which one is being acknowledged in order to be able to store the right values and send the next binding update with the correct *sequence number.*

Mobile nodes continue to retransmit the binding update (increasing the sequence number for each retransmission) to the home agent until an acknowledgment is received. However, such retransmission is done based on an exponential back-off algorithm. When using this algorithm, the mobile node doubles the time between retransmissions every time a binding update message

| 8 bits | 8 bits | 8 bits | 8 bits |
|---|---|---|---|
| next header | header len | MH Type | |
| checksum | | status | K / reserved |
| sequence number | | lifetime | |

**Figure 3–4** *Binding acknowledgment message format.*

is sent. The initial timeout value depends on whether the binding update sent was the first one to the home agent or a refresh of an existing binding. The first binding update would require more time, since the home agent performs DAD for the mobile node's home address (possibly 1 to 2 seconds). If the binding update were sent to refresh an existing binding, the initial timeout period is set to a default value of 1 second. The retransmission with exponential back-off (1 second, 2 seconds, 4 seconds, and so on) continues until the mobile node receives a binding acknowledgment or until a maximum timeout value (256 seconds) is reached. If the maximum timeout value is reached, a mobile node's implementation may repeat the entire algorithm or select another home agent using the Dynamic Home Agent Address Discovery (DHAAD) mechanism (described in section 3.2.7).

This retransmission algorithm is needed to minimize congestion in the network or unnecessary waste of limited bandwidth. The assumption behind this approach is that binding acknowledgments were not received because the binding update was lost (most likely due to congestion) or due to a home agent failure. Hence the need for back-off to avoid increasing the potential congestion.

After receiving the binding acknowledgment from the home agent, the mobile node needs to ensure that such information is stored to be able to refresh the binding before it expires. For this purpose, the mobile node maintains a data structure called the *binding update list*. This is equivalent to the binding cache maintained by the receivers of the binding update and contains the same information. Hence, the lifetimes in the mobile node's binding update list and the home agent's binding cache are synchronized (through the binding update and acknowledgment messages). Using the binding update list, the mobile node monitors the lifetimes of its bindings and refreshes them before they expire.

### 3.2.2 Refreshing Bindings

As described earlier, bindings are valid for a lifetime included in the binding update message. Mobile nodes should refresh bindings by sending another binding update before they expire or when the mobile node's care-of address changes. Mobile IPv6 allows the receiver of the binding update to request that the mobile node update its binding entry. This is done using the *binding refresh request*. This message requests that the mobile node update its current binding because it's about to expire. This message can be sent by the mobile node's home agent or a correspondent node communicating directly with the mobile node (section 3.3 talks about direct communication between the mobile and correspondent node).

The benefit of this message is not clear. Since the mobile node already knows (from its own binding update list) that it has a binding entry in its home agent's (or correspondent node's) binding cache, it should not need a reminder unless the mobile node's implementation contains bugs!

Like all Mobile IPv6 messages, the binding refresh request is carried inside the mobility header.

### 3.2.3 Why Reverse Tunneling?

So far we have discussed the need for tunneling from the home agent to the mobile node. To maintain the transparency of mobility to upper layers, *reverse tunneling* (from the mobile node to the home agent) is also needed. Recall that maintaining transparency to upper layers requires the mobile node to use the home address as the source address. If the mobile node sends packets directly to the correspondent node using the home address as a source address, ingress filtering causes the mobile node's packets to be dropped by the default router, since the source address is not derived from the foreign link's prefix. Ingress filtering problems are eliminated when reverse tunneling to the home agent is used. This is because the outer header (inspected by the default router in the foreign link) contains the mobile node's care-of address, which is topologically correct. When the home agent decapsulates the mobile node's packets, the source address in the inner header (the home address) belongs to the home agent's link's topology (prefix), and normal forwarding takes place.

### 3.2.4 Movement Detection

As discussed earlier, the mobile node should always inform its home agent of its current location (the care-of address). This is done using the binding update/acknowledgment messages. Consider a mobile node moving from one link to another; its home agent continues to forward packets to its previous care-of address until the mobile node updates it about its movement. This results in packet losses, as illustrated by Figure 3–5. Therefore, it is important for the mobile node to update the home agent as soon as movement to a new link

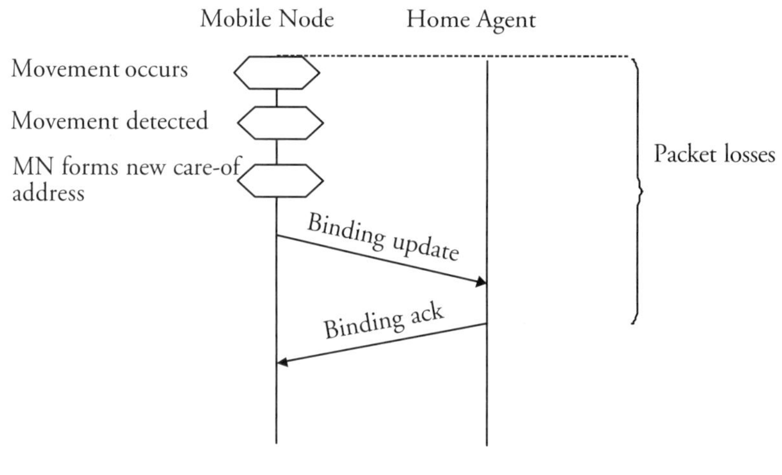


Figure 3–5 *Movement detection and associated packet losses.*

is detected. However, the mobile node needs to configure a new care-of address before it can update its home agent. This can be efficiently done using stateless address autoconfiguration. Hence, the following actions are required from the mobile node:

1. Detect movement.
2. Form a new care-of address.
3. Inform the home agent by sending a binding update containing the new care-of address.

To minimize packet losses, the first two steps need to be executed as quickly as possible. We now discuss some of the steps taken by Mobile IPv6 to accelerate movement detection.

Humans are aware of their physical movement from one location to another by simply looking around and noticing the changes in the surroundings, street names, addresses, and buildings. In a similar manner, a movement within the Internet topology implies a change of addresses. Each link has at least one unique prefix that identifies it and is advertised by one or more default routers on that link. Hence, if a mobile node notices a change in the link's prefix based on a new router advertisement, it can use that as a hint indicating that topological movement may have occurred. However, it is important to note that a new prefix advertised does not necessarily imply movement. The Neighbor Discovery specification allows routers to advertise more than one prefix on the same link. Furthermore, having one prefix option appear while another one disappears does not imply that the old prefix is no longer valid. Routers may be trying to reduce the size of a router advertisement by advertising a subset of the options whose timers are about to expire; this is perfectly legal according to Neighbor Discovery. Therefore, a mobile node can only be certain about movement when two events have taken place:

1. A new prefix has appeared on link, and
2. The current default router has disappeared.

The first step is achieved when the mobile node receives a router advertisement containing a new prefix option. However, the amount of time that passes before the reception of the router advertisement clearly depends on the frequency of the advertisements. To minimize this time (and the resulting packet losses), Mobile IPv6 relaxes the minimum interval between router advertisements to 0.05 seconds (as opposed to 3 seconds in standard Neighbor Discovery). In addition, the specification defines a new *advertisement interval* option (see Figure 3–6) for inclusion in router advertisements. This option informs mobile nodes about the maximum interval between router advertisements in milliseconds.

A mobile node receiving this option with an *advertisement interval* value of 0.1 seconds would realize that it may have moved if the router advertisement was not received on time and send a router solicitation for a new advertisement.

After noticing that a new prefix has appeared on link, the mobile node must check if it still shares a link with its default router. At this point another complication is raised: routers typically use their link-local address as a source address when sending router advertisements, which is only unique within the link. Therefore, different network operators may configure routers with the same link-local address, causing movement detection to fail (i.e., the mobile node would think that it is still attached to the same router). For this reason, the *R* flag was added to the prefix option (see section 2.6.1.2). The *R* flag informs the receivers of the option that the option includes the global address of the router (i.e., the last 64 bits are set to the router's interface identifier). Since

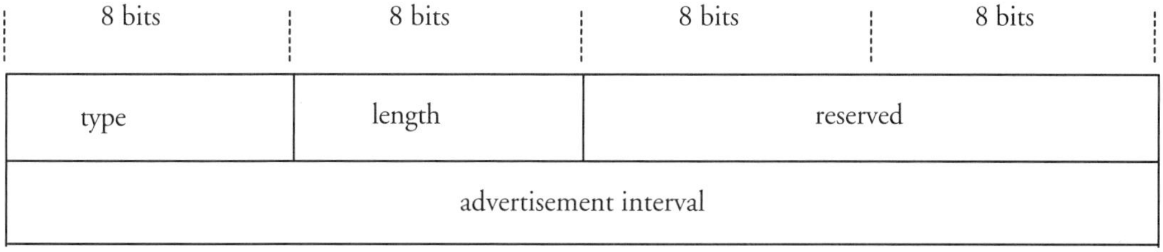

| 8 bits | 8 bits | 8 bits | 8 bits |
|---|---|---|---|
| type | length | reserved | |
| advertisement interval | | | |

Figure 3–6 Advertisement interval option.

the global address is globally unique, the mobile node knows when it has changed default routers.

However, the *R* flag does not solve the problem for all configurations. Consider Figure 3–7, where a mobile node is located on a link that has more than one default router, router_1 and router_2, connected. IPv6 specifications do not require all routers on a link to advertise the same prefixes. That is, router_1 may advertise one prefix and router_2 may advertise another prefix. If the mobile node initially configures its care-of address based on router_1's advertised prefix, then later receives an advertisement from router_2, it might think that it moved when in fact it did not. The *R* flag would not help, since each router has a different interface identifier. There are two ways to avoid this scenario:

1. Configure all routers on a link to advertise the same prefixes.
2. Design a new option that should be added to the router advertisement (e.g., link identifier option). This option could include a globally unique address that identifies the link. All routers on a link would have to include this option regardless of the prefixes they advertise. This would allow movement detection to work when several routers are connected to a link.

Another issue to consider when addressing movement detection is the possibility of gaining hints from lower layers. This can be discussed in light of Figure 3–8.

In Figure 3–8 a mobile node moving from Cell_1.2 to Cell_2.1 might get an indication from lower layers that an "intercell" movement has taken place before it detects its movement from router advertisements. While such an indication does not necessarily imply a topological movement, it can prompt the

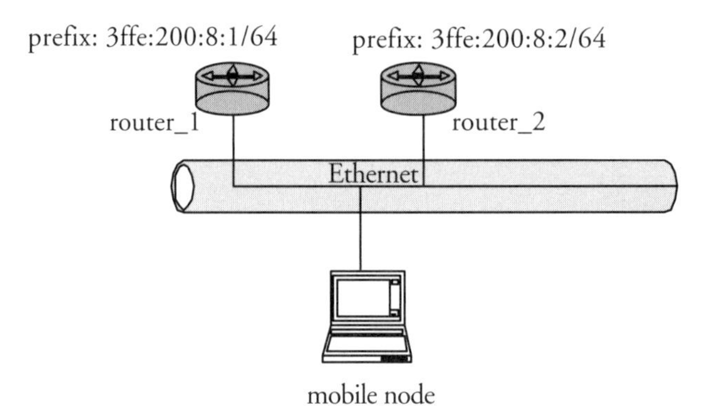


**Figure 3–7** *Different routers advertising different prefixes on the same link.*

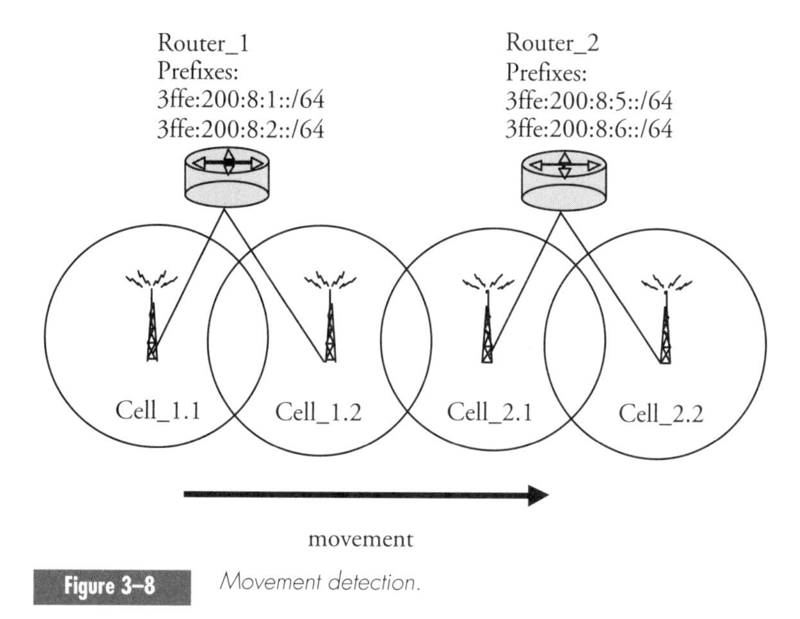


**Figure 3–8** *Movement detection.*

mobile node to solicit for a router advertisement. Generic lower-layer hints should not replace the need for movement detection on the IP layer; however, they can provide useful optimizations. In Figure 3–8, a mobile node might receive a lower-layer indication when moving from Cell_1.1 to Cell_1.2 or when moving from Cell_2.1 to Cell_2.2. Those cells are attached to the same router. Therefore, movement between those two cells does not constitute topological movement and should not invoke IP layer mobility. Clearly a link layer may be integrated with the IP layer so as to provide specific triggers that indicate IP layer movement (e.g., when a link is configured such that intercell movement is always associated with IP movement). However, this a special optimization (i.e., not a generic assumption that can be made in an IP layer protocol, which applies to all link layers).

### 3.2.5 Returning Home

We have discussed the operation of the protocol between the mobile node and its home agent when the mobile node is located in a foreign network or is moving between two foreign networks. When the mobile node returns to its home network, it needs to inform the home agent that it should stop receiving its packets on its behalf. To achieve this, the mobile node must send a binding update to the home agent with a lifetime of zero and a care-of address (source address) equal to the mobile node's home address, indicating that the home agent should no longer receive the mobile node's traffic. Here we face a chicken-and-egg problem: the mobile node needs to configure its home addresses to be able to send the binding update. However, the home agent still assumes that it should defend the mobile node's home addresses, which would cause DAD to fail and prevent the mobile node from configuring its home address. The mobile node also needs to learn the home agent's MAC address (on link layers with MAC addresses, like Ethernet). This is done by sending a *neighbor solicitation* with a *target address* field set to the home agent's global IP address. The destination address is set to the home agent's *solicited node multicast address* and the source address is set to the unspecified address (::). This message looks like a DAD message, causing the home agent to respond with a neighbor advertisement to the *all-nodes multicast address* (since the solicitor's address was not provided) with its unique address and its link-layer address.

To avoid this problem, the mobile node sends a binding update with its home address as a source and the home agent's address as a destination. This is done without performing DAD on the mobile node's home address (an exception to address autoconfiguration). The mobile node then waits for an acknowledgment from the home agent. As soon as the home agent processes the binding update, it will no longer defend the mobile node's home address. Therefore, it will be able to send a binding acknowledgment to the mobile node's home address. Note that in order for the home agent to send the binding acknowledgment, it may need to send a neighbor solicitation to the mobile node's home address. The mobile node must respond to this solicitation from the home agent, despite not performing DAD on its home address—that is, another exception from normal address autoconfiguration. This process is shown in Figure 3–9.

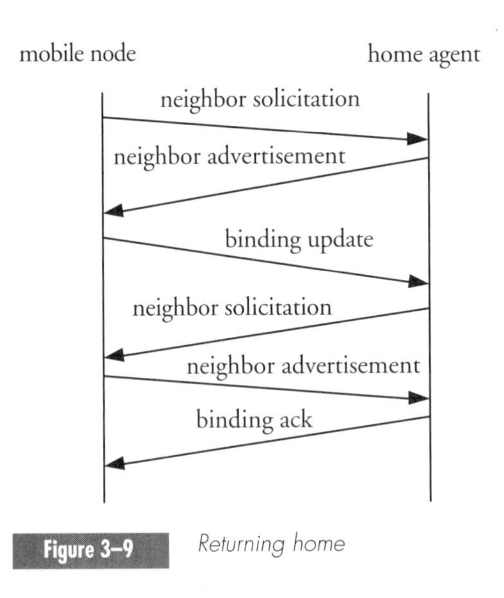


**Figure 3–9** *Returning home*

When the binding acknowledgment is received, the mobile node sends a neighbor advertisement to the *all-nodes multicast address* with the *O* flag set. The purpose of this advertisement is to inform all nodes that they should send traffic directly to the mobile node (i.e., override the home agent's earlier proxy advertisements). To minimize the probability of such advertisement being lost, the mobile node may send it more than once. At the end of this procedure, the mobile node should receive all its traffic from other nodes on-link directly without passing through the home agent.

### 3.2.6 Source Address Selection in Mobile Nodes

Mobile IPv6 provides transparent mobility to upper layers by providing a fixed home address to applications, independent of the mobile node's location within the Internet topology. To achieve this, an anchor point within the home network (the home agent) is needed to provide the forwarding service to the mobile node's care-of address.

However, in some cases there may be a need to provide a care-of address to applications. For instance, a mobile node would typically need to communicate with a DNS located in the same network (a foreign network). If the mobile node uses a home address as a source, all DNS queries and replies would pass through the home agent. This can cause unnecessary delays. Furthermore, since such queries represent a very short-lived connection, during which the mobile node is not expected to move, the mobile node is likely to be able to use its care-of address to communicate directly with the DNS. The DNS is only one example, and there may be many types of short connections that can use the care-of address successfully (e.g., Web browsing). The choice between the use of home and care-of addresses as a source address is left to the discretion of the implementation. However, the use of a care-of address for connections that are known to be short (one way of knowing this is by looking at well-known port numbers for applications, e.g., DNS) would provide better performance.

It is also possible to add some information in the sockets API to allow applications to request the type of address needed for their connections. However, an implementer cannot assume that all applications will have such intelligence (to request a care-of address or a home address). Therefore, a good default source address selection algorithm is needed (see [2] for further details).

### 3.2.7 Dynamic Home Agent Discovery

One of the details that was bypassed in our earlier analogy with the mail-forwarding service was how to locate the post office. One would need to locate the post office to be able to request a forwarding service. In a similar manner, a mobile node needs to locate a suitable home agent. The home agent's address can be configured in the mobile node (in a nonvolatile memory to survive reboots). However, this approach has some drawbacks. Addresses change, or the home agent may fail or simply get overloaded if too many mobile nodes register with it. To avoid these problems, Mobile IPv6 provides a *Dynamic Home Agent Address Discovery* mechanism that allows mobile nodes to discover home agents' addresses. In addition to address discovery, DHAAD allows home agents to share the load between them, in cases where multiple home agents are located on the same link, by utilizing a *preference* parameter. This parameter is included in a new option, the *home agent information option*, which is included in router advertisements sent by home agents and shown in Figure 3–10.

The option includes a *preference* field with a default value of zero. Larger values indicate higher availability of the home agent. This option can only be included if the router is also a home agent (i.e., when the *H* flag is set in the router advertisement).

The *home agent lifetime* is used to indicate how long a router can serve as a home agent. This field must never be set to zero. If a router cannot serve as a home agent, it should clear the *H* flag in the router advertisement and stop advertising this option.

When other home agents receive this option, they store the home agent's address, its preference, and its lifetime. If a home agent sends router advertisements that do not include this option, its preference is assumed to be zero. Hence, every home agent on a link keeps a list (the *home agents list*) containing an IP address of each home agent on the link and its preference. Home agents can change their preference values dynamically and communicate them in router advertisements, depending on their availability or other load-related parameters.

The next step is to communicate the information in the home agents list to mobile nodes; this is done using DHAAD messages. DHAAD messages are carried in two ICMP messages and consist of a *DHAAD request* and *DHAAD reply* message. The mobile node sends a message requesting a list of global IP addresses for possible home agents on its home link. The message is sent to the *home agents' anycast address*; the anycast address is formed by appending the special interface identifier reserved for home agents' anycast addresses (see

| 8 bits | 8 bits | 8 bits | 8 bits |
|---|---|---|---|
| type | length | reserved | |
| home agent preference | | home agent lifetime | |

Figure 3–10 Home agent information option.

section 2.5.4) to the prefix of the home link. Hence, the message will be delivered to one of the home agents on the home link. Since every home agent keeps a home agents list for all home agents on-link, the receiver of the DHAAD request message would be able to send a DHAAD reply message containing a list of IP addresses for the different home agents on-link, as shown in Figure 3–11. The home agents' IP addresses are arranged in order of preference, with the most preferred home agent on top of the list. If two or more home agents have the same preference, their order is randomized between different messages. Hence, some equality can be achieved between different home agents that have equal preferences.

The *identifier* field includes a number sent from the mobile node in the DHAAD requested and echoed in the DHAAD reply sent from the home agent. The actual value has no meaning, but it allows the mobile node to keep a record of its attempts to discover a home agent in cases where the mobile node had to retransmit the DHAAD request because the original message was lost.

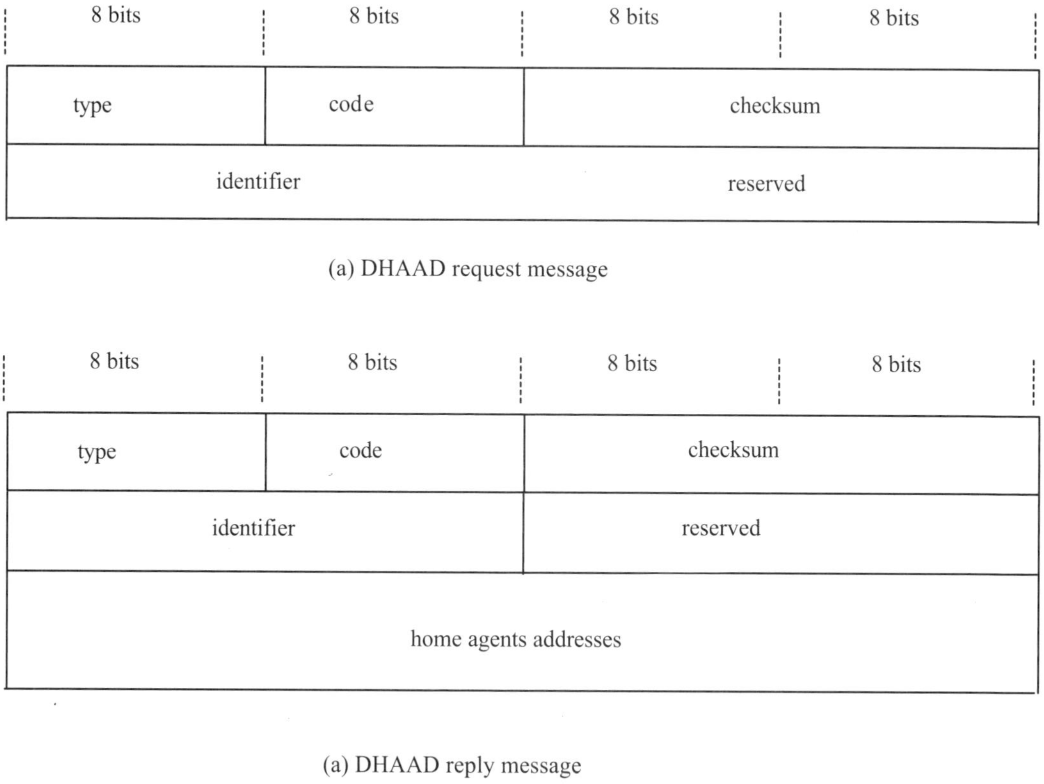

| 8 bits | 8 bits | 8 bits | 8 bits |
|---|---|---|---|
| type | code | checksum | |
| identifier | | reserved | |

(a) DHAAD request message

| 8 bits | 8 bits | 8 bits | 8 bits |
|---|---|---|---|
| type | code | checksum | |
| identifier | | reserved | |
| home agents addresses | | | |

(a) DHAAD reply message

Figure 3–11 DHAAD request and reply messages.

Clearly DHAAD does not completely eliminate manual configuration in the mobile node. Mobile nodes still need to know the home link's prefix in order to construct the home agent's anycast address. However, this mechanism is needed to allow load-sharing between different home agents on the same home link.

Why can't we rely on anycast addresses only for load-sharing? Why do we require home agents to maintain a list of all other home agents on-link? The simple answer is that anycast addresses may allow a router to pick a different anycast destination for each packet, but that does not necessarily mean that the load will be shared fairly between home agents. Home agents are routers that can have very different capabilities, including capacity, processing power, and throughput. Furthermore, not all home agents will experience the same amount of traffic from the mobile nodes that they serve. Therefore, even if all home agents were the same make and model, they could still experience different amounts of traffic from mobile nodes. Hence, load-sharing based on anycast routing alone is not sufficient. The preference value is still needed to give a more accurate view of the home agent's load.

#### 3.2.7.1 DETECTING CHANGES ON THE HOME LINK

Apart from discovering a home agent's address, mobile nodes often need to be updated with any changes on the home link. For instance, the home link may be renumbered with new prefixes, or new prefixes may be advertised in addition to the existing ones. These additional prefixes may imply new home addresses for the mobile node. It is important for the mobile node to detect such changes and detect new home addresses or deprecate the old ones (in case of renumbering). Neighbor Discovery allows nodes to detect such changes when located on the link in question. However, Neighbor Discovery messages cannot be sent beyond a link. For this reason, Mobile IPv6 defined two new messages: *Mobile Prefix Solicitation (MPS)* and *Mobile Prefix Advertisement (MPA)*. The first message is sent from the mobile node to its home agent, while the second is sent from the home agent to the mobile node. Just like the router solicitation and advertisement messages, these messages are carried in ICMP headers (with different ICMP types from those used in Neighbor Discovery). However, Mobile IPv6 requires that these messages are authenticated by IPsec. This is done to prevent a Bad Guy located between the mobile node and its home agent from intercepting these messages and modifying them in a malicious manner (e.g., setting a prefix's lifetime to zero, which would deprecate all addresses formed by the mobile node based on that prefix). More discussions on the security of these messages follow in Chapter 5.

#### 3.2.7.2 WHAT IF THE HOME AGENT FAILED?

The home agent is crucial for the mobile node's reachability and session continuity. It maintains state about the mobile node's location, possibly for long periods of time. Like all nodes that keep state, the home agent reduces reliability.

If the home agent fails, the mobile node becomes unreachable. Furthermore, any connections being tunneled through that home agent are lost. It might be possible to transfer the binding cache between home agents to allow other agents to take over when the home agent fails, but no protocol is currently defined to solve this problem.

When the home agent fails, a mobile node does not receive any indication. The mobile node, however, loses any connections routed through the home agent (if it had any connections with correspondent nodes). If no connections are going through the home agent, the mobile node becomes unreachable without knowing it. In any case, the mobile node eventually tries to refresh its binding with the home agent. If no response is received after several attempts, as described earlier, the mobile node can perform DHAAD and attempt to register with a different home agent.

This approach does not lead to quick detection of home agent failures. However, the use of route optimization (described in section 3.3) reduces the reliance on the home agent. In addition, further developments in Mobile IPv6 may result in a protocol that transfers state between home agents to avoid its being a single point of failure. One example of an existing standard can be found in [5], which is used for IP routers' redundancy.

### 3.2.8 Challenges Associated with Plug and Play for Mobile Nodes

We now know how a mobile node can discover a suitable home agent and remain updated about any changes in the home link. While DHAAD provides a dynamic mechanism for detecting changes on the home link (including the possibility of a home agent failing), such mechanisms still require some manual configuration inside the mobile node. The mobile node needs to know at least one of its home link's prefixes to be able to send the DHAAD request to the home agents' anycast address. The knowledge of a home link's prefix essentially gives the mobile node one of its home addresses (except for saving 64 bits of storage, there is no point in configuring a prefix and not the interface identifier, i.e., the entire home address). Furthermore, there is a need for a security association between the mobile node and its home agent to be able to secure the *Mobile Prefix Solicitation* and *Mobile Prefix Advertisement* messages. There is also a stronger need for mobile nodes to secure binding updates to the home agent, as described in Chapter 5.

You can imagine scenarios where a mobile node is out of coverage for long periods of time due to lack of wired or wireless coverage (or simply because you're trying to enjoy a holiday without being reminded of how much work you will do when the holidays are over!). If the home link were renumbered during such period, a mobile node would not be able to contact its home network (which was renumbered) and, as a result, would not be able to send a binding update to its home agent. Hence, removing the reliance on manual configuration would allow for better resilience to changes in the home link. However, removing such dependency is not a trivial task. In this section we discuss the challenges associated with removing manual configuration.

Achieving pure plug and play for mobile nodes would entail the following:

1. The mobile node discovers its own home address.
2. The mobile node uses the knowledge of the previous step to construct the home agents' anycast address and perform DHAAD.
3. The mobile node selects a home agent and sends a binding update.

DHAAD was described earlier and does not pose any special requirements. However, step 1 poses a more difficult challenge, as shown below. Step 3 requires mutual authentication between the mobile node and its home agent. This is described in detail in Chapter 5.

#### 3.2.8.1 HOW CAN A MOBILE NODE DISCOVER ITS HOME ADDRESS?

To remove the dependency on manual configuration of the home address, mobile nodes need to have an abstract identifier that can be mapped to a home address, for example, by storing that identifier in a database whose lookup would show the corresponding address. Such database already exists on the Internet: the DNS. Hence, one could configure a mobile node with its domain name. When starting the mobile node, it could query the DNS for the IP address corresponding to its own domain name and obtain its home address, then proceed with steps 2 and 3 above. However, DNS lookups are not usually secure. Bad Guy may intercept a DNS lookup and act as a DNS, providing the mobile node with a false home address. Currently the IETF is developing a DNS security protocol (DNSSEC) [3], which can allow mobile nodes to authenticate the DNS replies to ensure their authenticity. However, DNSSEC is currently not widely deployed and will take some time to get to wide deployment. In the worst-case scenario, Bad Guy can deny the mobile node access to its home agent by returning a false home address in the DNS reply. This is a form of a Denial of Service (DoS) attack, as it denies the mobile node access to its home agent. However, Bad Guy cannot pretend to be a home agent, since binding updates between the mobile node and its home agent are authenticated.

The next challenge with this approach is to be able to authenticate binding updates to the home agent. To avoid manual configuration, mobile nodes need to be able to establish security associations with the home agent in a dynamic manner. This requires the use of a key exchange mechanism (e.g., the Internet Key Exchange -IKE- protocol). Chapter 5 discusses these issues in more detail.

#### 3.2.8.2 CAN MOBILE NODES GENERATE A RANDOM HOME ADDRESS?

The concept of a permanent home address is extremely useful for mobile nodes that need to be reachable. However, some users may be concerned about being tracked when using the same home address all the time. For instance, someone snooping traffic near the home link may be able to track a home address (which can be attributed to human if the machine's owner is known) and find out which correspondent nodes it communicates with (e.g., Hesham.example.com reads news on newschannel.com and uses mailserver.net for his email service). This may lead to undesirable profiling of users, which raises privacy concerns.

RFC 3041 attempts to solve this problem for IPv6 nodes by providing an algorithm that they can use to generate random numbers and use them as interface identifiers. It also assumes that these addresses will be relatively stable (e.g., usable for several days). If a mobile node generates a home address, sends a binding update to its home agent, and then goes out of coverage until its binding expires, another mobile node may use this address with the same home agent, resulting in binding update failure due to home address duplication. The probability of this happening is low, but it can be significant for a large home agent and a bad random number generator in the mobile node. Hence, such address may not be stable for the period required by RFC 3041. This may not be a significant problem if the mobile node is ready for such change (i.e., does not allow applications to use this address after receiving a notification from the home agent that the binding update cannot be accepted) and can generate another address. However, another problem might be that a socket was already open with the previously generated home address before the mobile node went out of coverage. If an application tries to reuse that socket, it will fail, since the address no longer exists. In this case the application would need to open a new socket.

A more significant problem is due to the reachability requirement on the home address. For a home agent to accept a binding from a mobile node, it needs to make sure that this address is not allocated to another mobile node on a permanent basis. Simply checking that no other binding exists in the home agent's binding cache is not sufficient, as the real owner of the address may not have registered with the home agent yet (e.g., it is turned off or out of coverage). In chapters 4 and 5 we see how Cryptographically Generated Addresses (CGAs) can help with solving this problem.

### 3.2.9 Can a Mobile Node Have More than One Home Agent?

Mobile nodes can have several home addresses. The binding for each home address must be stored in **one** home agent only. A mobile node cannot send two different home registrations simultaneously (binding updates with the *H* flag set) to two different home agents. If it did, two home agents would attempt to defend the same home address on the same link, which would not work.

On the other hand, a mobile node can have several home addresses with one or more home agents. Whenever the mobile node moves, it must send a binding update for each home address to the corresponding home agent. That is, the same procedures described above are done for each home address. After the binding is established, the mobile node has several tunnel entry points, one for each home address.

### 3.2.10 Virtual Home Links

We saw how a mobile node can register and deregister a binding with its home agent. When creating a new binding, the home agent needs to perform DAD and subsequently defend the mobile node's home addresses and possibly its link-local address (if the *L* flag were set). Currently, there is no deployed mechanism defined to secure Neighbor Discovery messages (more on that in chapters 4 and 5). Moreover, the mechanisms currently being developed in IETF for this purpose have not yet provided a way of securing *proxy neighbor advertisements*.

It is possible for Bad Guy, sharing a link with the home agent, to launch a DoS attack by pretending to own the mobile node's home address and therefore cause the binding establishment to fail, rendering the mobile node unreachable. These problems can be avoided if the home link is a virtual link—that is, if the home prefix is not associated to any physical link. The home prefix can be configured on the home agent and mobile nodes only. The home agent would not send any router advertisements for this prefix. Consequently, the mobile node would never "return home"; it would always move from a foreign link to another. In this scenario, DAD would be done much quicker than if there were a physical link associated with the home prefix. The home agent would only need to make sure that the home address is not allocated to another node. Furthermore, no proxy neighbor advertisement would be sent on the wire, since no one else is located on the home link except mobile nodes and the home agent. Therefore, Bad Guy cannot launch DoS attacks on the mobile node's home address.

This configuration is possible with Mobile IPv6, but clearly not mandatory. The Mobile IPv6 specification does not restrict the home link configuration, but home agent implementers and network operators implementers are certainly free to allow for virtual home links, as they do have significant benefits.

## 3.3 Route Optimization

Going back to our post office forwarding service analogy once more, consider the case where you are planning to communicate with a certain person or organization on a regular basis after changing your address temporarily to another country. For example, you might be receiving a monthly credit card bill that needs to be paid within one week. If forwarding the bill to the new location takes a week, it is likely that your payment will be overdue and possibly cost more. The logical solution for this problem would be to inform the credit card company of your new address and avoid the additional delays.

Mobile nodes experience similar problems. Routing packets through the home agent adds additional delays. This can be clearly seen in Figure 3–1. As shown in Figure 3–12, the worst-case scenario would occur if the mobile node and the correspondent node share the same link. On the other hand, the best-case scenario (least possible delay) would occur if the correspondent node were on the mobile node's home link. However, in all cases there will be some additional delay when compared to direct communication.

Another problem with forcing traffic through the home agent is the introduction of a single point of failure in the network. That is, while it may be possible to eventually detect the failure of a home agent, its failure would cause the mobile node to lose all ongoing connections. In the future this problem may be avoided at the cost of introducing new protocols to synchronize binding cache states between home agents. However, it would be beneficial to remove this dependency for the general case. Another argument for eliminating the home agent from the path between the mobile node and correspondent node is that such routing uses more network bandwidth than direct communication and therefore makes inefficient utilization of the network bandwidth. Hence, if all mobile nodes' traffic is forwarded through the home agent, operators would need to be careful about the location of the home agent and the capacity of the links attached to it, which adds a burden on network design.

Route optimization is about routing packets between a mobile node and a correspondent node, using the shortest possible path (as it is normally done between two communicating hosts relying on normal routing). The mobile node is aware when packets are routed through the home agent when it receives tunneled

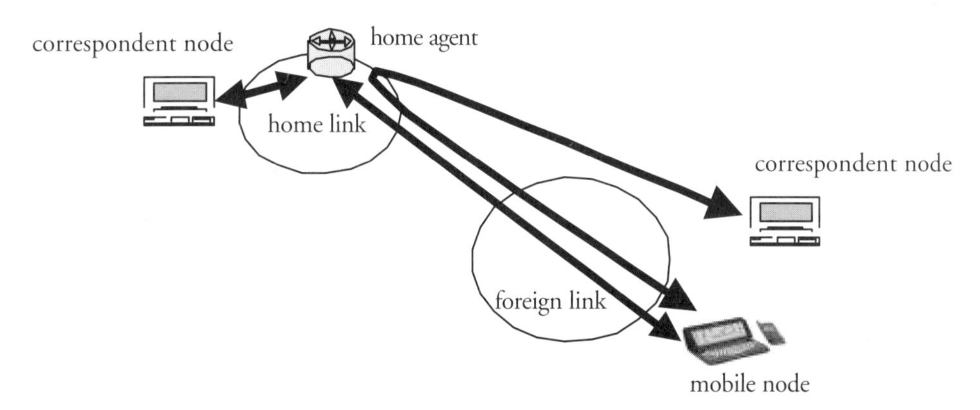


**Figure 3–12** *Worst-case and best-case scenarios for routing through the home agent.*

packets addressed to its home address. Depending on the nature of communication (long or short term), the mobile node can decide whether it should attempt to optimize the route between itself and the correspondent node. How does a mobile node know whether the communication with a correspondent node will last for a short or a long time? There is, unfortunately, no concrete answer for this question. Obviously, a user may configure the mobile node to treat some addresses in a special way, but this is far from being a realistic solution; most users know nothing about IP addresses or IP in general. Also, this does not cover communication with other nodes of which the user had no prior knowledge. A "smart" mobile node's implementation may make some assumptions about the type of applications used between the mobile and correspondent node (e.g., Web browsing usually involves short-lived connections), the duration of a connection, the RTT, and the amount of data being sent between the two nodes to decide whether route optimization is needed. On the other hand, using a simpler approach, a mobile node's implementation may also decide to always use route optimization or never use route optimization.

When a mobile node receives a packet tunneled from the home agent, it must decide whether route optimization is needed. If so, the mobile node informs the correspondent node of its current location. This is done using the same binding update message shown in Figure 3–2. The correspondent node maintains a binding cache similar to the one maintained by the home agent. However, a binding update sent to a correspondent node must not set the *H*-, *K*-, or *L*-bits, as they are only usable when communicating with the mobile node's home agent. Route optimization is shown in Figure 3–13.

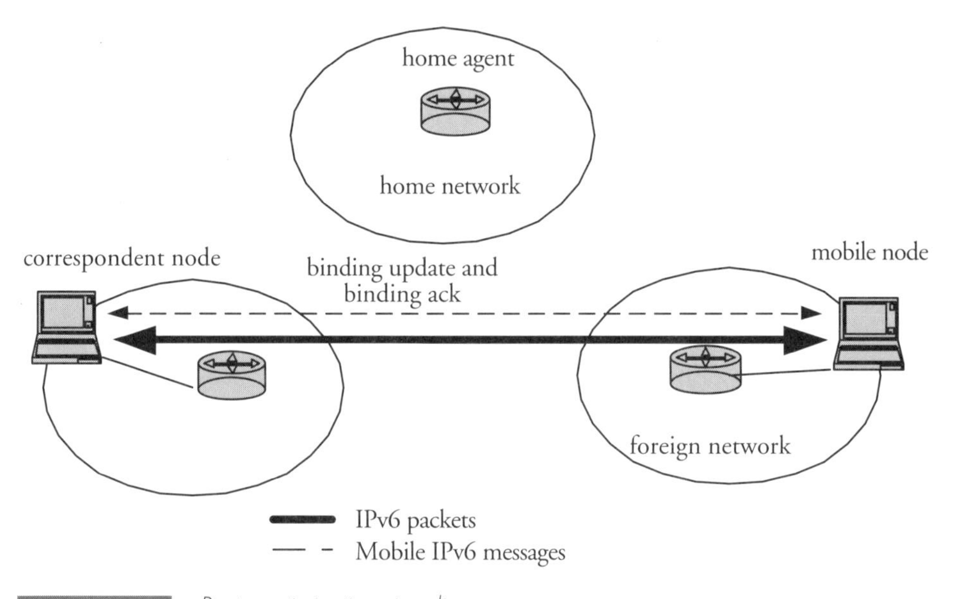


**Figure 3–13** *Route optimization signaling.*

When a correspondent node receives a binding update from a mobile node, it creates a new entry in the binding cache or updates the existing one with the new location of the mobile node. Following this step, the correspondent node can communicate directly with the mobile node by sending packets to the mobile node's care-of address.

The aim of the binding update is to achieve two goals: first, allow packets to be sent directly between the mobile and correspondent nodes without going through the home agent, and second, maintain ongoing connections in the meantime by allowing applications to keep using the home address as a source address (in the mobile node) and destination address (in the correspondent node). The first step can be achieved if both nodes ensure that the packets transmitted contain the mobile node's care-of address in the source and destination fields when being sent from and to the mobile node respectively. This is needed to allow Internet routing to deliver packets directly to the mobile node (for packets sent to the mobile node). In addition, this allows the mobile node to send outgoing packets with a topologically correct address. However, to maintain ongoing connections, the mobile node's home address must somehow be included in packets sent to and from the mobile node to allow the IP layer (in the mobile and correspondent nodes) to present it to upper layers. To do this, Mobile IPv6 defines two messages, a new *routing header type 2* and a new destination option called the *home address option*, that are included in packets sent to and from the mobile node respectively. We first discuss the home address option, then the new routing header. The security considerations for route optimization are discussed in detail in Chapter 5.

## 3.3.1 Sending Route Optimized Packets to Correspondent Nodes

When the mobile node sends a binding update to a correspondent node, it needs to indicate the home address for which the binding is sent. The home address is included in the home address option, which is included in the destination options extension header. After accepting the binding update, the mobile node's home address is stored in the correspondent node's binding cache with the rest of the contents of the binding update. If the *A* flag is set in the binding update, the correspondent node sends a binding acknowledgment to the mobile node. After receiving a binding acknowledgment, the mobile node updates its binding update list to include the information sent in the binding update to the correspondent node. This includes the correspondent node's IPv6 address, the mobile node's home and care-of addresses, the sequence number used, and the lifetime for the binding.

After successfully installing a binding in the correspondent node's binding cache, the mobile node uses the home address option in every packet sent that includes data from an application using the home address as a source and communicating with the correspondent node (destination). The home address option is essentially a disguised form of tunneling. When a correspondent node receives a packet containing a home address option, it replaces the source address in the packet's header with the address included in the home address option before passing the packet to upper layers, which effectively makes the packet appear to be received from the mobile node's home address when viewed by upper layers. Hence, mobility is kept transparent to upper layers.

It is important to note that the mobile node includes the home address option only in packets sent directly to correspondent nodes (i.e., to which a binding update was sent and accepted). This is done for security reasons that are discussed in Chapter 5. In order for mobile nodes to know which correspondent nodes have a binding cache entry for the mobile node (and hence will accept the home address option), the mobile node checks the content of the binding update list before sending packets. The binding update list contains an entry for each correspondent node's address. Each entry contains the details of the binding update sent to the correspondent node: the lifetime, sequence number, home address, and correspondent node's address, which is used as a search key for each entry. The authentication credentials for binding updates are also included in the binding update list. Finally, the binding update list contains a flag, set by the mobile node, to indicate whether a mobile node should attempt route optimization with a particular correspondent node in the future. The aim of this flag can be explained in light of one of the requirements on Mobile IPv6: backward compatibility. Since not all hosts on the Internet are expected to support Mobile IPv6, some correspondent nodes might not understand the mobility header at all and would send an ICMP error ("host unreachable") informing the mobile node that the mobility header is not supported. Hence, the mobile node can avoid sending future binding updates to those correspondent nodes by setting the flag mentioned above. How long should this entry be kept? This is left up to the mobile node implementer and primarily depends on the size of memory allocated for binding update lists.

Figure 3–14 shows how the home address option is added in the mobile node's IP layer implementation before sending the packet to the correspondent node.

Figure 3–14 shows three steps: first, the original packet as expected by the application is formed, then the home address option is added in the destination options header. This option includes the care-of address. We intended to show this step first to clarify exactly how the packet is processed. The reason for including the care-of address in the home address option is to allow any operations on the packet (between constructing the packet and sending it), like IPsec processing, to assume that the source address is in fact the home address. This is done to avoid changing, for example, an IPsec security association based on the mobile node's source address every time the mobile node changes its care-of address.

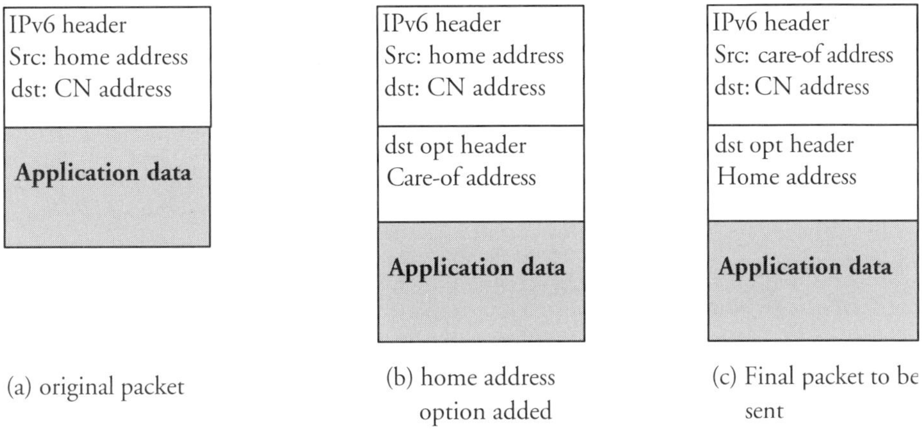

IPv6 header
Src: home address
dst: CN address

Application data

(a) original packet

IPv6 header
Src: home address
dst: CN address

dst opt header
Care-of address

Application data

(b) home address option added

IPv6 header
Src: care-of address
dst: CN address

dst opt header
Home address

Application data

(c) Final packet to be sent

**Figure 3–14** *Adding the home address option before sending a packet.*

When the packet is ready to be sent, the content of the home address option is swapped with the source address field. Hence, the packet leaving the mobile node will contain the care-of address in the source address field and the home address option in the destination options extension header.

When the packet is received by the correspondent node, the same operations are done in the reverse order. Hence, the packet seen by upper layers in the correspondent node will look like the original packet (seen in step a) in Figure 3–14.

### 3.3.2 Receiving Route Optimized Packets from Correspondent Nodes

When a correspondent node sends packets to a mobile node for which it has a binding cache entry, it must include a new routing header (with a *type* field set to 2). This routing header is identical to the one shown earlier in Chapter 2, with the exception that it can only include the mobile node's home address, that is, the number of segments cannot be larger than 1. The reason for this restriction becomes clear in section 5.1.2. When receiving the packet, the mobile node processes the routing header. This results in replacing the destination address in the packet (care-of address) with the address in the routing header (the home address). Since this address also belongs to the mobile node, the mobile node is essentially forwarding the packet to itself. Following this step the packet is passed to upper layers with the mobile node's home address in the destination address field, hiding the address change from upper layers.

### 3.3.3 Acknowledging Binding Updates Sent to Correspondent Nodes

The process of route optimization involves three distinct steps:

1. Detecting that packets are tunneled by the home agent.
2. Sending a binding update to the correspondent node.
3. Sending packets directly to the correspondent node and including the home address option in those packets.

The final step can only be done after the binding update is received and processed by the correspondent node. If the mobile node sends packets directly to correspondent nodes, including the home address option, when the correspondent node has not accepted the binding update, the correspondent node discards them. This act causes packets to be lost and consequently disrupts communication. Hence, the mobile node needs to ensure that the binding update was received and accepted by the correspondent node before routing packets directly to the correspondent node's address. To ensure the acceptance of the binding update, the mobile node can request an acknowledgment by setting the *A* flag in the binding update. If the binding is accepted, the correspondent node responds with a binding acknowledgment containing the appropriate status. After receiving the binding acknowledgment, the mobile node can be sure that the correspondent node will accept packets containing the home address option.

### 3.3.4 What if the Correspondent Node Failed?

The mobile node should perform route optimization signaling in a reliable manner to ensure that no packets are lost due to the premature inclusion of the home address option. However, it is possible for a correspondent node to lose the information in its binding cache (e.g., due to a reboot, which causes a node to lose the contents of its volatile memory that includes the binding cache, or because the cache exceeded the allocated memory and some entries were deleted) some time after accepting a binding update. The mobile node would not be aware of such reboot and would continue to send packets containing the home address option. The correspondent node receives these packets, checks if it has a binding cache entry corresponding to the home address, but finds none and consequently drops the packet. To inform the mobile node of such failure, the correspondent node can send a *binding error* message indicating that no binding cache entry exists for this home address. The binding error message may also be sent due to other errors in the mobility header. This message (see Figure 3–15) is sent to the source address in the packet (i.e., the mobile node's care-of address).

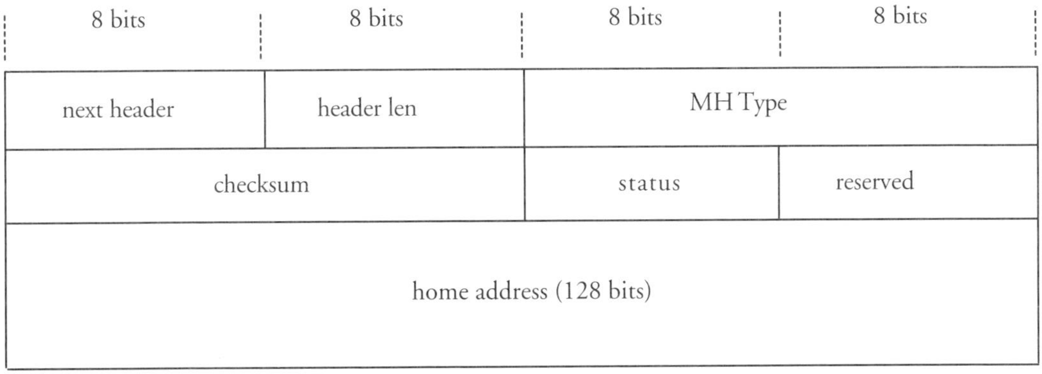

| 8 bits | 8 bits | 8 bits | 8 bits |
|---|---|---|---|
| next header | header len | MH Type | |
| checksum | | status | reserved |
| home address (128 bits) | | | |

Figure 3–15 *Binding error message.*

The binding error message has a *type* field value of 7. The *status* field has two possible values (and more may be defined in future): when set to 1 it indicates that there is no binding cache entry associated with the home address in the binding error message. This may be the case when the contents of the binding cache are lost or the binding update for this home address was never received. A value of 2 indicates that the type of mobility header received (i.e., the message included in the mobility header) is not known to the correspondent node. This value ensures that new mobility signals can be added in the future, in a backward-compatible manner. That is, new functions can be added without affecting legacy nodes, as they will inform mobile nodes if they are unable to support such functions.

### 3.3.5 Why Not IP in IP Tunneling for Route Optimization?

The basic idea behind route optimization is to hide the mobile node's home address inside the packets (using the routing header and the home address option) to avoid breaking ingress filtering (when the packets are sent by the mobile node) or to route packets to the right location in the topology (when they are sent by the correspondent node). However, the same effect could have been achieved by using IP in IP tunneling between the mobile node and the correspondent node. Following the binding update processing, a tunnel can be established between the mobile and correspondent nodes. Packets originating from the mobile node would contain the mobile node's home address in the source address field within the inner header; the destination address would be the correspondent node's address. The outer header would contain the mobile node's care-of address in the source address field and the correspondent node's

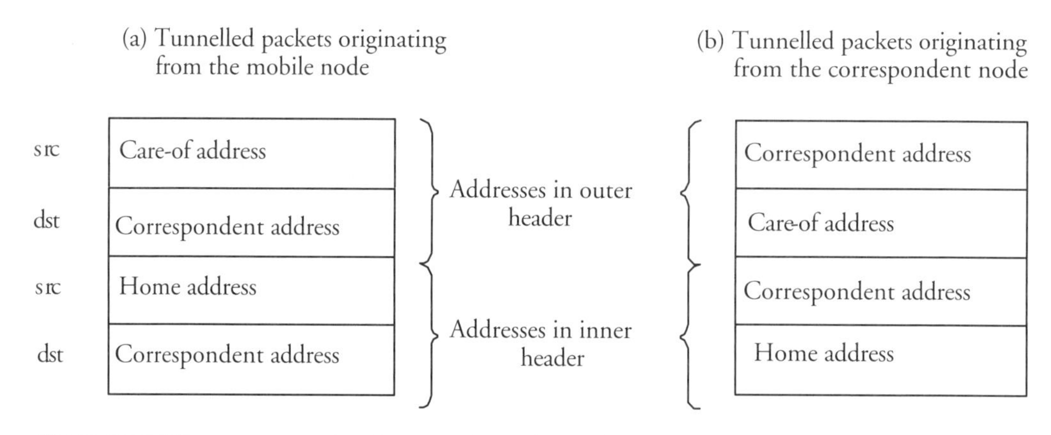


**Figure 3–16** *Using tunneling between the mobile node and the correspondent node.*

address in the destination field. The opposite can be done for packets originating from correspondent nodes, as shown in Figure 3–16.

While this approach has the benefit of using existing mechanisms (tunneling) instead of defining new ones (i.e., the routing header and the home address option), it has the disadvantage of using more network bandwidth when compared to using the routing header and the home address option. This is due to having four addresses (i.e., two IP headers) in the tunnel when compared to three when either the home address option or the routing header is used. In [1] it was proposed to compress the redundant address in the tunnel (the correspondent node's address). However, this approach was proposed at a later stage of Mobile IPv6's development and was viewed as a "good approach that came too late."

## 3.4 What if the Mobile Node Failed?

The mobile node maintains state about the nodes with which it has bindings in its binding update list. The loss of such information results in having different sets of information in the mobile node, its home agent, and correspondent nodes. If the mobile node reboots and loses this information, it immediately attempts to update its home agent. To update the current binding, the mobile node needs to use a sequence number larger than the one used in its last binding update message. Since the mobile node is not aware of the last value used, it picks a random value. If the chosen value is lower than the last one received by the home agent (or correspondent node), a binding acknowledgment is received with the status value set to 135, indicating that the sequence number is incorrect and providing the last used sequence number to the mobile node. The mobile node can then attempt to send the binding update with an appropriate value. Obviously, if the mobile node picks a random sequence number higher than the last used number, the binding update will be accepted the first time.

## 3.5 Site-Local Addresses and Mobile IPv6

One of the problems with site-local addresses is that a mobile node has no way of knowing whether it is in its home site or another one. That is, if a mobile node moves from one link to another, it may or may not have left its home site (the IPv6 site that was the home agent is located). There is no information in the router advertisement that tells nodes how large the site is. This will cause some confusion in the mobile node regarding its ability to use site-local care-of addresses. We can illustrate this confusion by demonstrating a couple of different scenarios where site-local addresses are used.

The first scenario to consider is a home site, including the mobile node's home agent, which uses site-local addresses only. A mobile node in this site will have a site-local home address. When the mobile node moves away from its home link, it does not know whether it has moved to a new site or within the same site. However, it does know that it moved from its home link. The mobile node then forms a site-local care-of address and attempts to send a binding update to its home agent. If the mobile node is in fact in the same site, the binding update is received by the home agent, and Mobile IPv6 will work. However, if the mobile node is in a different site, two different outcomes are possible:

- Another node in the same site is configured with the same site-local address as the home agent; therefore, this node will receive the binding update from the mobile node. However, since binding updates and acknowledgments are protected by IPsec there is no chance of confusion in this case; the node that receives the binding update will silently discard the packet.
- No node in this site is configured with the home agent's site-local address; therefore, the mobile node will receive an ICMP error "Destination unreachable."

From this discussion we can conclude that sites configured only with site-local addresses allow mobile nodes to move within the site; however, once the mobile node moves to another site, normal routing practices will inform the mobile node that its home agent is not reachable. We can also see that the mobile node's care-of address' scope should be equal to, or larger than, its home address' scope. In general, mobile nodes should use site-local care-of addresses only if no global care-of address exists.

## 3.6 A Communication Example

In this section we use the knowledge gained from this chapter to draw a simplified flowchart for the mobile and correspondent nodes' operation. This flowchart can be compared to the one used in Chapter 2 for normal IPv6 hosts. We do not consider the binding update security details in this section. Security issues are discussed in detail in chapters 4 and 5. Our flowchart for the mobile node's sending implementation is shown in Figure 3–17.

When the mobile node's IP layer implementation receives upper-layer data, it will have already supplied the home address as a source address for applications. Hence, an IP header can be constructed with the home address being the source address and the correspondent node's address in the destination address. Next, the binding update list is checked to see if the mobile node has already sent a binding update to the correspondent node and get the care-of address from the correspondent's node's entry. If it has, the mobile node can include its home address in the home address option. However, for now the mobile node places its care-of address in the home address option instead. We see why this is done when we analyze the following steps in the flowchart. The mobile node then checks its binding cache to see if the correspondent node has sent it a binding update. Note that *correspondent node* refers to a functional entity in an IPv6 node and does not exclude a correspondent node from being a mobile node as well. If the correspondent node had sent a binding update to the mobile node, a binding cache entry is found. Hence, the mobile node constructs a routing header type 2 and places the correspondent node's care-of address inside it.

Assuming that the mobile node does not need to add any other extension headers, the next step is to check whether IPsec is needed. This is done by checking the SPD to see whether the packet should be protected. When IPsec is used, the destination options header containing the home address option needs to be placed before the IPsec header (AH or ESP). This is done because IPsec identifies a security association in the SAD by *selectors*. Selectors include the source and destination addresses of the nodes sharing the security association. To avoid changing the security association every time the mobile node moves, it is best to use the home address (not the care-of address) as a selector identifying the mobile node. When the packet is received by the correspondent node, the headers are processed in the order they appear in the packet; therefore, IPsec headers are processed last, before the upper layer. Hence, an IPsec implementation in the correspondent node always sees the home address in the source address field of a received packet. Now we can see the reason for keeping the mobile node's home address in the source address field and its care-of address in the home address option until IPsec has added its header. We need to perform the IPsec operations on the header seen by the correspondent node after verifying the IPsec header. That is, the packet needs

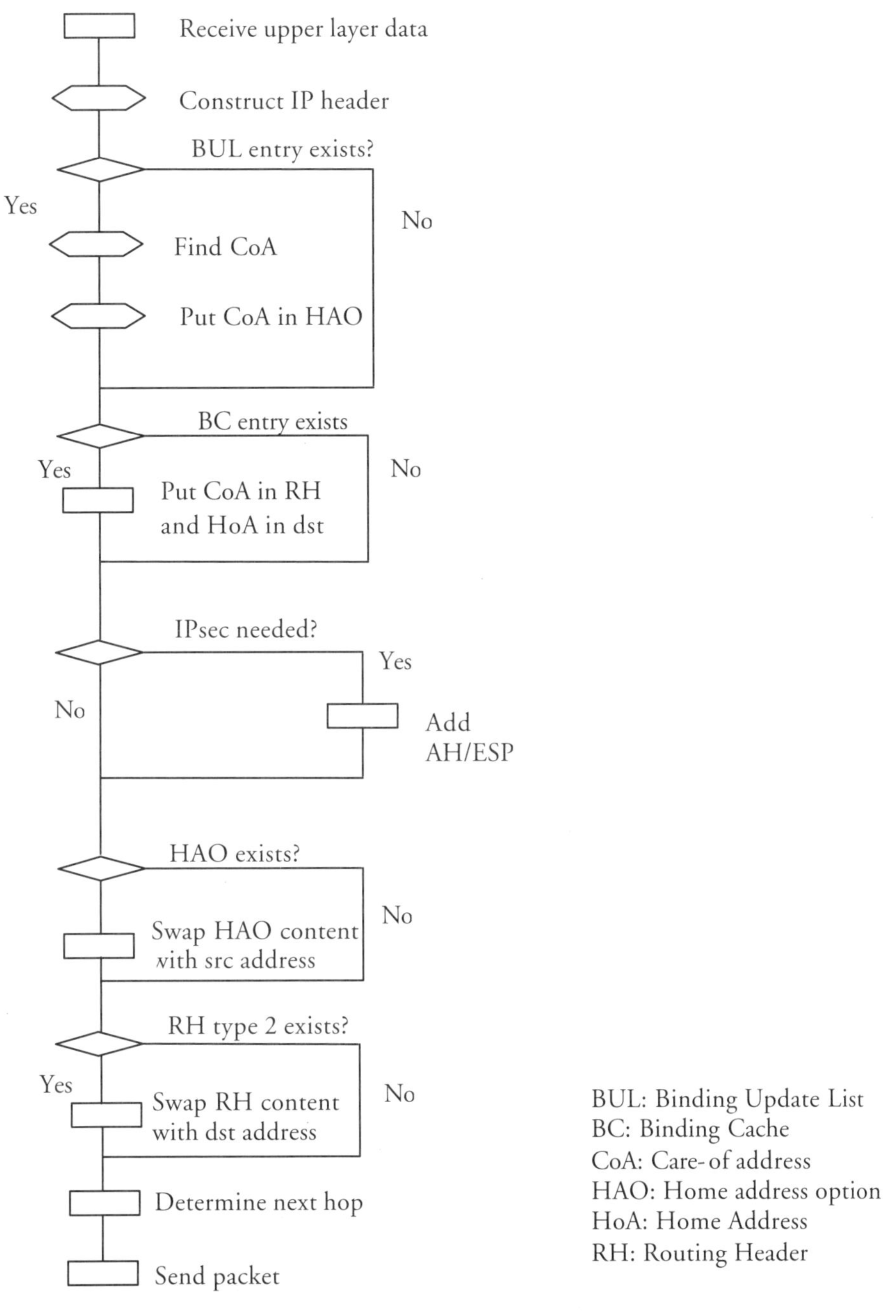


**Figure 13–17** *Simplified flowchart for mobile node operations (sending packets).*

to look exactly the same when passed to the correspondent node's IPsec implementation as it did when passed to the mobile node's IPsec implementation. Otherwise, to IPsec, the packet will seem to have been modified and will be dropped.

After IPsec processing is done, the mobile node swaps the contents of the home address option and the routing header with the source and destination addresses respectively. The next hop is determined based on the earlier discussion in Chapter 2 and whether the packet is tunneled to the home agent or sent directly to the correspondent node. As discussed in Chapter 2, addresses are tied to interfaces. When the mobile node is on a foreign link, its home address is associated with a *tunnel interface* to the home agent. When it moves back home, the tunnel is deleted and the home address is associated with a physical interface. The tunnel interface can be treated by the IP layer implementation as another physical interface that is one hop away from the home agent. Hence, if the packet contains the home address in the source address field, it is immediately tunneled to the home agent; otherwise, it is sent on the interface associated with the care-of address.

Now let's consider the steps taken by the correspondent node when it receives packets from the mobile node, as shown in Figure 3–18.

When the correspondent node receives a packet, it processes the headers in the order that they were sent. If a routing header type 2 is included, and provided that there is one address in it and that it is the correspondent node's home address, it swaps that address with the destination address in the packet; otherwise, the packet is dropped. The node then processes other extension headers. In Figure 3–18 we skip this processing until it gets to the destination options header. If a home address option is found, the binding cache is searched for a corresponding entry. If nothing is found, a binding error message is sent to the source address in the IPv6 header. If the correspondent node does not implement any Mobile IPv6 messages, then it does not understand the home address option and sends an ICMP error message to the source address in the packet. If a binding cache entry is found for this home address, the home address is swapped with the care-of address in the source address field.

If an IPsec header is included, it is verified as usual. Note that the packet would now look exactly the same when seen by IPsec implementations at both ends. Finally, the information is passed to upper layers.

## 3.7 Summary

This chapter described the operation of Mobile IPv6, starting with the role of each element: the mobile node, the correspondent node, and the home agent. We also saw how mobile node–home agent and mobile node–correspondent

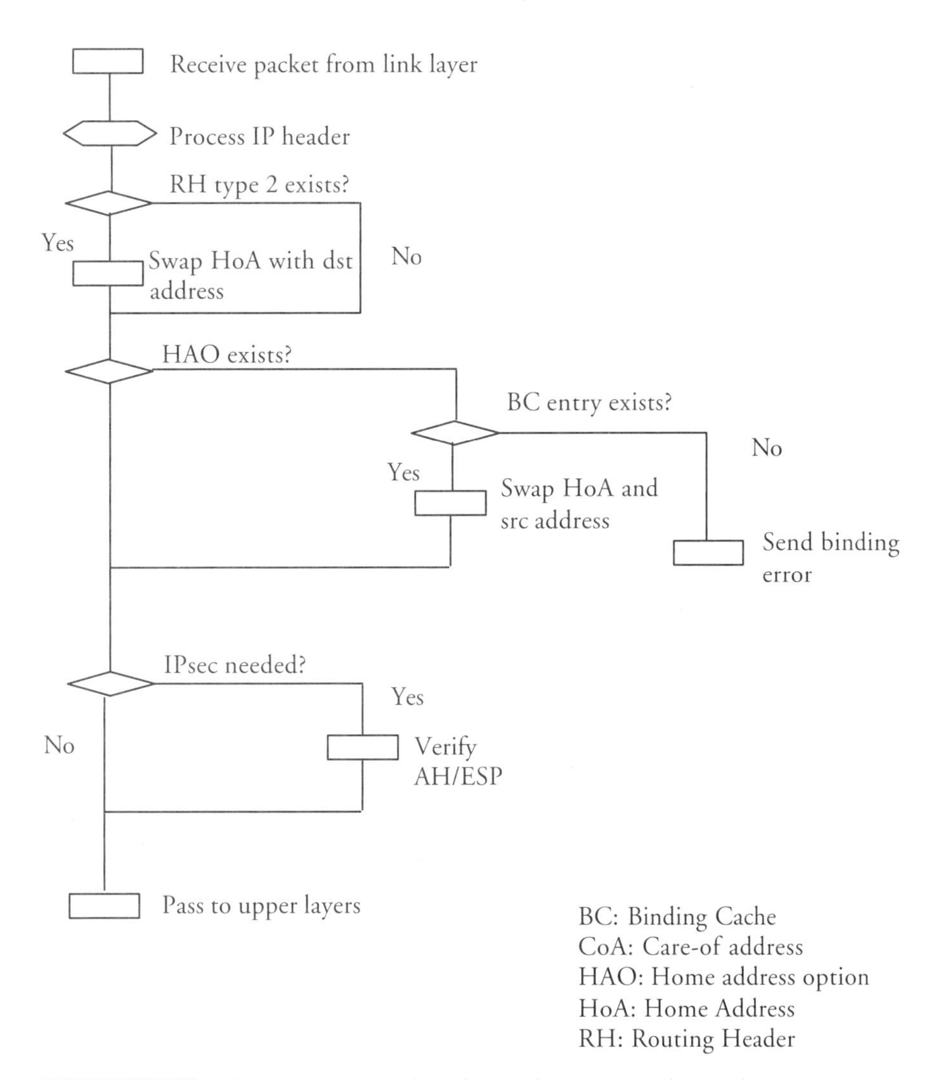


**Figure 13–18** *Processing received packets at the correspondent node.*

node signaling is performed. Two different communication scenarios between the mobile and correspondent nodes were described: communication through the home agent and direct communication (route optimization). The processing within each node was described for both communication scenarios.

We also discussed the various failure cases for the mobile node, correspondent node, and home agent; and we saw how the protocol can cope with each one. Finally, the knowledge gained in this chapter was used to build simplified flowcharts that describe the mobile and correspondent nodes' operations.

The next chapter introduces a new topic: security, an interesting topic required to set the scene for the Mobile IPv6 security discussion.

## Further Reading

[1] Deering, S., and B. Zill, "Redundant Address Deletion when Encapsulating IPv6 in IPv6," draft-deering-ipv6-encap-addr-deletion-00, work in progress, November 2001.

[2] Draves, R., "Default Address Selection for Internet Protocol version 6 (IPv6)," RFC 3484, February 2002.

[3] Eastlake, D., "Domain Name System Security Extensions," RFC 2535, March 1999.

[4] Johnson, D., C. Perkins, and J. Arkko, "Mobility Support in IPv6," draft-ietf-mobileip-ipv6-24, work in progress, June 2003.

[5] Knight, S., et al., "Virtual Router Redundancy Protocol," RFC 2338, April 1998.

[6] Nordmark, E., "MIPv6: From Hindsight to Foresight?" draft-nordmark-mobileip-mipv6-hindsight-00, work in progress, November 2001

FOUR

# Introduction to Security

During the design phase of Mobile IPv6, it became evident that security was a major component that needed to be considered for most messages used between mobile nodes, home agents, and correspondent nodes. In particular, securing route optimization messages presented a serious challenge to the Mobile IP working group. We introduce some basic security concepts to promote a better understanding of the Mobile IPv6 security mechanisms and the fundamental reasons that led to the current design. In Chapter 5, we analyze the security threats and requirements for Mobile IPv6 and present the selected mechanisms for Mobile IPv6 security.

## 4.1 What Is Security and Why Is It Needed?

The Internet is becoming popular for many applications that go beyond the traditional ones, like email and Web browsing. It is being used to streamline business transactions, banking, government departments, educational applications, and several other communication scenarios. The proliferation of such applications and many others has raised the need for security—more so than before. Within the context of computer networking, security is the science of protecting information and devices on a network from being misused by unauthorized users. This includes disclosure, modification, and destruction of information as well as unauthorized use of network resources, such as denying service to legitimate users.

For a complete understanding of security, we consider the following aspects:

- *Integrity checking*: Ensuring that data cannot be modified en route to its destination without being detected.
- *Confidentiality*: Transforming data in a way that makes them understood by authorized entities only.
- *Data origin authentication*: Ensuring that the received data came from the claimed sender.
- *Nonrepudiation*: Ensuring that the sender cannot deny being the origin of the received data. Nonrepudiation is a strong case of data origin authentication.
- *Entity authentication*: Verifying someone's or something's identity. Verification of a claimed identity can be bound to the integrity of the received messages from that identity, providing data origin authentication. That is, there is usually no point in attempting to authenticate an identity if one cannot ensure that the messages received, if any, from such identity are not themselves authentic. The granularity of identity authentication depends on whether a unicast or multicast communication is taking place. For instance, in a group setting, it might make sense to protect the integrity of the message without the need to identify the exact source of the data, except that it is "someone in the group."
- *Authorization*: Allowing someone or something to perform a certain action. Authorization takes place after authentication.

Different cryptographic functions are used to address these security aspects. We introduce these functions in the following sections.

## 4.2 Authentication

In this section, we are primarily concerned with entity authentication. In the following sections, we see other cases where data origin and message authentication are needed. Entity authentication is the process of verifying an entity's identity. This is typically done to permit someone or something to perform a task. Several examples of authentication can be observed in our daily lives. A customer must produce some identification at the bank to be able to withdraw money from a bank account; such identification must prove that the person standing at the counter is truly who he claims to be. Similarly, when landing at a foreign airport, a traveler must show a passport to be allowed to enter the country. In these examples, the credential used for authentication is usually designed in a way that makes it difficult to forge. However, other authentication systems rely on the use of a *secret* given to a certain entity to use for proving its identity (simply because it knows the secret). Password-based schemes fall into that category.

Some authentication systems are stronger than others. For instance, banks usually send cards and PIN codes to their clients by mail, typically in two separate mails. This method is clearly vulnerable to a malicious mailman who might intercept the legitimate clients' mail and gain access to their bank accounts (luckily, this is not a common scenario). A *strong authentication* system ensures that authenticators and messages of the actual authentication protocol are not exchanged in a manner that makes them vulnerable to being hijacked by an intermediate malicious node or human. That is, the information used to generate a proof of identity should not be exposed to anyone other than the person or machine that it is used to identify. Also, passively observing one successful execution of the authentication protocol should not enable a third party to perform subsequent successful authentications in the name of the legitimate user. In the following sections, we show different cryptographic techniques used to design strong authentication systems.

## 4.3 Authorization

Authorization is the tollgate at which a system decides whether or not a certain entity should be allowed to perform a requested task. This decision is made after authenticating the identity in question—hence, the tight coupling between authentication and authorization. When considering an authentication system for a particular application, it is crucial to understand the type of identifier required to provide a certain level of authorization. For instance, a passport provides an appropriate identifier to authorize a person to enter a country. However, a passport alone is unlikely to authorize the same person to enter a company's secure building. A specific company card is typically required to gain access. The importance of selecting the right identifier becomes clearer when we discuss Mobile IPv6 security in Chapter 5.

## 4.4 Confidentiality, Integrity Checks, Nonrepudiation, and Replay Attacks

The requirements on a security system depend on the nature and content of the messages in question. Confidentiality is needed when the message sent contains sensitive material that should not be read by others and therefore must not be sent in a comprehensible format. For example, if Bob sends a message to Alice that contains his bank account details, he must make sure that no one else can read this information, thus requiring that he encrypt the message before sending it. Encryption algorithms are used to hide such information to achieve confidentiality.

On the other hand, if Bob sends a letter to Alice asking her to transfer $1,000 from his account to Debora's account, Alice needs to make sure that the letter actually came from Bob. To know this, two requirements must be satisfied: Alice must ensure that the letter was not modified by someone else—for example, that no unauthorized person changed the amount. This is where *message integrity* becomes important. In addition, Alice must ensure that the message came from Bob and that no one else could have sent this letter. Why is the latter precaution needed? First, Alice needs to be sure that it is not Bad Guy who tries to empty Bob's account by faking messages. Second, suppose Bob changes his mind later about the money because he discovers that he is in financial trouble and he needs the $1,000 that he earlier transferred to Debora. He might claim that he never asked for such a transfer to take place and that Alice placed that order without his knowledge. However, if Alice received the original message signed by Bob, he would not be able to make such a claim. Therefore, in this scenario, signatures can achieve nonrepudiation (i.e., prevent Bob from denying that he sent the letter). In the context of digital communications, *Digital signatures* can achieve *nonrepudiation*.

Bad Guys can also launch *replay attacks* by resending packets previously sent by someone else through the network. If a packet contains some interesting information, like a password for a certain machine, Bad Guy could store it, then send it later, hoping he would be given access to the machine he sent it to. These attacks can be avoided by including a *sequence number* stored by both ends, which is incremented whenever a new message is sent. Another possibility is to use a *time stamp*—a field in the message that indicates the time it was sent. However, time stamps require a certain level of synchronization between the communicating nodes. Sequence numbers are usually preferred to time stamps. Replay attack prevention for Mobile IPv6 messages is discussed further in Chapter 5.

## 4.5 Cryptography

Cryptography is a science that uses mathematical techniques to achieve confidentiality (encryption), integrity, nonrepudiation, and authentication. Encryption is the science of communicating in a manner that prevents others from understanding the content of the communication. Encryption is an important area in cryptography; other areas include techniques to achieve integrity protection, nonrepudiation, and so on. Encryption is done by scrambling the contents of a message in an apparently random manner, but in a way that makes it possible for the legitimate receiver to unscramble it and restore its original content. Cryptographic techniques were used long before the Internet was developed. In fact, the Caesar cipher (a ciphering algorithm) is allegedly named after Julius Caesar, the Roman emperor who is said to have used it.

Cryptographers are mathematicians who develop complex cryptographic algorithms. Cryptanalysts try to evaluate these algorithms by attempting to break them, using various techniques. In some of our examples, the line that divides attackers from cryptanalysts might get blurry. However, our intention is to explain how cryptographic mechanisms work, and we do not wish to label all cryptanalysts as Bad Guys.

The words *ciphering* and *encryption* are both used in the literature to describe the operation of scrambling an original readable message, called *plaintext,* into an unreadable (encrypted) one, *ciphertext*. In this book, we use the word *encryption* to describe this process. We also use the names Alice and Bob to refer to two parties authorized to use the network and trying to communicate in a secure manner.

To ensure confidentiality when sending packets through a network, the sender must follow a certain encryption algorithm ($A$) that encrypts the plaintext message ($m$) to produce ciphertext ($c$). The algorithm requires a certain key ($K_e$) that is used during the encryption process. When receiving the message, the receiver applies the same algorithm, using a decryption key ($K_d$), to obtain the original message ($m$). This process is shown in Figure 4–1. Hence, for any message $m$:

$$A\ (K_{e,} m) = c$$

$$A\ (K_d,\ c) = m$$

There are three cryptographic functions: Secret key, Public key, and Hash functions. These three functions differ in the way they use keys and in their outputs. Secret key cryptography uses the same key at both ends, public key cryptography uses two different keys, and hash functions are *one-way transformations* that do not use keys. All three functions are discussed in this chapter.

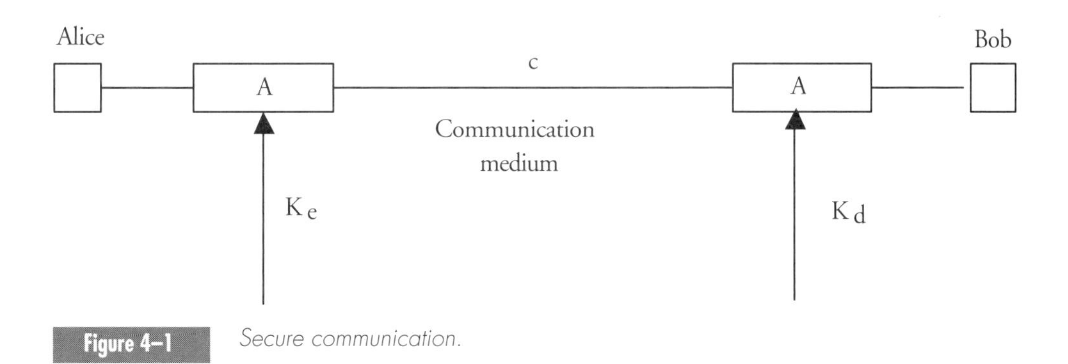


**Figure 4–1** *Secure communication.*

### 4.5.1 Encryption Algorithms and Keys

An encryption algorithm converts a plaintext input into a *random* output. By random we mean that the output of the algorithm should look like it was randomly generated and should not have recognizable patterns. One recognizable pattern could be that a change in the least significant bit of the input causes a change in the seventh bit of the output. This would make it easier for an attacker to be able to extract the original plaintext. In addition to producing an apparently random output, an algorithm must produce an output that (together with the key) uniquely defines the input. This is needed to be able to decrypt the message at the receiving end. If the same ciphertext were produced for different plaintexts, the decryption of the ciphertext could lead to nondeterministic results and defeat the point of encryption in the first place.

An algorithm is applied together with a key, as shown in Figure 4–1. The combination of an algorithm and a key would provide a unique encryption rule. That is, given a plaintext input $m$, an algorithm $A$, and a key $K$, an encrypted message $c$ can be produced. However, given the same input message and algorithm, but a different key, say $K1$, a different output, $c1$, would be produced. A cryptographic algorithm's input is the plaintext and the key. Changing either of these variables would normally result in a different ciphertext. The confidentiality of the information depends only on the confidentiality of the key. Confidentiality can be achieved by maintaining a well-known (published standard) algorithm and a secret key (known only to the correspondents). This is analogous to a lock; the design of the lock is not a secret, but the key is only kept by the lock's owner. This allows lock manufacturers to reuse the same lock designs and simply have a different key for each one. On the Internet, this property allows different machines to communicate securely without needing different encryption algorithms for different correspondents.

Breaking algorithms, or guessing the keys, might be achieved in different ways, many of which are described in [7]. In this section, we briefly discuss some of these methods to show some of the basic requirements on cryptographic algorithms.

The least sophisticated and most time-consuming attack is the *brute-force* attack, also known as *exhaustive search*. This involves guessing keys until the right one is found—that is, until it decrypts the message. Clearly, knowing that the message is correctly decrypted depends on several factors:

- The language in which the message was sent.
- The length of the message. If the original plaintext were too short (e.g., three letters), it would be difficult to know for sure when the plaintext was obtained: Is the word "but" correct? Or is it "hut," or "nut"?
- The expected content of the message (e.g., this might be known through the knowledge of a port number in the IP packet, which tells us what application is being used).

In addition, the ability of this method to succeed depends largely on the key length. If the key were only 8 bits long, this could be done very quickly. However, if the key were 56 bits long (as is the case for the Data Encryption Standard, DES), this exercise would take much longer. If guessing the key would take 30 years and cost $10 million, when the actual value of the information is worth $1,000 and the content is useful for only 10 days, then clearly this is not a good way to go. Hence, a cryptographic algorithm is successful if it ensures that the original plaintext cannot be retrieved for the lifetime during which the data is valuable. Today, 128-bit keys are recommended for symmetric keys. *Asymmetric keys* (used in public key cryptography) are considerably longer, as shown later in this chapter.

A simpler attack would involve knowledge of the original plaintext being sent (e.g., through out-of-band mechanisms like human spying). If the attacker knows that Alice will send a message to Bob saying, "Please buy 12000 shares," he can somehow compare the encrypted text with the original plaintext and attempt to break the encryption scheme. Such an attack is referred to in [7] as the *known plaintext* attack. This attack is executed in order for the attacker to know the encryption key, since the attacker clearly knows the plaintext.

Encryption algorithms vary with the type of cryptographic system. Public key cryptography uses different algorithms, from secret key cryptography, due to the nature of the keys being used. In the following sections, we discuss both systems, starting with secret key cryptography.

## 4.5.2 Secret Key Encryption

Consider Figure 4–2. A message $m$ is encrypted using algorithm $A$ and encryption key $K_e$. At the receiving side, the message is decrypted with $A$ and $K_d$. When secret key encryption is used, $K_e = K_d$. That is, the same key is used for both encryption and decryption. For this reason, secret key encryption is also

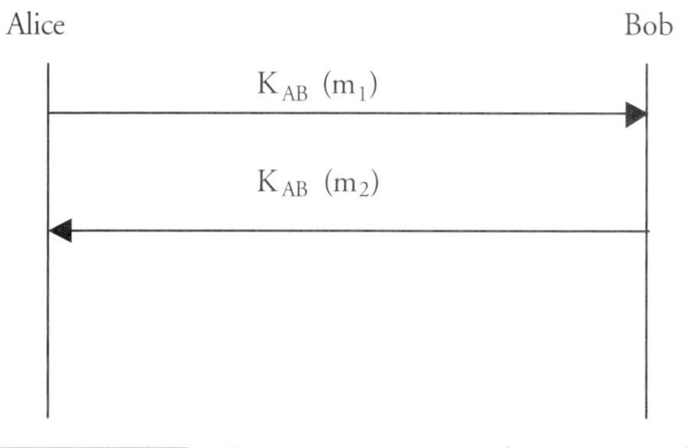


**Figure 4–2** *Encryption using secret key cryptography.*

called *symmetric key* encryption (since the same key is used for encryption and decryption). If Alice knows that she will, at some point in the future, communicate with Bob, they can agree on a key to use for their secure communication. Obviously, the key should be sent in a secure manner (so that it cannot be intercepted by Bad Guy); the agreement on such key would need to happen in an out-of-band fashion. For instance, Alice might give the key to Bob when they meet, or, if there is another secure channel available, they can send the key over it. Alternatively, a person trusted by both Alice and Bob may take the key from Alice and give it to Bob. Typically, a network administrator would manually configure the key on both Alice's and Bob's devices.

Following their key exchange, Alice can send a message to Bob encrypted with their secret key $K_{AB}$. Bob would then be able to decrypt this message with the same key and retrieve the plaintext. The same process can be done when Bob is sending a message to Alice, as shown in Figure 4–2.

Examples of secret key encryption algorithms include the DES, 3DES (DES performed three times using two or three different keys), and the International Data Encryption Algorithm (IDEA). When Alice and Bob agree on the secret key for securing their communication, they also need to agree on the secret key algorithm that they will use. This is part of establishing a security association, as discussed in section 4.5.6.

#### 4.5.2.1 MUTUAL AUTHENTICATION WITH SECRET KEYS

So far we have seen how Alice and Bob can use their secret key to encrypt sensitive messages and ensure confidentiality. In addition to confidentiality, Alice wants to ensure that she is talking to Bob. Depending on the particular communication scenario, Bob may also want to make sure that he is talking to Alice. Figure 4–3 shows one way of achieving mutual authentication between Alice and Bob.

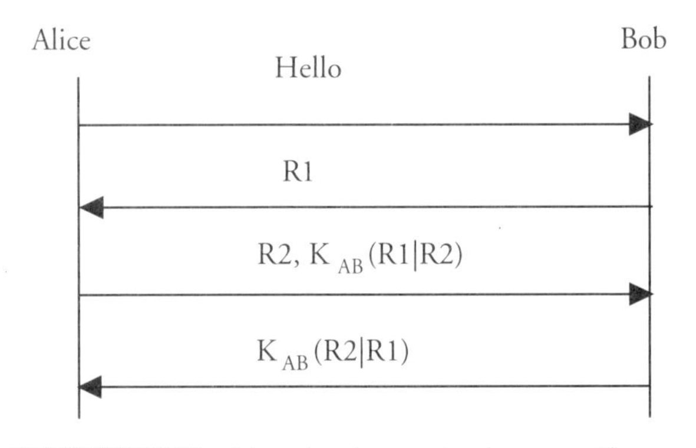


**Figure 4–3** *Mutual authentication between Alice and Bob.*

In this example, Alice is initiating communication with Bob. Bob generates a random number, $R1$, and sends it to Alice. If Alice can encrypt that number with their secret key, Bob can be sure that he is talking to Alice, since no one else knows the key. However, since Alice wants to authenticate Bob, she includes another random number, $R2$, in addition to the encrypted concatenation of $R1$ and $R2$ ("|" means concatenate). The reason for encrypting both $R1$ and $R2$ is that it allows Bob to verify that the sender of $R2$ knows $K_{AB}$—that is, Alice. If $R2$ were not encrypted, Bad Guy could have intercepted the encrypted message and pretended to be Alice. When Bob receives this message, he can verify that he is communicating with Alice. To authenticate himself to Alice, Bob concatenates $R2$ and $R1$, encrypts them, and sends them to Alice. Note the reverse order of concatenation used to ensure that the message is different from the previous one and therefore could not have been replayed by Bad Guy. Finally, Alice and Bob can proceed with their normal communication.

#### 4.5.2.2 LIMITATIONS OF SECRET KEY CRYPTOGRAPHY

While secret key cryptography seems like a simple way to communicate securely, its use is limited by the need to exchange keys securely. Alice and Bob would need to know that they will communicate with each other at some point in the future, agree on keys over a secure channel, and then start communication. Clearly, such a mechanism is not suited to spontaneous communication with previously unknown correspondents. For instance, if Howard left his contact details with Alice (and he was previously unknown to her) and asked her to contact him, she would not be able to do so securely, since they have not agreed on a secret key. This restriction represents a scalability issue for secret key cryptography, since it cannot be used for communication between any two random correspondents. The use of trusted third parties can alleviate this problem, but they have to share the same trusted third party.

In addition to the scalability limitations, secret key cryptography usually does not provide nonrepudiation in a convenient way. Suppose Alice has asked Bob (via confidential email) to sell 1,000 of her shares at the current market price of $10. At the same time, the share price dropped to $5. Alice could claim that she never ordered this sale and could sue Bob. If secret key cryptography is used for Alice's order, Bob cannot prove that Alice actually made the order, since he also has the same key and Alice could claim that he generated the message himself. For nonrepudiation to be provided, we need to prove that no one except the sender could have generated the message, which is not possible when symmetric keying is used.[1].

[1] Strictly speaking, it is possible, but again involves a common trusted party and non-shared secret keys.

### 4.5.3 Public Key Cryptography

Unlike secret key cryptography, public key cryptography uses two different keys: a public key and a private key. When used to gain confidentiality, the public key is used for encryption and the private key is used for decryption. The opposite is done when public key cryptography is used for authentication. Both uses (confidentiality and authentication) are described in the following sections. Public key cryptography is also called asymmetric cryptography due to its use of two different keys for encryption and decryption. A public key—as the name suggests—is public and may be known by anyone. On the other hand, a private key is known to only one entity and should never be revealed to anyone else at any time. Hence, an important requirement of public key cryptography is that the knowledge of one key cannot be used to derive the other.

Public key cryptography is based on the belief that given a message encrypted with $k_1$ and algorithm $A$, there is a very small set of values (in practice, often only one value), $k_2$, that can be used with $A$ to decrypt this message. In addition, $k2$ (or any other admissible decryption key) is hard to derive from $k1$. While this is not mathematically proven, it is believed that it is computationally infeasible to prove otherwise. This makes the case for a strong cryptographic function. Hence, there is a need for two keys in this cryptographic function. The generation of these keys involves a lot of computational power and is often based on fundamental Modular arithmetic theorems, Euclid's algorithm, and the Chinese remainder theorem.[2] While we do not present these theorems and their proofs here, we present some mathematical terms and facts, without proof, in order to make the RSA example in the next section easier to understand.

First we need to understand the concept of *relatively prime* numbers. A *prime* number is a positive integer that can be divided only by itself and 1 evenly (i.e., producing whole integers); for instance, 1, 2, 3, 5, and 7 are all prime numbers, while 10 is not because it is divisible by 2 and 5. On the other hand, two numbers are relatively prime if there is no number greater than 1 that divides **both** evenly. For instance, 3 and 10 are relatively prime. On the other hand, 8 and 10 are not relatively prime because they are both divisible by 2.

The next concept to be introduced is modular arithmetic. Modular arithmetic uses positive integers that are less than a number, $n$, to perform addition and multiplication. The result of such operations is then divided by the number $n$, and the remainder is the final result of the operation: modulo $n$, or mod $n$. For instance, $3 + 9 \bmod 10 = 2$. Similarly, $3 \times 9 \bmod 10 = 7$.

Finally, we state the following as a fact without proof: if two prime numbers, say $a$ and $b$, are multiplied, producing the number $n$, the number of integers that are *relatively prime* to $n$ equals $(a - 1)(b - 1)$. The notation $\phi(n)$

[2] Readers who are not familiar with Modular arithmetic or wish to have a deeper understanding of the mathematics behind public key cryptography should consult [7] and/or [20] for a comprehensive and detailed explanation of the background.

(called *totient*) refers to the number of integers that are *relatively prime* to $n$. For instance, $\phi(10) = 4$; that is, there are four integers less than 10 that are *relatively prime* to 10 (i.e., 1, 3, 7, 9). It turns out that for any number $y$ within the set of numbers 1, ..., $\phi(n)$, such that $y$ is relatively prime to $\phi(n)$, it is possible to pick a number, $c$, such that $y \times c = 1 \bmod \phi(n)$. Hence, $y$ and $c$ are *multiplicative inverses* mod $\phi(n)$. It also turns out that $x^y \bmod n = x^{(y \bmod \phi(n))} \bmod n$, where $x < n$. This is generally true if $n$ is a prime number or the product of two distinct prime numbers. It therefore also holds that

$$(x^y)^c \bmod n = x^{y \times c} \bmod n = x^{(y \times c \bmod \phi(n))} \bmod n = x^1 \bmod n = x$$

Let's illustrate this by considering the following example: the number 77 is a product of two distinct prime numbers, 7 and 11. Therefore, $\phi(77) = 6 \times 10 = 60$. That is, there are 60 numbers less than 77 that are relatively prime to 77. From the 60 numbers 1, ..., 60, we can pick the numbers 7 and 43 as *multiplicative inverses* mod 60. That is, $7 \times 43 \bmod 60 = 1$ (301 decimal). We can also see that $(x^7)^{43} \bmod 77 = x^{7 \times 43} \bmod 77 = x^{(301 \bmod 60)} \bmod 77 = x^1 \bmod 77 = x$.

Based on the knowledge we gained in the above crash course, we describe in the next section the fundamental steps needed to generate a public/private key pair (using RSA, the most popular public key algorithm, as an example) to illustrate the complexity of such operation.

#### 4.5.3.1 AN RSA EXAMPLE

We illustrate how to generate a public/private key pair by showing an example based on the RSA algorithm (named after its inventors, Rivest, Shamir, and Adleman). In this algorithm, the following steps must be taken:

1. Find two large prime numbers, say $p$ and $q$ (around 200 digits each). These numbers must not be disclosed to other entities.
2. Multiply those two numbers and call the result $n$.
3. $\phi(n)$ contains all the numbers that are relatively prime to $n$. $\phi(n) = (p - 1)(q - 1)$. To generate a public key, find a number $e$, which is relatively prime to $\phi(n)$. The public key is $(e|n)$.
4. To generate a private key, using Euclid's algorithm, find the multiplicative inverse for $e$ (call it $d$) such that $de \bmod \phi(n) = 1$. The private key is $(d|n)$.
5. Now, given a message $m$, it can be encrypted by raising it to $m^e \bmod n$. The encrypted message can be decrypted by raising it to $d$—that is, $(m^e)^d \bmod n$. Since $e$ and $d$ are multiplicative inverses, $m^{ed} \bmod n = m$—the original message.

The above algorithm seems simple enough at first glance; however, some of the steps involve computationally complex and intensive operations. The first step involves finding two very large prime numbers and then multiplying them. Finding these large (over 200 digits) prime numbers is not a trivial task. The numbers would need to be tested to see if they are in fact prime. Luckily, there are quite efficient tests that can be done to provide a high degree of certainty about whether or not a number is prime.[3] However, it is important to note that achieving higher certainty would involve more computation. On the other hand, a public/private key pair is expected to have a long lifetime; therefore the heavy computation involved in the key generation may not be a substantial obstacle for many devices.

In addition to testing prime numbers, encrypting messages using public key cryptographic algorithms can be computationally cumbersome. As shown in step 5 above, it involves raising a large number to the *e*th power. For this reason, a secure channel may utilize both symmetric (for encryption) and asymmetric (for authentication and nonrepudiation) cryptography to get confidentiality, authentication, and nonrepudiation. Public key cryptography is typically not used for encrypting large messages. A typical application is to securely transfer a relatively short symmetric key between Alice and Bob, and then rely on that symmetric key for "bulk" encryption.

#### 4.5.3.2 ENCRYPTING MESSAGES USING PUBLIC KEY CRYPTOGRAPHY

Consider a scenario where Alice and Bob need to communicate confidential information about a secret business deal. Alice and Bob each have a key pair, and they both know each other's public keys. When Alice sends a message to Bob, she needs to encrypt it with Bob's public key. Since only one private key can invert this operation and retrieve the original message, Bob will be the only one able to understand the contents of such message. Figure 4–4 illustrates this process. $PK_{bob}(m_1)$ is used to describe that Bob's public key was used to encrypt the message $m_1$.

When Bob replies to Alice, he encrypts his message, $m_2$, with Alice's public key $PK_{alice}$. This way he can ensure that only Alice will be able to understand the contents of the message he sent.

#### 4.5.3.3 AUTHENTICATION WITH PUBLIC KEY CRYPTOGRAPHY

Let's assume Bob and Alice don't care about the confidentiality of their messages, but they think it's important to ensure that the messages are not modified en route. In this case, when Bob communicates with Alice, he could encrypt messages using his own private key. The only key that can decrypt the signature of this message is Bob's public key. That is, anyone can decrypt and verify this message (since the public key can be known by anyone). However, this is sufficient to ensure that Bob sent the message because he is the only one who knows the corresponding private key.

[3] Again, see [7] for the details of these tests.

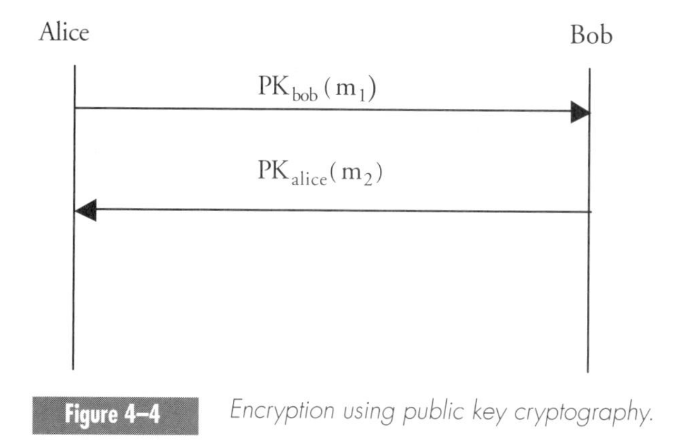


Figure 4–4 *Encryption using public key cryptography.*

#### 4.5.3.4 DIGITAL SIGNATURES

In Figure 4–4, we showed how public key cryptography can be used to encrypt messages between Alice and Bob. This ensures that no one except Alice and Bob know the contents of the messages sent from Alice to Bob, and vice versa. But this does not assure Bob that the messages were not modified after being sent by Alice. Bad Guy could modify the message sent by Alice and send false information to Bob. This might be difficult if the content of the messages coming from Alice depends on the messages sent from Bob. For example, if Bob sent a message saying, "Should we book a trip to Hawaii?" and Bad Guy (pretending to be Alice) replies with, "What a nice day this is," Bob is likely to realize that something is wrong. However, there are other examples where Bad Guy can modify Alice's reply information with undesirable consequences. We stated in the previous section that entity authentication can be achieved by public key cryptography. However, such authentication still involves the computationally intensive encryption of large messages. Digital signatures allow Bob and Alice to authenticate messages while reducing the amount of computation involved.

Digital signatures depend on hash functions, which are described in section 4.5.4. Hash functions are one-way functions that take large inputs and produce a fixed-size output. The size of the output depends on the algorithm (typically between 100 and 200 bits). Given a particular output, the original message cannot be computed.

A digital signature is similar to a human fingerprint in the sense that it is extremely difficult to forge. If Alice wanted to ensure that Bob could verify that the received messages actually came from Alice and were not modified en route, she would have to sign the messages before sending them to Bob. Alice's signature is a number that is generated based on two inputs: her private key and the contents of the message. To do this, Alice follows these steps:

1. Compute a hash of the message.
2. Encrypt the hash with her private key, and include the encrypted hash in the message sent to Bob.
3. When Bob receives the message, he performs step 1.
4. Bob decrypts the encrypted hash using Alice's public key. If he gets the same value as produced in step 3, then he knows that the message was not modified en route and that it came from Alice.

Since the above algorithm provides a "fingerprint" that depends on the content of the message, Bad Guy cannot simply modify the message or copy a fingerprint and put it on another message.

In addition to authentication, digital signatures allow for nonrepudiation. If Alice sends a signed message to Bob, she cannot later claim that someone else impersonated her, since the signature was based on her private key, of which no one else has knowledge.

#### 4.5.3.5 CERTIFICATES

So far we have discussed how a public/private key pair can be used to provide confidentiality, authentication, integrity, and nonrepudiation. It is now clear that the use of a public/private key pair that can satisfy these requirements relies on three important assumptions:

1. For every public key, there is only one private key, and vice versa.
2. A public key can be known by anyone. A private key is known only to the owner of the public key.
3. Messages encrypted with a public key can be decrypted only with the corresponding private key, and vice versa.

The next question is, How can Alice know Bob's public key? A trivial answer would be that Alice can simply ask for it and Bob will give it to her. After all, it is public knowledge and therefore can be sent as plaintext. The problem with this approach is that Bad Guy can intercept Alice's request or Bob's reply, and send his own public key to Alice. Since public keys do not contain any information about Bob's identity, Alice would innocently use Bad Guy's public key to encrypt messages that would effectively be sent to Bad Guy. Bad Guy would also be able to decrypt the messages and send replies if necessary, since he does have the right private key needed for decryption. This attack is shown in Figure 4–5.

The problem here is not that the keys were forged or an encryption algorithm was broken; the problem is that Alice has no means of associating a

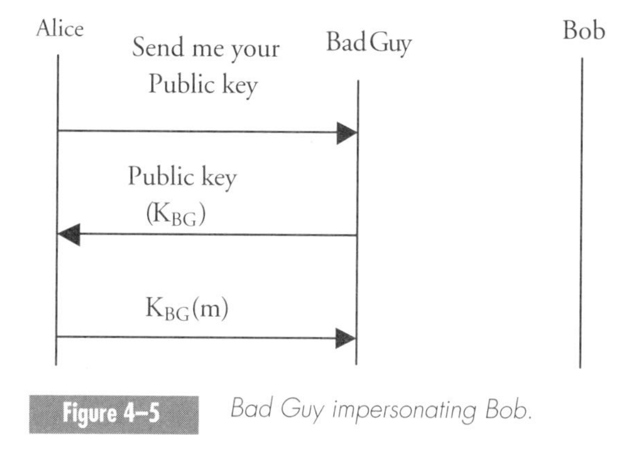


Figure 4–5 *Bad Guy impersonating Bob.*

public key with Bob. Therefore, there is no way of checking who owns that public key.

This impersonation attack can be avoided by using public key certificates. A public key certificate is a digitally signed statement by a trusted entity—a Certificate Authority (CA)—publishing some information about another entity (e.g., Bob). The certificate is signed by the CA's private key. Essentially, the certificate binds an identifier to a public key pair. The identifier can be a person's name, an organization, an email address, or an IP address. Today's widely used standard for public key certificates is the X.509 standard [6]. It is used by several protocols, including Privacy Extended Mail (PEM), Secure Socket Layer (SSL, known as Transport Layer Security, TLS), Secure HTTP (S-HTTP), and the Public Key Cryptography Standard (PKCS). There are three different versions of the X.509 certificate standard; the fields contained in X.509 certificates are shown below with reference to their support in different versions:

- version
- serialNumber
- signature
- issuer
- validity
- subject
- subjectPublicKeyInfo
- issuerUniqueIdentifier (optional; supported in versions 2 and 3 only)
- subjectUniqueIdentifier (optional; supported in versions 2 and 3 only)
- extensions (optional; supported in version 3 only)
- signature

The *version* identifies the certificate's version and defaults to 1. The *serialNumber* is the certificate's serial number. The *signature* field (deceptively named) in fact identifies a value corresponding to the algorithm used to generate the signature in this certificate (e.g., a number corresponding to the Digital Signature Standard, DSS). The *issuer* is the name of the CA. The *validity* field contains the starting and expiry time of the certificate. The syntax used to define this field assumes that CAs will agree on a global time zone that can be used as a reference point. When a certificate is no longer valid, it is considered to be *revoked*. CAs advertise revoked certificates in a Certificate Revocation List (CRL). Revocation lists are upgraded regularly; the frequency of these updates is also included in the CRL. This allows entities to verify whether a public key for one of their correspondents is still valid and how often they need to check a CA's CRL. A more "paranoid" approach is to do an online check of the status of the certificate each time it is to be used.

The *subject* is the name of the entity being certified by the CA (e.g., Bob Black), and the *subjectPublicKeyInfo* contains the public key of the subject. The *issuerUniqueIdentifier* is an optional field that can be used to further identify the CA. The *subjectUniqueIdentifier* is a field that provides a unique identifier for the subject; this is useful in cases where the subject's name is not unique within an organization or a CA. The *extensions* field describes possible future extensions that can be added to public key certificates. The *signature* field contains the CA's signature computed over the rest of the information, using its own private key.

Clearly, the CA's private key is extremely important and must be kept very safe. The private key is stored in a tamper-proof box, referred to as the Certificate Signing Unit (CSU). A CA may also change its public/private key pairs (e.g., every year) to ensure that the private key remains unknown.

Let us now reconsider the communication scenario shown in Figure 4–5. If Bob sends Alice his certificate, she can verify the CA's signature by obtaining the CA's public key. Once this is known, the signature can be verified. When the CA's signature is verified, Alice would trust that the subject's public key included in the certificate actually belongs to Bob and communicate with him securely.

The next question is, How does Alice know the CA's public key? This can be manually configured on Alice's machine. However, if Alice wishes to use public key cryptography with many different (and sometimes previously unknown) correspondents from different parts of the world who have certificates from different authorities, manual configuration becomes difficult. In some cases, Alice would not even know the CA and might not want to trust it. To increase the scalability of this mechanism, CAs have a hierarchical structure. Figure 4–6 shows a hierarchy of CAs. In this hierarchy, CA1, CA3, and CA5 trust each other. CA5 is the highest node in the hierarchy; certificates from CA3 are also signed by CA5 (as well as CA3). Certificates from CA1 are signed by both CA3 and CA5.

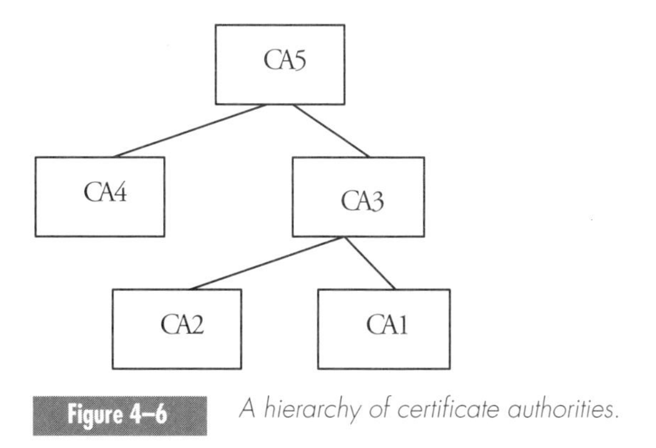


**Figure 4–6** A hierarchy of certificate authorities.

Hence, Alice does not need to be aware of CA1 (which could be Bob's CA) if she is aware and trusts either CA3 or CA5. Since Bob's certificate is signed by all the CAs above CA1 in the hierarchy, only one CA needs to be trusted (CA5), and one public key would need to be configured in Alice's machine.

### 4.5.4 Hash Functions, Message Digests, and Message Authentication Codes

*Hash functions* and *message digests* are synonyms for a class of one-way functions. The reason for calling them one-way functions is that given the output of their operations, the input cannot be efficiently computed. For example, let $h(m) = x$ be the output of a hash function $h$ with an input message $m$; given $x$, the original message $m$ cannot be efficiently computed. By "efficiently computed" we mean that an attacker would need to try a large number of messages to guess the input.

The requirements on hash functions are similar to those mentioned earlier for encryption algorithms. That is, the output of these functions should seem random, with approximately half the bits set to the opposite values of the other half. Furthermore, changing one bit of the input should result in a completely different output; That is, there should be no statistical trends observed when comparing two inputs with only one bit changed. The only way of finding a message that will produce $h(m)$ should be to try many random messages and compare the output against $h(m)$. Hence, reducing the probability of finding a message that produces the same output hinges on making it very difficult to compute the necessary number of random messages needed in order to find it.

A hash function takes a variable-size input and produces a fixed-size output—let's call it $y$. Given this fact, it is clear that there is an infinity of messages that, when hashed, could produce the same output. If $y$ is too small, then it will be easy to try all possible values that will produce $h(m)$. Suppose that $y$ is 32 bits; this means that one would need to try approximately $2^{32}$ different messages to produce a particular output. Or, put differently, given a certain output, the probability of finding another input that would give the same output is $2^{-32}$. On the other hand, one would need to try approximately $2^{16}$ different messages to find two messages that will produce the same hash.[4]

Note the difference between the two cases above and when they can become relevant. That is, finding two different messages that produce the same output becomes a problem when an attacker is trying to generate two different messages that would produce the same hash, that is, that would look like they have the same content. On the other hand, finding a message that, when hashed, produces a particular output is relevant when an attacker is trying to modify an existing message without making it obvious to the receiver that the message was modified.

The most commonly known hash functions are the Secure Hash Algorithm (SHA-1) and MD5 (MD stands for Message Digest). The output of all of these algorithms is 128 bits or more.

#### 4.5.4.1 USING HASH FUNCTIONS FOR AUTHENTICATION

In section 4.5.2.1, we saw how Alice and Bob can authenticate each other by using their secret key, a secret key encryption algorithm (like 3DES), and a random number generator. Figure 4–7 shows that mutual authentication can also be achieved by replacing the secret key algorithm with a hash function.

The first two messages are identical to those in section 4.5.2.1. After receiving $R1$, Alice computes $h(K_{AB} | R1 | R2)$. She concatenates $K_{AB}$, $R1$, and $R2$,

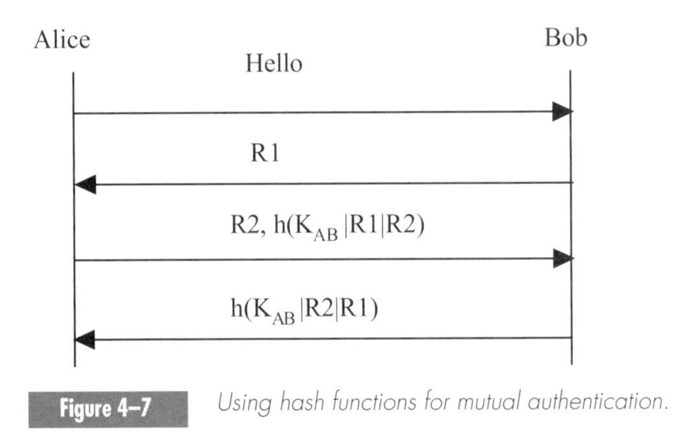


**Figure 4–7** *Using hash functions for mutual authentication.*

[4] This is known as the Birthday paradox. The proof is not discussed in this book, but can be found in [7].

then hashes them as a single message. Alice then sends $h(K_{AB}|R1|R2)$ along with $R2$. When Bob receives the message, he won't be able to reverse the hash function; however, since he knows $R1$, $R2$ (sent as plaintext), and $K_{AB}$, he could generate $h(K_{AB}|R1|R2)$ and verify whether the message came from Alice. Following this, Bob calculates $h(K_{AB}|R2|R1)$ and sends it to Alice, who is able to verify that the message came from Bob and is finally glad to continue their normal communication.

If we replace $R1$ and $R2$ with a message $m$, $h(K_{AB}|m)$ can be included at the end of this message when sent from Alice to Bob. When Bob receives the message, he can recalculate the hash function to prove that the message is authentic (i.e., has not been modified by Bad Guy). Hence, $h(K_{AB}|m)$ can be referred to as the Message Authentication Code (MAC). For various security reasons, however, it is recommended to authenticate using $h(K_{AB}|m|K_{AB})$ or even more elaborate constructions. A MAC is also called the integrity check value (ICV) or the message integrity code (MIC).

HMAC [12] is a mechanism designed to seed a hash function with a secret key (and hence sometimes called a keyed hash function) and use the hash output for message authentication (hence the name hashed MAC). HMAC is independent of the actual hash function being used. When used with SHA-1, it is denoted HMAC SHA-1. HMAC MD5 denotes a keyed MD5 hash function using the HMAC algorithm.

It is worth noting that the above message authentication does not provide nonrepudiation, which is provided with digital signatures. Hence, the use of this mechanism is usually limited to cases where providing nonrepudiation is not needed. As seen earlier, random numbers (like R1 and R2) play a key role in cryptography. The following section introduces random numbers and cookies.

### 4.5.5 Nonces and Cookies

A *nonce* is a random number that is used only once. Nonces can also be sequence numbers, but this would require nodes to store them in a nonvolatile memory to avoid losing them when they crash. Nonces can be used to avoid replay attacks, since by definition they are used only once. If a sequence number is not used, a large enough random number can be used, as it is unlikely to be generated twice. Another possibility is to use time stamps, but that is often inconvenient, as it requires clock synchronization. Mobile IPv6 uses nonces to secure binding updates, as explained in Chapter 5.

A cookie is also a large number that may contain some information like a transaction identity between two nodes or programs. Cookies are used between HTTP servers and clients. For instance, some Web sites send cookies to clients when they connect to the server. The cookie may contain information about the client itself or what the user did on that server. When the client reconnects to the server, it sends the cookie back, hence acting as an identifier for the client and keeping state about it.

## 4.5.6 Establishing a Security Association

In all our earlier examples, we assumed that Alice and Bob somehow knew each other's public keys, or a shared secret, and used them to authenticate each other or for encrypting sensitive messages. We also assumed that they knew what encryption algorithm or hash function should be used. Furthermore, we assumed that both parties knew the different parameters relevant to particular algorithms (e.g., RSA). Such knowledge represents a Security Association (SA), which is an agreement between two entities that allows them to know all the parameters necessary in order to communicate securely. These include keys, algorithms and their parameters, policies related to the type of data being secured, and the lifetime of this agreement.

In the absence of manual configuration, SAs need to be established between two nodes in a dynamic manner. The Internet Security Association and Key Management Protocol (ISAKMP) defines a framework that includes a number of parameters and secure handshakes that allow nodes to agree on an SA in a secure manner. The Internet Key Exchange (IKE) uses a hybrid of the mechanisms defined in ISAKMP [13], Oakley [17], and SKEME [11] to exchange keys and establish an SA between two nodes in a secure manner. In this section, we present an overview of IKE's operation and use that as an example to show how an SA can be established without manual configuration.[5]

IKE can work between two nodes, using public or secret key cryptography. To limit the time in which keys are exposed on a communication channel, IKE does not use existing keys (secret or public keys already configured in nodes) for securing communication between different applications. Instead, it creates an SA between the two nodes, which uses different keys from their "permanent" public or secret keys. To do this, the protocol is divided into phase 1 and phase 2, as shown in Figure 4–8. Each phase has different modes of operation. That is, each phase is needed to achieve a certain goal, but this can be done in different ways (*modes*). In phase 1, the two nodes create a secure channel in order to secure the communication that will take place in phase 2. This can be done using public keys and certificates, as discussed earlier, or using a preconfigured secret key for mutual authentication and securing the channel. Phase 1 has two different modes: *main mode* and *aggressive mode*. Aggressive mode involves fewer messages than main mode, but this results in fewer options that can be exchanged for the secure channel (e.g., fewer attributes are allowed for the SA parameters due to the rules for message construction). Phase 2 uses the secure channel created in phase 1 to negotiate an SA; this includes the protocol that will be used to protect communication (e.g., ESP or AH), the type of data that will be protected (e.g., TCP, UDP, or all packets), the lifetime of the SA,

[5] IKE is a complex protocol that deserves more than one chapter to discuss it; however, our intention is to provide an overview of the protocol. A detailed understanding would require reading RFCs 2408 and 2409.

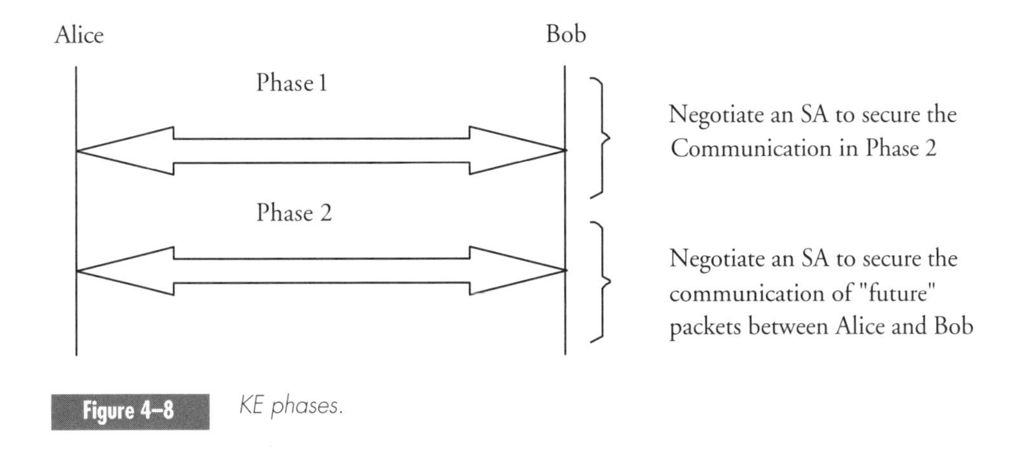


**Figure 4–8** *KE phases.*

and the material that can be used to generate the keys. The exact method used to generate keys from the key material provided by IKE is left up to the protection mechanism used later to protect the communication. For instance, if ESP is used to protect traffic, it is the responsibility of the ESP implementation to take the key material provided by IKE and use it to produce keys.

Following the IKE exchange, two nodes would have enough material to generate two different keys. Each key is used to protect information in one direction only. For instance, if Bob is communicating with Alice, two keys are generated: $K_{Bob\text{->}Alice}$ and $K_{Alice\text{->}Bob}$. When Alice sends a message to Bob, she can encrypt it with $K_{Alice\text{->}Bob}$; Bob can decrypt it with the same key. On the other hand, when Bob sends a message to Alice, he can encrypt it with $K_{Bob\text{->}Alice}$ and Alice should decrypt it with the same key. Hence, symmetric cryptography is used with unidirectional SAs, as seen in Figure 4–9.

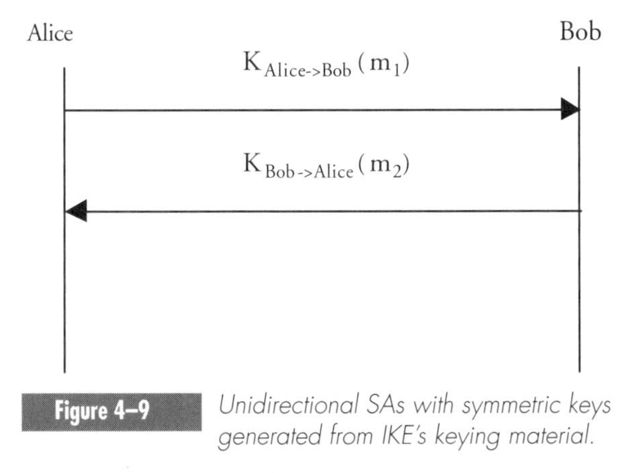


**Figure 4–9** *Unidirectional SAs with symmetric keys generated from IKE's keying material.*

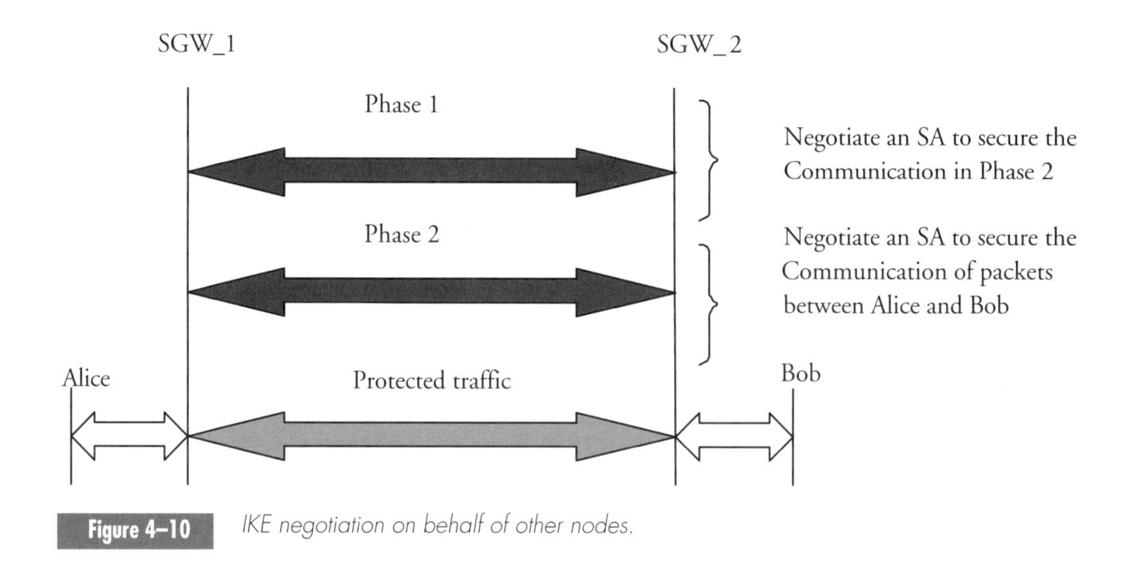


**Figure 4–10** *IKE negotiation on behalf of other nodes.*

IKE allows two nodes to negotiate an SA for two other nodes, using a proxy mode. In phase 2, either the initiator or responder can include an identity other than itself to be used for the SA being negotiated. The original intention of this option was to allow two different security gateways to negotiate an SA on behalf of other nodes within their domain (see Figure 4–10). Interestingly enough, this feature will be useful when we consider Mobile IPv6 security.

In this figure, SGW_1 and SGW_2 are security gateways, which are devices (routers) that check packets (on their egress or ingress interfaces) to see if they have particular security requirements and apply the corresponding protection mechanisms on these packets if necessary. In this scenario, they may wish to negotiate an ESP tunnel to allow Alice and Bob to communicate securely. This scenario may be needed if Alice and Bob are working in two different sites that belong to the same organization (e.g., a multinational company or a university with several campuses). Such organization may not need to secure its network internally but may need to secure traffic between its different campuses, which communicate over the Internet. This can be done as shown above by essentially creating a secure Virtual Private Network (VPN) using *tunnel mode* ESP or AH.

## 4.5.7 Cryptographically Generated Addresses

In section 4.5.3, we saw how Alice and Bob can communicate over a secure channel using public key cryptography. Furthermore, we saw how they can authenticate each other's identities and public keys by using X.509 public key certificates that bind a certain subject to a public key.

In some cases, there might be a need to bind a certain IP address to a public key. Consider the following example. Suppose Alice wishes to talk to Bob and she knows that they share a link. Alice would send a neighbor solicitation message (described in Chapter 2) to resolve Bob's address. The reply to this message binds Bob's IP address to a link-layer address. This allows Alice to send the following packets to Bob's link-layer address. But it is possible that Bad Guy, instead of Bob, replies to the solicitation message and causes Alice to send her traffic to him. Note that we cannot run IKE in this case to authenticate Bob, since sending the first IKE message would require a neighbor solicitation message. This leaves us with the chicken and egg problem described in [1]. That is, we need IKE for authentication, but IKE would need address resolution, which is not secure. This can also be expressed differently: IKE may be used to secure Neighbor Discovery messages, but to send an IKE message, we need to send a Neighbor Discovery message (for address resolution) in an insecure manner, which defeats the purpose of securing it! It turns out that this is not the only case where one node needs to prove that another node is assigned a particular address; Mobile IPv6 is a perfect example, as described in Chapter 5.

Cryptographically Generated Addresses (CGAs) are designed to bind the IP address of a node to its public key and consequently prove that a node is assigned a particular address. A node may use CGAs in the following manner:[6]

1. Generate a public key pair.
2. Compute $h(PK|context)$ where $h(m)$ is a hash of $m$, and $PK$ is the node's public key. We use *context* to indicate that another value is concatenated with the $PK$ and used as an input to the hash function. The content of this extra parameter depends on the context in which CGAs are used. Two different applications for CGAs and their corresponding context will be discussed.
3. Include 64 bits of the output of the hash function in the interface identifier field.
4. Set the $u$- and $g$- bits in the interface identifier to zero, indicating that the interface identifier is not globally unique.
5. Whenever proof of address ownership is needed for a particular message, the message would include the node's $PK$ and would be signed by the node's private key.

When a receiver gets the message (step 5), it can easily verify that the sender had the private key corresponding to the public key included in the message. It can also verify that the sending node owns the address, provided that *context* is also included in the message, hence allowing the hash to be verified.

[6] This is a simplified description for the generation of a cryptographically generated address. More details can be found in [2].

An attacker wanting to break this mechanism would need to try approximately $2^{62}$ different public keys to be able to find the private key used to sign this message (only 62 bits from the hash function are left in the interface identifier after setting the *u*- and *g*-bits). This is considered secure enough for the type of applications it may be used for. For instance, if it takes an attacker one year to find the private key corresponding to the hash of *PK* in the interface identifier and steal a node's address (e.g., respond to neighbor solicitations sent to that node), by that stage a node might have left the link or changed its public/private key pair. On the other hand, it would take the attacker approximately $2^{31}$ attempts to be able to generate two different random strings that produce the same hash output. This may be considered a problem when protection against repudiation is needed. However, it is worth noting that just because an attacker can find two different strings that can produce the same hash, it is highly unlikely that he will be able to find the right public/private key pair used to sign the entire message sent from the CGA.

Apart from the fact that CGAs solve an important problem, which is not conveniently solved by the traditional application of public cryptography mechanisms, one of their main advantages is that they do not require any infrastructure support. Only the sender and the ultimate receiver need to support this mechanism. This feature makes CGAs an easy solution to deploy. The same cannot be said about traditional public key cryptography mechanisms, which rely on CAs. This is illustrated in the following section.

#### 4.5.7.1 CAN WE SOLVE THIS PROBLEM WITH X.509 CERTIFICATES?

In theory, X.509 certificates can be used to solve the proof of address ownership problem. However, there are serious limitations to this approach. Let's consider how certificates can be used to solve this problem, and then observe the limitations of this approach.

Using certificates to prove address ownership would require a node to include its address in a certificate. For instance, the *subject* field or the *subject UniqueIdentifier* field could contain an IP address. Since the certificate is signed by a CA, a correspondent could possibly verify (given the CA's public key) that the node owning that certificate also owns the address inside it. This would work, but let's consider the implications of this approach. Certificates are not meant to change on a regular basis; they can last for more than one year. Furthermore, certificates take some time to issue. On the other hand, addresses can change regularly, especially for mobile nodes. How can we update certificates for nodes moving at 50 km/hr in a car or a bus (i.e., considering the change to the node's care-of address, not home address)?

Another problem with this approach is that it combines two different parameters that have different roles and are obtained from two different entities: IP addresses and certificates. Certificates are issued by CAs, while IP addresses are provided by network operators or ISPs. A CA may not belong to an ISP (in many cases they don't). Adding the IP address to the certificate can make it difficult to manage certificates when a network is renumbered, for example. This would require some cooperation between the ISPs and CAs or some notification from ISPs to different nodes to inform them that their certificates are about to be revoked due to renumbering. The difficulties with this approach are clearly tied to the fact that it requires infrastructure support; that is, support from entities other than the two communicating nodes.

### 4.5.8 Firewalls and Application Level Gateways

The mathematics behind cryptography can solve many security problems. However, sometimes there is a need for physical protection of networks. This is especially true when strong cryptographic mechanisms are not applied in all communication or when they are applied incorrectly. Firewalls are special routers located at the edge of a network (toward the Internet) to protect the network from unauthorized or unwanted traffic in both directions (incoming and outgoing). To do this, firewalls must be configured with certain policies. For example, they can be configured to not forward packets containing a particular source address or a range of addresses. This simple form of firewall is often called a *filtering router*.

Firewalls can be more sophisticated; they can filter on more than just addresses. For instance, a firewall may be configured to not allow any ICMP traffic.

Application Level Gateways (ALGs) are applications that run on computers located between two end users, one inside the protected network and another somewhere on the Internet. ALGs terminate connections between the two end hosts and act as proxies. That is, in order to send a file from inside the network to the outside, the host application would need to establish a connection with the ALG and send the file to it; the ALG would in turn establish another connection to the intended end host and relay the same file. ALGs combined with firewalls present a typical strategy used by enterprise network administrators today to protect their networks from external attacks. An ALG used with firewalls is shown in Figure 4–11.

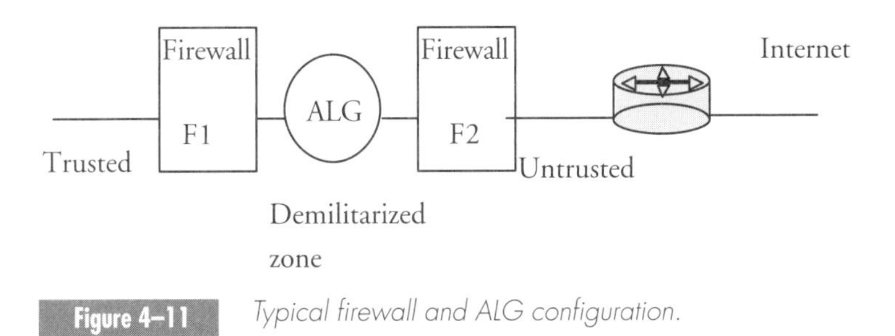


Figure 4–11 *Typical firewall and ALG configuration.*

In Figure 4–11, the trusted interface is connected to the internal (e.g., enterprise) network, and the untrusted interface is connected to the public Internet. The zone between F1 and F2 is called the *DeMilitarized Zone* (DMZ). A network administrator can add rules in firewalls F1 and F2 to make sure that they only allow traffic destined to the ALG. Therefore, anyone on the trusted or untrusted sides can reach the ALG. This is a typical configuration for enterprise networks.

In this manner, all communication between the trusted and untrusted domains, regardless of its origin, will have to go through the ALG. The immediate question is, Does the ALG support all kinds of applications? The answer is no. Typically, enterprise networks allow only email and Web (HTTP) traffic. In fact, one of the advantages of the ALG is that it gives the network administrator control over which applications can be used to communicate with the outside world.

Why do we need firewalls or ALGs? If everything is perfectly secured inside the network, we probably don't need them. However, this may never be the case. New applications are released regularly, and they have vulnerabilities that we gradually learn. Therefore, practically, it is too difficult to assume that the network is perfectly secure against any external attack. If one mistake is made and a particular application becomes vulnerable, an administrator would need to fix this mistake in all of the enterprise's machines. It would be much simpler to do this in one machine (ALG). Furthermore, network administrators are required to ensure that people in the trusted domain (perhaps disgruntled employees) do not voluntarily provide information to the other side of the firewall. For these reasons, as a general rule, enterprise networks never connect to the Internet without a firewall.

In a public access network, it is difficult to draw a line between a trusted and an untrusted domain. For instance, machines connected to a public shared wireless link can attack other machines on the same link or routers and servers in the network. In this scenario, a centralized firewall would not help. Firewalls or filtering routers would be needed on access routers and individual users' machines.

## 4.6 Summary

This chapter introduced the basic security principles needed to understand Mobile IPv6 security. We discussed the different security aspects and defined each one. Cryptography was then introduced; public and secret key cryptography are the main types of cryptography used today. Secret key cryptography is the most intuitive because the same key is used by both ends. Public key cryptography relies on using different keys for encryption and decryption. Both authentication and encryption can be achieved with either public or secret key cryptography.

Hash functions are very useful for authentication and nonrepudiation (when used to generate digital signatures). The use of hash functions is shown in Chapter 5.

The importance and use of certificates with public key cryptography were also described in the context of X.509 certificates. In addition, we saw that certificate authorities are organized in a hierarchy to allow different correspondents (associated with different CAs) to identify each other in a secure manner.

This chapter built on the knowledge gained earlier to show how a security association can be established between two nodes. An example was shown based on the Internet Key Exchange (IKE) protocol.

Finally, firewalls and ALGs were discussed. These tools are essential for any enterprise network. Therefore, it is important to understand their basic operation in order to appreciate why they were considered during the design of Mobile IPv6 security.

Chapter 5 builds on the knowledge presented here by using it to show how Mobile IPv6 security works.

# Further Reading

[1] Arkko, J., "Effects of ICMPv6 on IKE," draft-arkko-icmpv6-ike-effects-02, work in progress, March 2003.

[2] Aura, T., "Cryptographically Generated Addresses (CGA)," draft-ietf-send-cga-05, work in progress, February 2004.

[3] Cheswick, W. R., S. M. Bellovin, and A. D. Rubin. *Firewalls and Internet Security: Repelling the Wily Hacker*, 2nd ed. Addison-Wesley, 2003.

[4] *Eastlake, D., S. Crocker, and J. Schiller, "Randomness Recommendations for Security," RFC 1750, December 1994.*

[5] Harkins, D. and D. Carrel, "The Internet Key Exchange (IKE)," RFC 2409, November 1998.

[6] ITU-T recommendation, X.509, "Public-key and attribute certificate frameworks," March 2000.

[7] Kaufman, C., R. Perlman, and M Speciner. *Network Security: PRIVATE Communication in a PUBLIC World*. Prentice Hall, 1995.

[8] Kent, S., and R. Atkinson, "IP Authentication Header," RFC 2402, November 1998.

[9] Kent, S., and R. Atkinson, "IP Encapsulating Security Payload," RFC 2406, November 1998.

[10] Kent, S., and R. Atkinson, "Security Architecture for the Internet," RFC 2401, November 1998.

[11] Krawczyk, H. "SKEME: A Versatile Secure Key Exchange Mechanism for Internet." ISOC Secure Networks and Distributed Systems Symposium, San Diego, 1996.

[12] Krawczyk H., M. Bellare, and R. Canetti, "HMAC: Keyed-Hashing for Message Authentication," RFC 2104, February 1997.

[13] Maughan, D., M. Shertler, M. Schneider, and J. Turner, "Internet Security Association and Key Management Protocol (ISAKMP)," RFC 2408, November 1998.

[14] National Institute for Standards and Technology, "Secure Hash Standard," FIPS PUB 180-1, April 1995.

[15] Nikander, P., "Address Ownership Problem in IPv6," draft-nikander-ipng-address-ownership-00, work in progress, February 2001.

[16] O'Shea, G., and M. Roe, "Child-proof Authentication for MIPv6 (CAM)," *Computer Communications Review*, April 2001.

[17] Orman, H., "The OAKLEY Key Determination Protocol," RFC 2412, November 1998.

[18] Piper, D., "The Internet IP Security Domain of Interpretation for ISAKMP," RFC 2407, November 1998.

[19] Rivest, R., "The MD5 Message-Digest Algorithm," RFC 1321, April 1992.

[20] Stinson, D. R. *Cryptography Theory and Practice*. CRC Press, 1996.

FIVE

# Securing Mobile IPv6 Signaling

Mobility adds inherent security risks to those already in the Internet today. Some of these risks are introduced by the specific mobility protocol. Mobile IPv6 is a new protocol that attempts to do something that has not been done before on the Internet: redirect traffic between a mobile node and other correspondent nodes from one address to another. The signaling for such redirection is done between the mobile and correspondent nodes. To be able to design a protocol that avoids some or all of the security risks associated with it, we need to identify the types of threats specific to this protocol. Then we need to place requirements on the protocol to avoid some or all of these threats. In some cases, it is acceptable to have known threats associated with a protocol, provided that they are documented and understood. The output of the requirements study is used to test the protocol and see whether or not it conforms.

In this chapter, we focus on the security threats that result from the introduction of Mobile IPv6. We analyze different Mobile IPv6 messages and show how each one can be used by Bad Guy to produce undesired effects to the mobile node, correspondent node, and home agent. We then present the mechanisms used by Mobile IPv6 to secure its messages.

## 5.1 Why Do We Need to Secure Mobile IPv6?

Before we analyze the threats of Mobile IPv6's messages, we consider two different communication scenarios that are possible when Mobile IPv6 is used. Figure 5–1 shows the different cases.

A mobile node may tunnel its packets to the home agent, which in turn decapsulates and forwards them to the correspondent node. If route optimization were used (i.e., the mobile node sent a binding update to the correspondent

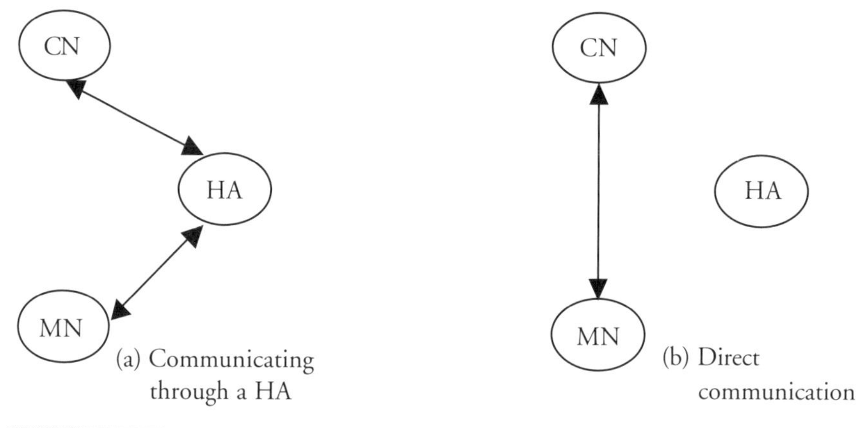


**Figure 5–1** *Communication scenarios with and without route optimization.*

node), the mobile node would send packets directly to the correspondent node after adding a *home address* option, as described in Chapter 3. The correspondent node would also send packets directly to the mobile node using a *routing header type 2* that includes the mobile node's home address. We need to analyze the types of attacks that Bad Guy can launch when he is on-path or off-path. An on-path attacker is one that can see packets going through a certain link between two nodes. For instance, an attacker can be on-path between the mobile and correspondent nodes if he is located at the mobile node's link, the correspondent node's link, or any link between the two where packets between the two nodes are routed. On the other hand, an off-path attacker is unable to see packets sent between the two nodes he is trying to attack. Now let us consider the different attacks that Bad Guy can launch using this protocol.

## 5.1.1 Using Binding Updates to Launch Attacks

The binding update is used to redirect traffic from one address to another. If not used carefully, it can have some detrimental effects on the mobile–correspondent communication. It should be noted that while our examples are focused on the mobile–correspondent node communication, they can all be applied to the mobile node–home agent binding updates.

### 5.1.1.1 STEALING TRAFFIC

If Bad Guy knows the mobile node's home address (which is not difficult, since it is public knowledge and may be stored in the DNS), he can send a binding update to a correspondent node to redirect the mobile node's traffic to himself, as shown in Figure 5–2.

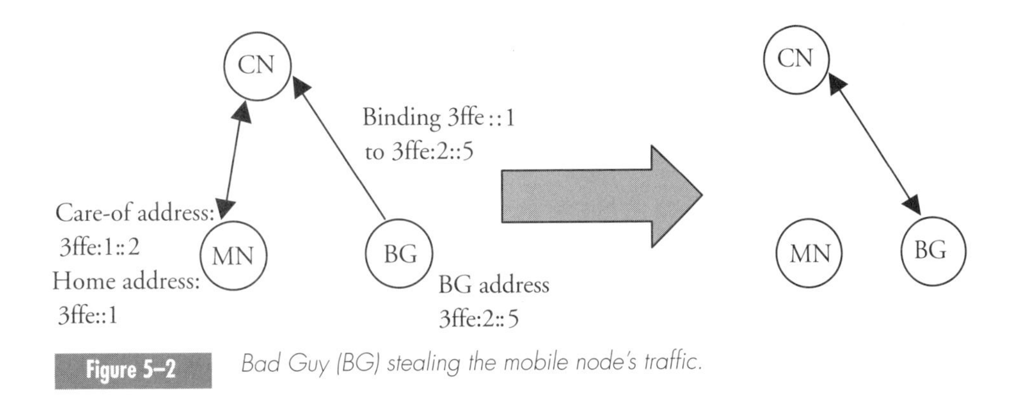


**Figure 5–2** *Bad Guy (BG) stealing the mobile node's traffic.*

In this example, Bad Guy steals the traffic originally intended for the mobile node and could possibly hijack the mobile node's connection with a correspondent node and pretend to be the mobile node for the rest of the connection. Note that in this case, Bad Guy is not on the path between the mobile and correspondent node.

#### 5.1.1.2 REFLECTION AND FLOODING ATTACKS

When Alice and Bob are communicating, it is perfectly acceptable that Alice send a packet to Bob, which causes Bob to send a reply back to Alice. However, if Alice sends a packet to Bob that causes Bob to reply to Angela, it is called a *reflection attack*. While it is perfectly acceptable for one node to communicate with another via a relaying agent, a packet sent from A to B should not cause B to **reply** to X. It is acceptable for B to **relay** the same message to X and for X to **reply** to A (note the distinction between relaying and replying), but it is unacceptable to have B replying to another node. Bad Guy (who need not be on-path) may use a binding update to launch a reflection attack. Figure 5–3 shows an example of a reflection attack.

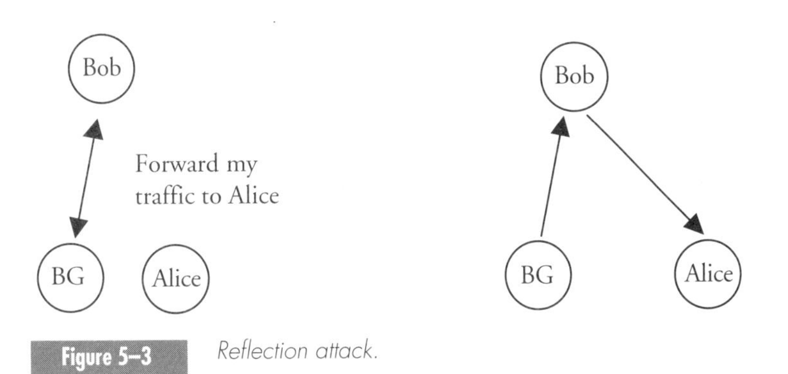


**Figure 5–3** *Reflection attack.*

In this case, Bad Guy can start a connection with Bob. After the connection starts, Bad Guy can send a binding update to Bob asking him to direct his traffic to another address (pretending that he moved), which is assigned to Alice. This would cause Bob to send all packets to Alice. If Bob is a server streaming video, for example, or Bad Guy was downloading a large file from Bob, this would cause Bob to **flood** Alice's link with information that she is not interested in. Flooding attacks are sometimes called **bombing** attacks.

#### 5.1.1.3 MAN IN THE MIDDLE (MITM) ATTACKS ON THE BINDING UPDATE

MITM attacks (Figure 5–4) involve Bad Guy on the communication path between two nodes. Bad Guy may interfere with the communication in different ways. The most obvious is to change the contents of a packet to cause some effects that were not intended by the original sender of the packet.

When Bad Guy is located on the path between a mobile and a correspondent node, he can modify the content of the binding update, possibly causing a reflection attack or hijacking of ongoing connection(s) between the mobile and correspondent nodes.

#### 5.1.1.4 DENIAL OF SERVICE ATTACKS (DOS)

DoS attacks can be done in several ways, with or without Mobile IPv6. The aim of a DoS attack is to deny service to a legitimate node. For example, an attacker may send many bogus messages to a particular server (e.g., asking to initiate a connection) to keep it so busy that it cannot respond to legitimate nodes. These kinds of attacks are very difficult to stop. Avoiding them usually involves having a Firewall that detects that too many requests are coming from a certain address and can drop all or some of them. However, this is not a real solution, since the attacker can use more than one address (i.e., distributed DoS, or DDoS, attacks).

Using Mobile IPv6, however, Bad Guy may use the binding update to deny a mobile node service from a particular correspondent node. This can be done by sending a binding update to the correspondent node asking it to forward packets addressed to the mobile node to another address (not the mobile node's care-of address). Bad Guy may also continue to refresh the correspondent node's binding cache to stop it from directing traffic to the mobile node's real location. Clearly the mobile node would try to rectify the situation by sending a binding update that includes the real care-of address. However, if Bad Guy happens to be on the path between the mobile and correspondent nodes (MITM), he can modify the packet to suit his attack.

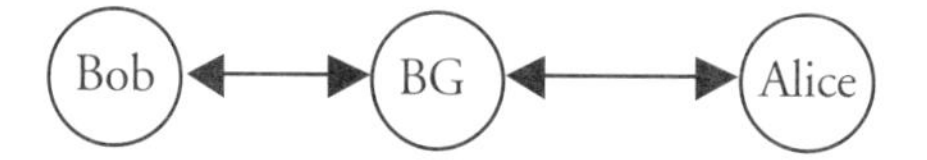


**Figure 5–4** *Man in the middle attack.*

Other types of DoS attacks can be launched against a correspondent node by sending a large number of bogus binding updates (i.e., bindings that do not necessarily correspond to real home addresses) that can fill up the correspondent node's memory allocated for storing binding cache entries. This would result in the correspondent node being unable to process new messages sent by real mobile nodes. The same attack can be launched against a home agent.

A DoS attack may also involve Bad Guy somehow compromising and controlling a router on the path between the mobile node and correspondent node or the home agent. In this case, Bad Guy might decide to simply drop the mobile node's packet, hence denying it service. However, this type of attack is not specific to Mobile IPv6. In fact there is nothing that can be done for it but make it impossible (or extremely difficult) for Bad Guy to be able to compromise a router.

### 5.1.2 Attacks Using the Routing Header and Home Address Option

The routing header type 2 is included in packets sent from the correspondent node to the mobile node; it includes the mobile node's home address. Therefore, the routing header allows the mobile node, upon receiving the packet, to place its home address in the destination field and effectively forward the packet (internally) to itself. However, what would happen if the address in the routing header were not the mobile node's home address? The mobile node would follow the same steps, except the packet would be forwarded to the new destination address (which does not belong to any of the mobile node's interfaces). If Bad Guy were communicating with a mobile node, he might put another address in the routing header, causing the mobile node to forward his packets to another node. This is not the same type of attack as the reflection attack described earlier, since the mobile node would not be replying to another node, but relaying Bad Guy's packets instead. However, this is still an undesirable effect for two reasons: first, Bad Guy might be using the mobile node as a relay to another node that, under normal circumstances, he is not authorized to communicate with (see section 2.2.2); and second, this use of the routing header type 2 was not the intention of the Mobile IPv6 protocol.

If Bad Guy were a mobile node instead, he might attempt to create a reflection attack similar to the one in section 5.1.1.2, using the home address option. In this case, Bad Guy could send a packet containing a home address option to a correspondent node. Since correspondent nodes replace the source address of the packet with the one in the home address option (to maintain transparency to upper layers), the reply to this message would go to the address included in the home address option. If this is not Bad Guy's address, another node (victim) will receive this information, which could possibly flood its link.

### 5.1.3 MITM Attacks on MPS/MPA

The Mobile Prefix Solicitation (MPS) and Mobile Prefix Advertisement (MPA) messages are used by the mobile node to learn about prefixes advertised on its home link. The MPS is sent by the mobile node and causes the home agent to send an MPA. The mobile node uses the information in the MPA to bind its home addresses to its current care-of addresses. Bad Guy can launch an MITM attack on these messages if he happens to be on the path between the mobile node and its home agent. Bad Guy can intercept packets and change their contents. Let's consider some of the undesirable effects:

- *Modifying lifetimes:* Suppose that the home network was being renumbered, or that for some reason, one of the prefixes on the home link was being removed. In this case, routers are likely to gradually reduce the lifetime on the prefix being removed to allow mobile nodes to pick the prefix with a higher lifetime. Eventually, a prefix will be deprecated and should not be used by the mobile node to form a home address. Suppose that Bad Guy saw one prefix (say prefix_x) with a lifetime of 10 minutes and another (say prefix_y) with a lifetime of 2 hours. He could swap the lifetimes of the prefixes. The mobile node would pick prefix_x, as it seems to have a longer lifetime. However, in 10 minutes prefix_x is going to be deprecated. When this happens, the mobile node will be unreachable.
- *Impersonating the home agent:* Bad Guy may intercept the mobile node's message (or simply read it) and generate his own replies. He could make up some prefixes or use prefixes that would cause the mobile node's traffic to be sent to his link. The mobile node would send a binding update to Bad Guy (thinking it is communicating with a home agent). If the binding acknowledgment were not authenticated, Bad Guy might simply reply to the mobile node, informing it of a successful binding. From this point on, all of the mobile node's traffic would be directed to Bad Guy. He might copy the packets and forward them, or he might drop the packets (DoS). Both actions will harm the mobile node and its correspondents, and should be avoided.

## 5.2 Requirements for Mobile IPv6 Security

We have now seen how dangerous Mobile IPv6 can be if not secured. The next step is to understand what type of security is needed and which threats are more relevant than others. Many of the requirements on Mobile IPv6 security are summarized by the IESG's recommendation to the Mobile IP WG: "Do no harm to the existing Internet." In other words, today's IPv4 Internet has known threats and is not perfect, but we ought not make it worse by introducing new threats. Hence, if a threat is known to exist in today's Internet, we do not have to solve it in this protocol (of course, solving it would be a bonus but is not required). From this high-level requirement, we can extrapolate that off-path attacks cannot be tolerated. MITM attacks can be done on today's Internet in some situations; however, off-path attacks like the ones described in section 5.1.1 should not be introduced by Mobile IPv6. That is, Mobile IPv6 signaling should be as secure as if the mobile node were located at the home link and no movement was involved.

Mobile devices vary in shapes and sizes; however, one of their common characteristics is that they are usually constrained by how much battery power they can have. Hence, the chosen cryptographic mechanism should not rely on computationally intensive algorithms (although this might be an optional feature for strong authentication, it should not be the only one).

In Chapter 4 we saw how the use of traditional public key cryptography needs Public Key Infrastructure (PKI) support—that is, hierarchies of certificate authorities—for proof of identity. Currently there is no global PKI on the Internet, and there is no expectation of global PKI deployment in the near future. This is mainly due to the difficulty involved in setting up trust between the different CAs. Mobile IPv6 should be deployed on a global scale; mobile nodes might be communicating with any random correspondent node on the Internet. Hence, a Mobile IPv6 security solution for mobile node to correspondent node signaling cannot assume the existence of global PKI.

On top of the general requirements discussed above, each function needs to have its own set of requirements.

## 5.2.1 Securing Communication Between Mobile and Correspondent Nodes

Messages between the mobile and correspondent nodes are needed for binding cache and binding update list management in the correspondent and mobile nodes respectively. The binding update is essentially acting as a redirection request, which if not secured can produce the undesirable result shown in section 5.1.1. Therefore, it is crucial that a correspondent node trust the mobile node to be able to accept this request. Let's consider what requirements this statement would generate.

### 5.2.1.1 HOW DOES A CORRESPONDENT NODE TRUST A MOBILE NODE?

Authorization is conditional. One might authorize a doctor to perform a certain surgical procedure because she can provide certain credentials that would make us believe that she is qualified for certain tasks. A heart surgeon is not generally trusted to perform plastic surgery, and an electrician is allowed to fix power problems but not problems related to a human's brain. All these people are trusted because they can prove that they meet a set of requirements that authorize them to perform their tasks (i.e., a trusted license). The same logic is needed when mobile nodes communicate with correspondent nodes. The important question is, What does a mobile node need to prove in order to be authorized to redirect packets from one address to another? Is a subject field of *Bob@example.com* in an X.509 certificate sufficient to authorize Bob to forward packets from `3ffe:1::1 to 3ffe::5`? The answer is no (in the general case, but of course if Bob is **known** to a correspondent, then this may be acceptable). If this were true, then we would be implying that Bob is a good person who would not do anything wrong, and if he does, we can sue him—which implies absolute trust and late detection of any breach of that trust, but does not **prevent** Bob from breaching the trust, which is our goal (prevention is better than late detection). The fact is, hosts on the Internet are allowed to forward traffic to themselves only. Hence, if a mobile node wishes to redirect traffic from one address to another, it needs to prove that it has such authority by proving that it is in fact assigned both addresses. This is the most important requirement for securing binding updates: proof of address ownership. The proof of home and care-of address ownership gives the mobile node the authority to request a redirection of traffic between those two addresses.

#### 5.2.1.2 MESSAGE INTEGRITY: AVOIDING MITM ATTACKS

Even if Bob can prove address ownership, Bad Guy might be on the path between Bob (mobile node) and Alice (correspondent node). He can modify the contents of the binding update message and the IPv6 header to serve his purposes. Hence, Alice needs to check whether Bob's message was modified en route. This requires binding updates to include some form of message authentication. The same requirement can be placed on the binding acknowledgment; a mobile node needs to ensure that the binding acknowledgment came from the correspondent node. Otherwise, Bad Guy might impersonate the correspondent node and reply with a fake binding acknowledgment. Hence, mutual authentication between the mobile and correspondent nodes is needed.

#### 5.2.1.3 AVOIDING DOS ATTACKS IN CORRESPONDENT NODES

The DoS attack mentioned in section 5.1.1.4 involves filling up the correspondent node's binding cache with unnecessary information. To avoid this type of attack, correspondent nodes must ensure that they do not maintain state per mobile node until the binding update is accepted. Hence, the protocol needs to be stateless from the correspondent node's viewpoint until an authenticated binding update message is received. This allows correspondent nodes to avoid the DoS attacks described earlier.

### 5.2.2 Securing Messages to the Home Agent

The same attacks on the binding update to correspondent nodes are relevant when a mobile node is communicating with its home agent. In addition, home agent impersonation (see sections 5.1.3 and 5.1.1.3) could be detrimental to the mobile and correspondent nodes. Hence, binding updates and acknowledgments and MPS/MPA exchanged between a mobile node and its home agent should be authenticated. Due to the importance of the home agent, it is important that mobile nodes' communication is secured using strong authentication. This is not a hard requirement, since the home agent(s) is known and mobile nodes are not expected to pick them at random, which is the case for correspondent nodes. There is no requirement for securing DHAAD messages; the reasons are shown in section 5.3.2.1.

### 5.2.3 Assumptions about Mobile IPv6 Security

We can certainly assume that any route optimization security solution that involves manual configuration is not useful as a generic solution. Hence, symmetric cryptography that requires manual configuration of keys in mobile and correspondent nodes is not a generic solution. A solution that requires no manual configuration for mobile–correspondent node bindings is needed. Furthermore, we can also assume that a solution that requires global PKI falls in the same—infeasible—category. We need a solution that requires no infrastructure support. Ideally, only nodes implementing Mobile IPv6 functions should be involved in securing the protocol between the mobile and correspondent nodes. On the other hand, a mobile node's home agent is a known node. Hence, manual configuration of security associations between the mobile node and its home agent is possible. This would allow for securing binding updates and MPS/MPA messages with strong authentication techniques. In addition, public keys can be used to secure binding updates between the mobile node and its home agent without requiring global PKI. The mobile node and its home agent can belong to the same trust domain; hence, only local PKI would be needed (e.g., a network operator may issue and sign certificates to its mobile nodes and home agents). Assuming a secure channel between the mobile node and its home agent also reduces the number of paths vulnerable to MITM attacks between the mobile and correspondent nodes. This reduces the number of locations that can be used by Bad Guy to launch attacks (see Figure 5–1).

## 5.3 Mobile IPv6 Security

Mobile IPv6 signaling utilizes several messages that require different levels of security. When dealing with the mobile node–home agent messages, it is possible to use manual configuration to set up an IPsec security association. However, this approach is not generally feasible between mobile and correspondent nodes. In the following sections we address each message and show how they can be secured.

### 5.3.1 Securing Binding Updates to the Home Agent

A home agent and a mobile node are expected to have prior knowledge of each other. Home agents could be provided by ISPs to serve their mobile subscribers. Subscribers typically have trusted credentials and a security relationship with their ISPs. The same type of relationship can be set up between mobile subscribers and their home agents. This would allow home agents to know the right credentials needed by mobile nodes to use certain home addresses. Furthermore, it would allow a home agent to impose certain actions on misbehaving mobile nodes, for instance, by terminating the home agent service provided to these nodes. This special relationship between a mobile node and its home agent allows us to relax the requirements on some of the threats mentioned earlier; specifically, reflection attacks shown in section 5.1.1.2 can be traced to the misbehaving mobile node. A responsible network manager would not want to have mobile nodes that use the network to flood other links in his network or in other networks on the Internet. For this reason, a mobile node and its home agent can be treated as a single entity when it comes to these attacks. A consequence to this assumption is that we do not require mobile nodes to prove ownership of their care-of addresses to their home agents.

It should be noted that home agents and mobile nodes are not required to have a subscriber–ISP relationship. A user might wish to have a home agent at home (in a house) that serves her mobile devices (mobile phone, PDA, laptop, etc.). The same assumptions are still valid for this case. If someone owns a home agent, there is no difference between using that home agent to flood other nodes' links on the Internet and using her mobile node to directly flood these links. The fact that packets are reflected makes no difference, since the attacker administers both entities (mobile node and home agent). Therefore, when sending a binding update to a home agent, the only things a mobile node needs to prove are that

- it has been allocated the home address (i.e., it is authorized to redirect traffic), and
- it is authorized to use this particular home agent to forward its packets (i.e., it was authorized to use the forwarding service).

To satisfy these requirements, an IPsec security association is needed between the mobile node and its home agent. IPsec uses IP addresses to identify security associations in its Security Association Database (SAD). Hence, by knowing the mobile node's home address and using that to look up a corresponding security association, a home agent can verify both points above in a simple way if the security association was manually configured. That is, since the network administrator configures a security association to protect the mobility header, then he is implicitly allowing a mobile node (which knows the secret key) to use this home agent to forward its traffic; and since each address has a single key associated with it, knowing that key is a proof that the mobile node has been allocated a particular home address.

If the security association were established dynamically, using public keys and certificates, the mobile node would need to provide credentials showing that it has been allocated this particular home address. This can be done in different ways. One way is to include the mobile node's home address in a certificate signed by a trusted CA. This might be the ISP's own CA or another one trusted by the home agent. Clearly, if the CA does not belong to the ISP, it needs to verify with the ISP (through an out-of-band method) that the ISP has allocated a particular home address to the mobile node. Alternatively, the association between a particular certificate and a corresponding home address can be manually configured in both the home agent and the mobile node. This adds more manual configuration in the home agent; however, it is not a significant amount. In both cases, the mobile node needs to know its home address. However, the latter method does not require any cooperation between the CA and the ISP (except, of course, the home agent must trust the CA). This is a significant advantage and involves less work when the home network is being renumbered.

The next issue is how to protect Mobile IPv6 messages between the mobile node and home agent using the established security association. IPsec has two different mechanisms, AH and ESP (discussed in Chapter 2). Both can be used to get authentication and replay protection. However, strong replay protection can only be achieved with dynamic keying. When preconfigured keys are used (secret key cryptography), only the sequence number in the binding update can be used. However, this would not provide strong replay protection (the sequence number rolls over and the same numbers are used again with the same key). The major difference between AH and ESP is that ESP provides confidentiality (encryption) while AH does not. Another difference is that AH protects the entire packet, including the IPv6 header; that is, the care-of address (in the source address) is protected. On the other hand, ESP only protects the headers below it; hence, the source address is not protected. Consequently, when using ESP, the *alternate care-of address* option must be included in the mobility header, which is covered by ESP (and should contain the same address as the source address field).

When considering the binding update and acknowledgment messages, we can see that the most important requirement is mutual authentication and message integrity. The mobile node and the home agent need to authenticate each other and ensure that the message was not modified by MITM. Both AH and ESP (for ESP, authentication must be enabled) can achieve this. Since binding updates are sent to set up a tunnel between the mobile node and the home agent (i.e., before the tunnel is set up), either AH or ESP can protect the binding update only when used in *transport* mode.

The IPsec security association should be based on the mobile node's home address. Therefore, when sending a binding update to the home agent, the mobile node needs to include its home address in the packet (remember that the source address is the care-of address). This is done by including the home address inside the *home address* option, which allows the home agent to look up the security association based on the mobile node's home address. Similarly, when sending a binding acknowledgment to the mobile node, the home agent needs to include a routing header type 2, including the mobile node's home address, to allow mobile nodes to find the right security association. An overview of the mobile node and home agent processing of binding updates is shown in Figure 5–5. Figure 5–6 shows an overview of the binding acknowledgment processing by the mobile node and home agent.

**Mobile Node**

Sending binding updates
Construct IPv6 header: src: home address dst:HA
Construct MH with BU
Construct HAO with CoA
Protect with IPsec
Swap the content of the HAO and src address
Send the packet.

**Home Agent**

Receiving binding updates
Swap the contents of the HAO and src address
Verify IPsec
Verify the MH and BU
BU contents correct?
Yes
Status = success.
Create/update binding cache entry
No
Status = Fail.
Prepare BA with Status value

BU = Binding Update
CoA = Care-of Address
HA = Home Agent
HAO = Home Agent Option
MH = Mobility Header

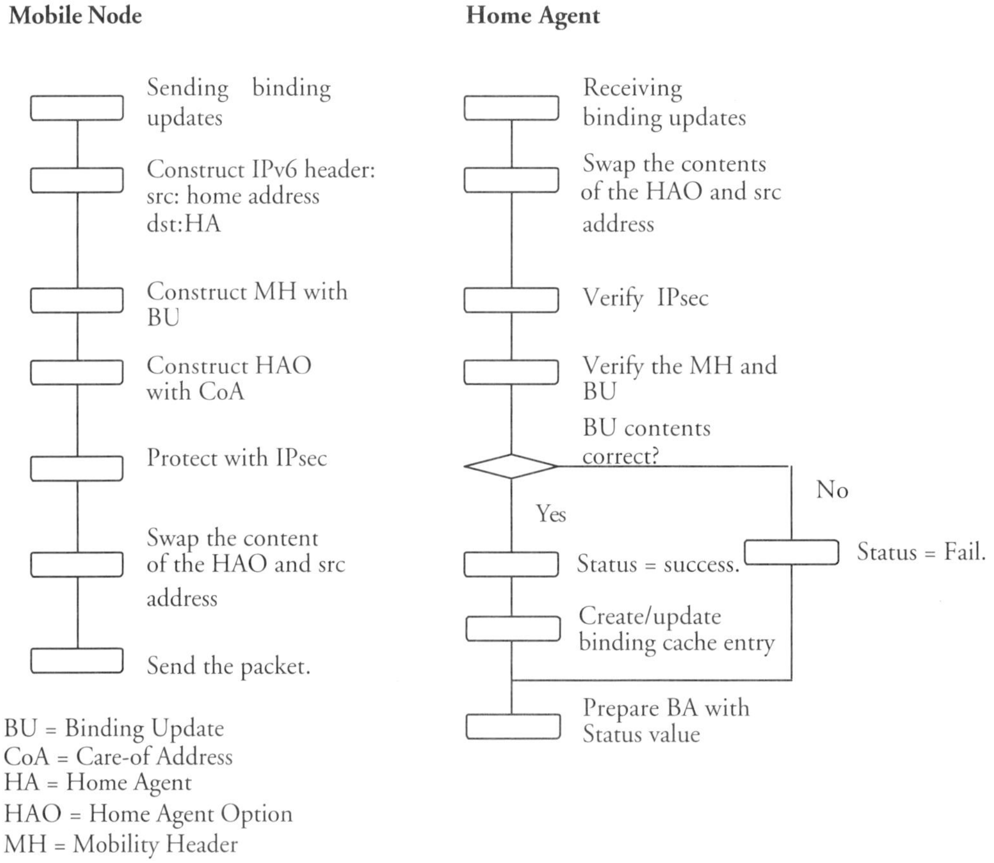

**Figure 5–5** *A simplified flowchart for sending and receiving binding updates by the mobile node and home agent respectively.*

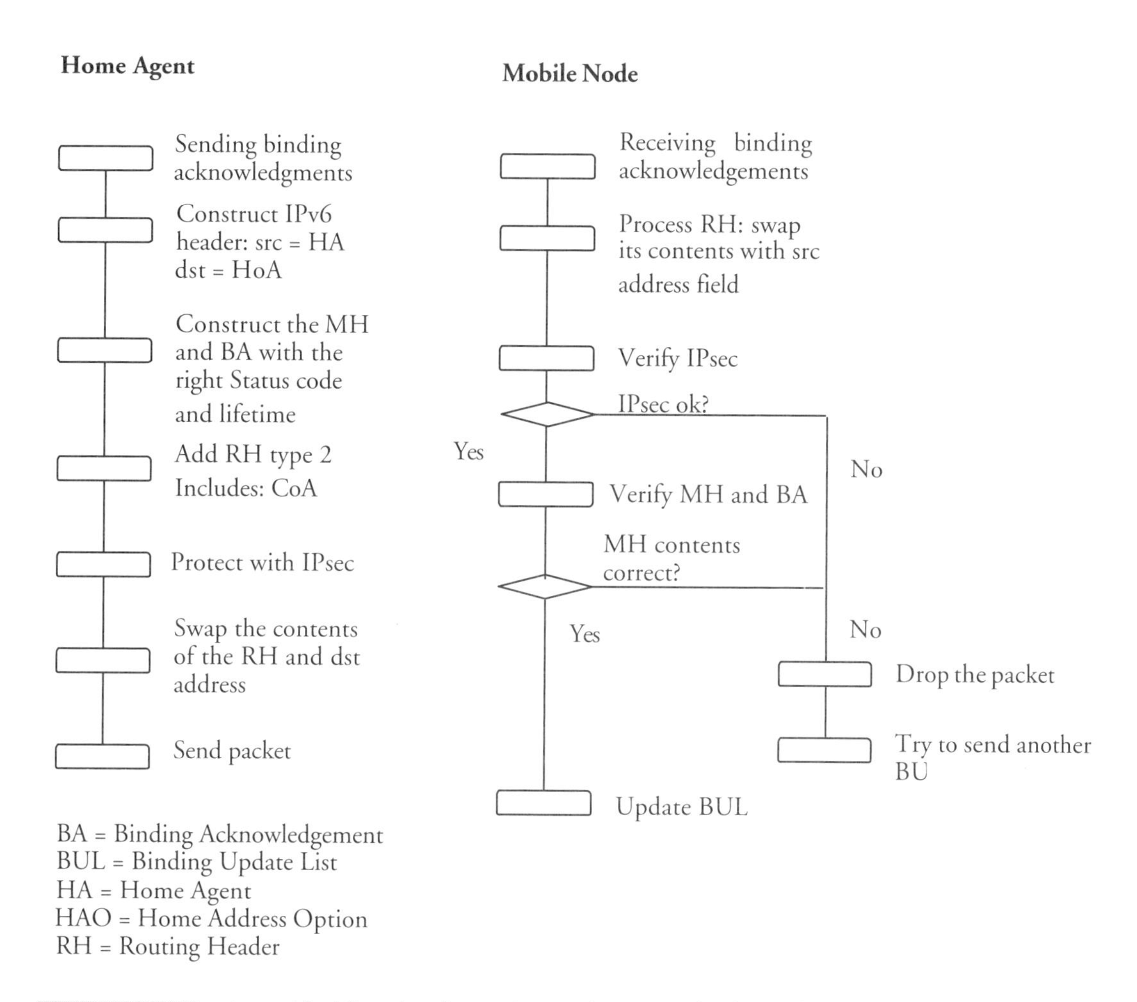


**Figure 5–6** *A simplified flowchart for sending and receiving binding acknowledgments by the home agent and mobile node respectively.*

Note that while the contents of the binding update list and binding cache are stored in volatile memory, the sequence number requires a different treatment when dynamic keying is not used (i.e., no IKE). The home agent needs to make sure that the last valid sequence number used is stored in nonvolatile memory to be able to survive reboots. Otherwise, the home agent will accept a replayed binding update message after a reboot, causing it to tunnel traffic to a potentially wrong address. This problem does not exist when dynamic keying is used, as a reboot results in losing the contents of the SAD (stored in volatile memory). Hence, a security association must be negotiated before a binding update can be accepted.

Note that when dynamic keying is used, it is possible for the mobile node to establish a security association with its home agent that requires renegotiation every time the mobile node moves. In this case, the *K* flag in the binding update (refer to Chapter 3) would be cleared. However, this implies that the mobile node needs to renegotiate a new security association whenever its care-of address changes (i.e., every time it moves). This case is not covered in the flowcharts shown in this section.

## 5.3.2 Securing Mobile Prefix Solicitations and Advertisements

The requirements and security mechanism for MPS and MPA are basically the same as those for the binding update/acknowledgment. The main difference is the configuration of the SAD and Security Policy Database (SPD); they clearly need to indicate ICMP for the protocol type, as opposed to the *mobility header* used for binding updates and acknowledgments.

### 5.3.2.1 WHAT ABOUT DHAAD?

Dynamic Home Agent Address Discovery allows a mobile node to discover home agent addresses while located on a foreign link. The actual messages do not need to be hidden; the home agent's address is not a secret. It is exposed in all packets tunneled to and from the home agent and cannot be hidden. Bad Guy might intercept packets, modify them, or reply on behalf of the home agent. However, none of this would cause traffic disruption, since the mobile node sends an authenticated binding update and expects an authenticated binding acknowledgment, as shown in the previous sections. The attacker cannot send an authenticated binding acknowledgment to the mobile node, and no harm can be done. The best that Bad Guy can do is launch a DoS attack, which is already possible today. Therefore, no Mobile IPv6-specific attacks can be launched using DHAAD.

## 5.3.3 Manual Versus Dynamic SAs Between the Mobile Node and Its Home Agent Configuration

Two types of keying mechanisms can be used when setting up a security association with the home agent: static and dynamic keying. In the static case, a network administrator configures a security association on the home agent and the mobile node; in the dynamic keying case (e.g., IKE), public key cryptography can be used with appropriate certificates, as discussed earlier. We now see how static security associations can be established.

IPsec security associations are unidirectional. That is, when establishing a security association between two nodes for bidirectional communication, one needs to configure a security association for each direction: IN (for traffic routed toward a node) and OUT (for traffic sent from the node). In our example, we set up a security association between the mobile node and its home agent to protect the *mobility header*. Each node contains the necessary information for both the SAD and the SPD. In this example, we use the same encryption and authentication keys for both directions of communication.

**Example**

Let's consider an example where the mobile node's home address is `3ffe:200:8:1:1234:5678::1`, its care-of address is `3ffe:ffff:2:1:1234:5678::1`, and its home agent's address is `3ffe:200:8:1::5`. To set up a security association for binding update protection, we need to configure the mobile node and the home agent with the necessary keys, Security Parameter Index (SPI), and the protocol to be protected.

The SPD configuration has the following format:

```
spdadd <source> <destination> <protocol> <dir> <protection>
<mode>
```

where `<source>` is the source address, `<destination>` is the destination address, `<protocol>` is the protocol to be protected, `<dir>` is the direction of communication, `<protection>` is the type of IPsec protection (ESP or AH), and `<mode>` specifies *transport* or *tunnel* modes. Note that the source and destination may contain a range of addresses, for instance, by specifying a prefix. Using this definition, let's see how this can be used to configure the SPD in the mobile node and home agent for binding update/acknowledgment protection using ESP transport mode. The following is the SPD configuration in the mobile node.[1]

```
#setkey -c

spdadd 3ffe:200:8:1:1234:5678::1 - 3ffe:200:8:1::5 mh -P out
esp/transport/require;

spdadd 3ffe:200:8:1::5 - 3ffe:200:8:1:1234:5678::1 mh -P in
esp/transport/require;
```

In English, this essentially implies two Boolean conditions:[2]

- If any outgoing packet has a source address of `3ffe:200:8:1:1234:5678::1`, a destination address of `3ffe:200:8:1::5`, and carries an extension header of type *mobility header*, it must (the "require" part) be protected by ESP in transport mode.
- If any incoming packet has a source address of `3ffe:200:8:1::5`, a destination address of `3ffe:200:8:1:1234:5678::1`, and carries an extension header of type *mobility header*, it must be protected by ESP in transport mode. If not, the packet must be dropped.

The same configuration for the SPD must be done for the home agent:

```
#setkey -c

spdadd 3ffe:200:8:1::5 - 3ffe:200:8:1:1234:5678::1 mh -P out
esp/transport/require;

spdadd 3ffe:200:8:1:1234:5678::1 - 3ffe:200:8:1::5 mh -P in
esp/transport/require;
```

Now let's consider how the SAD can be configured:

```
Add <source> <destination> <protection> <mode> <spi> <algconf>
```

The first four fields are the same as those mentioned earlier for the SPD. The `<spi>` is the SPI used by the IPsec header, and `<algconf>` is a set of parameters describing the keys and algorithms to be used for encryption, authentication, or both. The following is an example of an SAD configuration command in the mobile node (this is a continuation of the previous command, i.e., `setkey -c`):

```
add 3ffe:200:8:1:1234:5678::1 - 3ffe:200:8:1::5 esp 0x10001 -m
transport -E 3DES "maryhasalittlelamb121234"

-A hmac-sha1 "this is an mn example";

add 3ffe:200:8:1::5 - 3ffe:200:8:1:1234:5678::1 esp 0x10002 -m
transport -E 3DES "maryhasalittlelamb121234"

-A hmac-sha1 "this is an mn example";

EOF
```

In this example we configured the mobile node to use ESP in transport mode to protect any packets containing the mobility header. The packets will be encrypted with 3DES (key is `maryhasalittlelamb121234`) and authenticated with the message authentication code HMAC-SHA1 (key is `this is an mn example`). The switch `-E` indicates encryption algorithm; `-A` indicates authentication algorithm, and `-m` indicates mode.

Continuing with the home agent configuration, the SAD configuration is done as follows:

```
add 3ffe:200:8:1:1234:5678::1 - 3ffe:200:8:1::5 esp 0x10001 -m
transport -E 3DES "maryhasalittlelamb121234"

-A hmac-sha1 "this is an mn example";

add 3ffe:200:8:1::5 - 3ffe:200:8:1:1234:5678::1 esp 0x10002 -m
transport -E 3DES "maryhasalittlelamb121234"

-A hmac-sha1 "this is an mn example";

EOF
```

[1] We use an example configuration for a FreeBSD operating system (www.freebsd.org).

[2] There was no protocol type "mh" defined in FreeBSD at the time of writing because the Mobile IPv6 implementation was not complete yet. This (mh) was added for the purpose of illustration.

At this stage, the mobile node and home agent can send and receive encrypted and authenticated packets that contain binding updates and acknowledgments. To configure a security association for MPS and MPA messages, the same commands need to be done, with the exception that the protocol is ICMPv6 (as opposed to the mobility header). Note that this results in protecting **all** ICMPv6 messages between the mobile node and the home agent due to the inability to define higher granularity (to look for certain *types* of ICMPv6 messages) in the SPD (some, but not all, implementations may support a finer granularity). However, since there are no other likely ICMPv6 packets (except for ICMPv6 errors, which should be rare), the additional overhead of encrypting all ICMPv6 packets is minimal.

This configuration example clearly shows some of the reasons for processing the home address option and the routing header shown in Figures 5–5 and 5–6. The security association depends on the home address, which is also seen by applications as the source address of all packets. Therefore, IPsec needs to see this address in the source address field of the packet. If the security association were based on the care-of address, the mobile node would need to set up a new security association every time it moves, which is impossible with manual configuration and would cause unnecessary signaling when IKE is used. It would also cause additional delays before the mobile node can send an authenticated binding update.

### 5.3.4 Securing Binding Updates to Correspondent Nodes

Unlike the home agent, the correspondent node is not likely to have established any security relationship with a mobile node. A correspondent node can be a Web server, a news channel, or any random destination that the mobile node wishes to communicate with. This random node will not have the resources, or the desire, to keep track of mobile nodes' actions and report them if they misbehave. For instance, if a mobile node launches a flooding attack on another node through *www.bbc.co.uk* (e.g., downloading a large video file), we cannot expect *www.bbc.co.uk* to keep track of that or punish the mobile node for doing so. It is simply unrealistic to assume that correspondent nodes will police the mobile nodes' actions; it is much more realistic to assume that the protocol will prevent these attacks from taking place. Preventing the mobile node from stealing another node's traffic or directing traffic to a node other than itself can be done if the mobile node proves to correspondent nodes that it is authorized to redirect traffic from address A to address B; that is, the mobile node must prove that it **owns**[3] both addresses.

[3] We use the term "owns" loosely. Nodes do not own addresses; they are assigned addresses that can be changed at any time. Therefore, a host that "owns" an address at a particular moment in time is one that has rightfully configured that address on its interface, provided that the address is still valid.

#### 5.3.4.1 OVERVIEW OF THE RETURN ROUTABILITY PROCEDURE

Suppose that the correspondent node sent a message to the mobile node's home address and another one to its care-of address; and suppose that each one of those messages required a particular reply that depends on the content of the message. If the mobile node answers both messages correctly, then we can assume that it **received** both messages. Since the messages were sent to the mobile node's claimed home address and care-of address, then answering them correctly implies that the mobile node actually received both messages and therefore owns both addresses. This is basically how a return routability test works in a Bad Guy–free Internet. However, since this ideal Internet does not exist, we need to somehow authenticate these messages. We now show how a return routability test works and how it can be used to secure binding updates to correspondent nodes.

A correspondent node generates a key, $K_{cn}$, which can be used with any mobile node. In addition, the correspondent node generates nonces at regular intervals. For instance, a correspondent node may generate $K_{cn}$ when it boots, then generate a nonce every two minutes. The size of the nonce can vary depending on the correspondent node's implementation; however, the Mobile IPv6 specification recommends that nonce size be 64 bits. Nonces and $K_{cn}$ will be used to generate a security association between the mobile and correspondent nodes. Nonces are stored for some time in an indexed list. Each nonce will have a 16-bit index. The return routability procedure is shown in Figure 5–7.

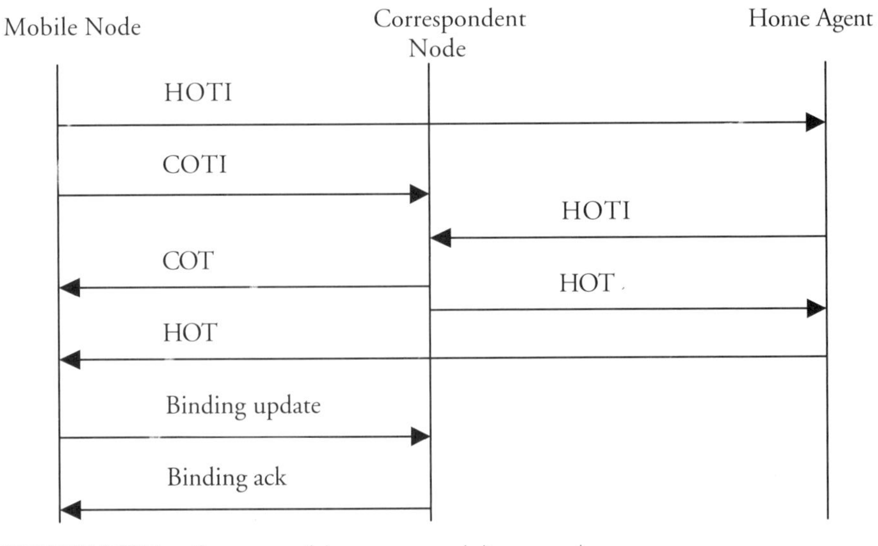


**Figure 5–7** *Operation of the return routability procedure.*

The essence of the return routability procedure is that the mobile node requests that the correspondent node test its ownership of the home and care-of addresses. This is done by sending two independent messages: the HOme address Test Init (HOTI) and Care-Of address Test Init (COTI). The correspondent node creates two tokens that only the correspondent node can create and sends one token to each address (home and care-of addresses) in two separate messages: HOme Test (HOT) and Care-Of Test (COT). The mobile node uses both of these tokens to create a key ($K_{bm}$) that can be used to authenticate a binding update message to the correspondent node. Since the correspondent node knows all the information needed to produce the key, it can reproduce it when the binding update is received, and so authenticate the message. The same key is used to authenticate the binding acknowledgment.

The HOTI message is sent by the mobile node to request a test of the home address. The source address used in the IPv6 header is the mobile node's home address, and the destination is the correspondent node's address. Hence, this message has to be tunneled to the home agent (since the home address is not topologically correct in the visited network), which decapsulates the message and forwards it to the correspondent node. The HOTI message is transported inside a mobility header type 1. This message contains a cookie (called *home init cookie*) generated by the mobile node and later returned by the correspondent node. The cookie is a random number that has no significance; it is included to ensure that the entity responding to the HOTI message has actually received it. This message is protected on the mobile node–home agent path by ESP in *tunnel* mode. When sent by the mobile node, the HOTI message has the following format:

```
IPv6 header
   src = care-of address
   dst = home agent
ESP header
IPv6 header
   src = home address
   dst = correspondent node
Mobility Header type 1
   Home init cookie
```

The home agent verifies the ESP header and forwards the internal message to the correspondent node. Note that in this case the home agent is not provided with a home address option in the outer header (unlike the binding update message) to use in order to locate the right security association in the SAD. In this scenario, the home agent's SPD is configured to treat the mobile node's care-of address as a security gateway address. The implication of this configuration is that the home agent can associate a security association entry in the SAD with a specific tunnel interface, identified by the mobile node's care-of address. Hence, the home agent will be able to identify the security association based on the interface it was received from. Why is this message (and the HOT message) treated differently by not including the home address option? The reason is that the binding update is sent before establishing the tunnel; therefore, no tunnel interface can be used to identify the security association.

Almost simultaneously, the mobile node can send a COTI message. The COTI message is sent from the mobile node's care-of address directly to the correspondent node. It is transported in a mobility header type 2. The message contains another random cookie (called *care-of init cookie*). The COTI cookie is a random number used to ensure that the responder to a COTI message has actually received the original (COTI) message. The content of the message is shown below:

```
IPv6 header
    src = care-of address
    dst = correspondent node
Mobility Header type 2
    Care-of init cookie
```

When the correspondent node receives the HOTI message, it generates a 64-bit *home keygen* token (the token generated based on the home address). The home keygen token is generated by taking the first 64 bits of the output of a message authentication code function using $K_{cn}$ and computed on the concatenation of the home address and a nonce generated by the correspondent node as follows:

```
Home keygen token = First(64, HMAC_SHA1(Kcn, home
address|nonce|0))
```

where `First(n, j)` represents the first `n` bits in `j`. `HMAC_SHA1(K, info)` means a hashed message authentication code (or a keyed hash) based on the SHA1 hash algorithm and uses `K` to key the function, which operates on `info`. The 0 is used to distinguish the home keygen token from the care-of keygen token, shown later. Hash functions and MACs were explained earlier in Chapter 4.

The correspondent node then constructs a HOT message and sends it to the mobile node. This message contains the home init cookie originally sent by the mobile node and the home keygen token. Since the correspondent node generates nonces frequently, it needs to be aware of the nonce used to generate a particular cookie. Nonces are stored in an indexed list. Therefore, a correspondent node only needs to know the index corresponding to a particular nonce to be able to generate the *home keygen token* again. The *nonce index* is included in the HOT message. This will be needed later by the correspondent node to authenticate the binding update. The content of the HOT message is

```
IPv6 header
    src = correspondent node
    dst = home address
```

```
Mobility Header type 3
    Home nonce index
    Home init cookie
    Home keygen token
```

The message will be intercepted by the home agent and tunneled to the mobile node's care-of address. A secure tunnel (ESP) is used to forward this message to the mobile node.

A similar operation is done when the correspondent node receives the COTI message. It generates a care-of keygen token, where

Care-of keygen token = First(64, MAC ($K_{cn}$, care-of address | nonce|1))

The nonce used in this operation might not be the same nonce used to create a *home keygen token*, depending on when the COTI message was received (the correspondent node might have generated a new nonce). Therefore, the nonce index should be sent to the mobile node in the COT message. The COT message is formed as follows:

```
IPv6 header
    src = correspondent node
    dst = care-of address
Mobility Header type 4
    Care-of nonce index
    Care-of init cookie
    Care-of keygen token
```

This message concludes the return routability procedure. At this point, the correspondent node has not yet stored any more information than it had at the beginning of this procedure: $K_{cn}$ and an indexed list of nonces. The correspondent node stores neither the home keygen token nor the care-of keygen token. When needed, these tokens can be regenerated, given the nonce indices originally used to generate them.

After receiving the HOT (tunneled from the home agent) and COT messages, the mobile node is in a position to generate a *binding management key*, $K_{bm}$. This is done as follows:

$K_{bm}$ = SHA1 (home keygen token|care-of keygen token)

The mobile node can now construct the mobility header used for the binding update message. The mobility header includes the binding update, a *nonce indices* option, and a *binding authorization data* option (shown in Figure 5–8). The nonce indices option contains the two indices received in the HOT and COT messages.

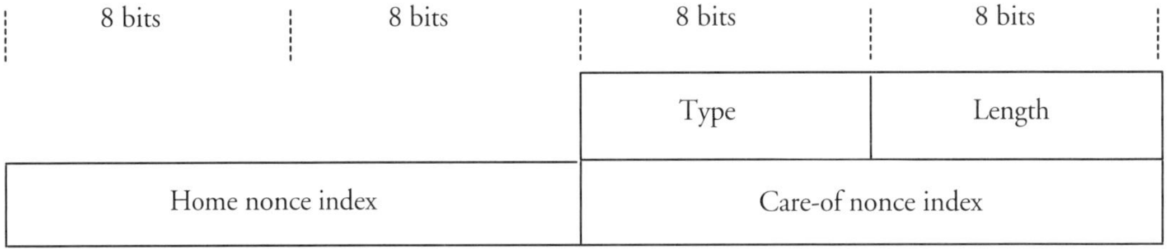

8 bits | 8 bits | 8 bits | 8 bits

Type | Length

Home nonce index | Care-of nonce index

(a) nonce indices optimum

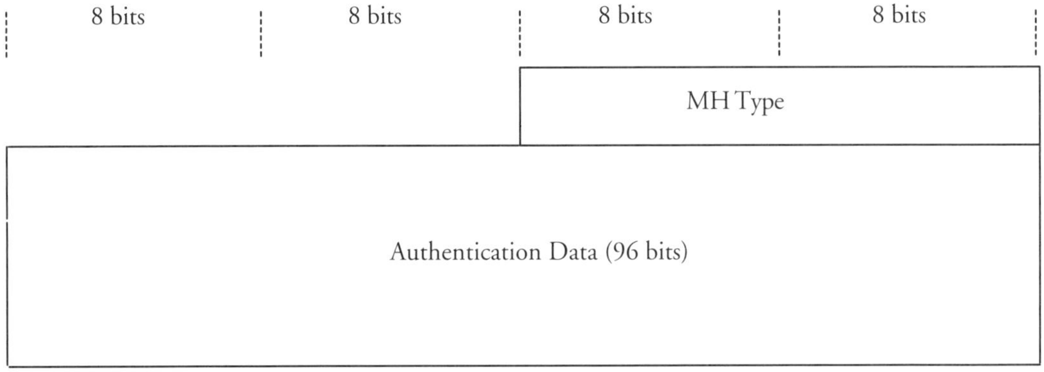

8 bits | 8 bits | 8 bits | 8 bits

MH Type

Authentication Data (96 bits)

(b) Authorisation data option

**Figure 5–8** *Nonce indices and binding authorization data options.*

The authentication data are calculated as follows:

Auth_data = First (96, MAC($K_{bm}$, Mobility_data)

where

Mobility_data = care-of address|*final dst*|Mobility header data

The mobility header data includes the content of the mobility header with the exception of the authorization data option itself. The `final dst` is the packet's final destination, that is, the correspondent node's address. If the correspondent node were also a mobile node, a routing header type 2 (containing its home address) would be included in the packet. Since the routing header is processed before the mobility header, the `final dst` field should contain that correspondent node's home address.

Since the correspondent node does not keep state for any mobile nodes during the return routability procedure, the mobile node needs to include its home and care-of addresses in the binding update. The home address is included in a home address option (in a destination options extension header), which precedes the mobility header. If the care-of address were different from the packet's source address, it should be included in the alternate-care-of address option; otherwise, the packet's source address is assumed to be the care-of address. In any case, the care-of address should always be the one used in the source address field of the COTI message; otherwise, the wrong care-of keygen token will be used to generate $K_{bm}$ when the binding update is received at the correspondent node.

After the binding update message is constructed, the mobile node sends it to the correspondent node with the following format:

```
IPv6 header
     src = care-of address
     dst = correspondent node
DST-options header
     Home address option
Mobility header type 5
     Binding update
     Nonce indices option
     [optional alternate-care-of address option]
     Authorization data option
```

Alternatively, if the correspondent node were also a mobile node, and it had a binding cache entry in the mobile node, the packet format would differ slightly:

```
IPv6 header
     src = care-of address
     dst = correspondent node
Routing header type 2
     Home address_cn
DST-options header
     Home address option
Mobility header type 5
     Binding update
     Nonce indices option
     [optional alternate-care-of address option]
     Authorization data option
```

where `Home address_cn` is the correspondent node's home address.

When the correspondent node receives the binding update, it looks into the nonce indices option and finds the corresponding nonces. The correspondent node will be able to regenerate $K_{bm}$ as follows:

1. Generate home keygen token: `First(64, MAC(`$K_{cn}$`, home address|nonce|0))`. The home address is taken from the home address option.
2. Generate care-of keygen token: `First(64, MAC(`$K_{cn}$`, care-of address|nonce|1))`. The care-of address is taken from the alternate-care-of address option when present; otherwise, the source address is used.
3. Generate $K_{bm}$: `Hash(`*`home keygen token|care-of keygen token`*`)`.
4. Calculate `Auth_data`: `First(96, MAC(`$K_{bm}$`, Mobility_data)`.
5. If `Auth_data` is equal to the content of the binding authorization data option, accept the binding update.

If an acknowledgment is requested, the correspondent node must send a binding acknowledgment. The binding acknowledgment should also contain the binding authorization data option in Figure 5–8. The contents of the binding acknowledgment message are shown below:[4]

```
IPv6 header
   src: correspondent node
   dst: care-of address
Routing header type 2
   mobile node's home address
DST-options header
      Home address option (if the correspondent node were
      also a mobile node)
Mobility header type 6
      Binding acknowledgment
      [optional binding refresh advice option]
      Authorization data option
```

where

`Auth_data = First(96, MAC(`$K_{bm}$`, Mobility_data),`

and

`Mobility_data = care-of address|`*`CN_addr`*`|Mobility header data`

`CN_addr` is the correspondent node's address. The *binding refresh advice* option informs the mobile node about the time when a new binding update is needed.

[4] There are exceptions to this format, which are described in the following section.

#### 5.3.4.2 REMOVING BINDING CACHE ENTRIES IN THE CORRESPONDENT NODE

The mobile node attempts to delete its binding cache entry in the correspondent node's binding cache when it moves to its home link to allow packets to be sent directly to its home address. In this case, the mobile node would not test a care-of address because it does not have one. Hence, only a home keygen token is needed; that is, only a HOTI/HOT exchange is needed.

When sending a binding update to remove a binding cache entry, the source address in the IPv6 header will contain the mobile node's home address, and the destination address will contain the correspondent node's address. $K_{bm}$ is calculated as follows:

$K_{bm}$ = SHA1 (Home keygen token)

The binding update message sent to the correspondent node will contain the following information:

```
IPv6 header
     src = home address
     dst = correspondent node
Routing header type 2 [optional if the correspondent
node were also mobile]
     Home address_cn
Mobility header type 5
     Binding update (lifetime set to zero)
     Nonce indices option
     Authorization data option
```

Note that no home address option or alternate-care-of address can be included in this message. In this scenario, the care-of nonce index is irrelevant, since $K_{bm}$ was formed based on the home keygen token. Therefore, this nonce index is ignored by the correspondent node.

When the correspondent node accepts the binding update, it sends a binding acknowledgment (if requested) with the following format:

```
IPv6 header
   src: correspondent node
   dst: home address
DST-options header
   Home address option (if the correspondent node were also a
   mobile node)
Mobility header type 6
   Binding acknowledgment
   Authorization data option
```

In this scenario, the routing header type 2 is not included in the binding acknowledgment. The authorization data is calculated based on $K_{bm}$, as shown in the previous section.

#### 5.3.4.3 MAINTAINING NONCES AND $K_{CN}$ CORRESPONDENT NODES

The correspondent node is the only node that can generate the home and care-of keygen tokens based on two secrets: $K_{cn}$ and a nonce. $K_{cn}$ is 160 bits, and the nonce may have an arbitrary size; however, 64 bits is the recommended size in the Mobile IPv6 specification. A 64-bit nonce is large enough to minimize the probability that it will be reproduced in the future (for the same mobile node). If $K_{cn}$ and the nonce stay constant for a long period of time, both tokens will remain the same, allowing Bad Guy to steal a legitimate mobile node's traffic. For instance, suppose that Bad Guy snooped a HOT message sent to the mobile node's home address (if he happened to be on the path between the correspondent node and the mobile node's home agent). He can later move to another location, send a COTI message to the correspondent node, then use the snooped token with that received in the COT message to direct the mobile node's traffic to his new location (see section 5.1.1.1). However, if tokens change regularly (because the nonce or $K_{cn}$ changes), this attack will eventually fail. Hence, it is wise to require the correspondent node to regularly generate a new nonce that can be used for token generation. It is also necessary that tokens have short lifetimes. When a mobile node moves, it can reuse the same home keygen token while it is valid (the care-of keygen token is clearly invalid because the address changed). After the token expires, the mobile node must run the return routability procedure again in order to send an authenticated binding update.

The Mobile IPv6 specification requires that a nonce have a maximum lifetime of 240 seconds. On the other hand, a token is valid for 210 seconds. Hence, a fast-moving mobile node may reuse its home keygen token to generate $K_{bm}$ for 210 seconds, and only a new COTI/COT exchange would be needed every time the mobile node moves within the 210 second period following the first binding update.

Since the correspondent node needs to know the nonce used to create the home keygen token, it must "remember" nonces for the lifetime of the token. Note that the difference between the nonce and token lifetimes is 30 seconds. Hence, a correspondent node can generate a new nonce every 30 seconds and accept bindings for the last eight nonces. Using this mechanism, only eight nonces would need to be remembered by the correspondent node at any given time (assuming that it uses the latest nonce to generate tokens for any new HOT/COT messages). If the correspondent node updates $K_{cn}$ regularly, it needs to know which $K_{cn}$ was used with a particular nonce to generate a token. Each one of the eight nonces stored should point to a particular $K_{cn}$. To simplify matters, the correspondent node could update $K_{cn}$ at the same time it creates a new nonce; this allows it to use the nonce index to identify a nonce and the corresponding $K_{cn}$.

If a correspondent node receives a binding update with nonce indices pointing to nonces that have expired, it should respond with a binding acknowledgment, including an appropriate error code in the status field. The following error codes can be used for this case:

136 Expired home nonce index
137 Expired care-of nonce index
138 Expired nonces (if both nonces expired)

Upon reception of one of these errors, the mobile node should restart the return routability procedure. Note that if one of these errors occurs, the correspondent node cannot authenticate the binding acknowledgment, as it would not share a valid $K_{bm}$ with the mobile node. Therefore, when these error codes are used, the correspondent node should not include an authorization data option in the binding acknowledgment.

#### 5.3.4.4 SECURING HOTI AND HOT BETWEEN THE MOBILE NODE AND THE HOME AGENT

The strength of return routability relies to a large extent on a secure tunnel between the mobile node and its home agent (this is analyzed in detail in sections 5.3.4.6 and 5.3.4.7). For HOTI/HOT messages, confidentiality is required, hence the need for protection using ESP. The HOTI/HOT messages are protected by ESP in tunnel mode. As we saw earlier, those messages do not require the addition of a home address option (when sent by the mobile node) or a routing header type 2 (when sent by the home agent). The SAD entry required for protecting both of these messages would point to a tunnel mode ESP protection, which includes the tunnel exit point. Hence, the IPsec implementation in the sending node would add the outer header used to send the packet. Similarly, when one of these messages is received, the receiver would look at the interface (tunnel interface) that it was received on, and based on this information, identify the right security association in the SAD.

How does a mobile node configure its SPD to ensure that outgoing HOTI packets are protected? The mobile node's SPD entry checks the source address in the packet (home address), then checks whether a mobility header is included in the packet. If both of these conditions are satisfied, then the packet is a HOTI (pardon the pun!). Fortunately, these conditions are sufficient regardless of the content of the destination address field (the correspondent node's address), which could not be preconfigured, as no one knows which correspondent nodes the mobile node might communicate with in the future.

When a home agent tunnels a HOT message to the mobile node, its SPD entry contains checks on both the destination address (mobile node's home address) and whether a mobility header is included. Again, these conditions are sufficient independently of the source address field (the correspondent node's address).

#### 5.3.4.5 SECURING BINDING REFRESH REQUESTS

*Binding refresh requests* are used to request a binding update from the mobile node. However, the mobile node is not required to respond to this message with a binding update.

The only potential vulnerability that can be introduced with binding refresh requests is a resource-exhaustion DoS attack. Bad Guy could send many binding refresh requests to trick the mobile node into believing that a return routability test and a consequent binding update is required. If this is done too many times, it can use processing power and memory in the mobile node. However, this kind of attack is not considered to be serious for two main reasons:

- The mobile node is not obliged to start a return routability test or send a binding update when this message is received. As shown earlier, the binding update and acknowledgment messages synchronize the binding cache and binding update list in the correspondent and mobile node respectively. Therefore, the mobile node should know when a binding update needs to be sent to a correspondent node or a home agent.
- The same effect can be produced without sending a binding refresh request. Bad Guy can simply request a connection with the mobile node and send many packets initially to make the mobile node think that it should optimize its path to the correspondent. Therefore, this message does not add new vulnerabilities to the existing ones on the Internet.

#### 5.3.4.6 WHY DOES RETURN ROUTABILITY WORK?

The return routability procedure implicitly assumes that the routing infrastructure is secure and trusted. Thus, it is appropriate to design a protocol to secure the binding update as long as it is no less secure than the underlying routing infrastructure. In other words, if a packet is sent to a particular destination, the routing system delivers it to that destination. If Bad Guy(s) compromise the routing infrastructure and manage to control one or more routers, several serious attacks can be launched independently of return routability procedures.

Another assumption made by return routability is that it is difficult for Bad Guy to be located on two different paths at the same time and receive both tokens needed to generate $K_{bm}$. This could happen if Bad Guy is sharing a link with the correspondent node; he would be able to see all of the return routability packets, construct a binding update message, send it to the correspondent node, and receive all of the correspondent node's traffic addressed to the mobile node. However, Bad Guy does not need to go through all this trouble to hijack the correspondent node's connections with the mobile node if he shares a link with the correspondent node; he can simply pretend to be a router by stealing the default router's link-layer address and sending a fake router advertisement to the correspondent node. Alternatively, he can send a Neighbor Discovery redirect message to the correspondent node requesting that all its traffic be sent to his link-layer address. Thus, an attacker sharing a link with the correspondent node can cause serious harm without Mobile IPv6; that is, Neighbor Discovery messages are the weakest link when an attacker is sharing a link with the correspondent node.

Since our main goal was to ensure that securing route optimization does not make things worse than they are in today's Internet, the above case can be ignored. However, it is worth noting that this type of attack will become significant as soon as a mechanism is devised to secure Neighbor Discovery messages. When this happens, the return routability procedure will become the weakest link.

Bad Guy can be located on the mobile node–correspondent node path. In this location, he would only be able to see the care-of keygen token, which would not allow him to construct $K_{bm}$ correctly to steal the mobile node's traffic.

Bad Guy might send a large number of HOTI and COTI messages to try to consume the correspondent node's resources in a way that makes it unable to process legitimate requests from real mobile nodes. The return routability procedure is designed to allow correspondent nodes to be protected from memory-exhaustion attacks; a correspondent node would only keep state when it receives an authenticated binding update from a mobile node. Clearly, this procedure cannot protect against an attacker aiming at using up the correspondent node's link bandwidth by sending a very large number of HOTI/COTI messages. However, this attack can be launched without return routability by simply sending a large number of bogus messages. It is worth noting, though, that the correspondent node can simply decide to not receive any HOTI/COTI messages if it detects that it is being attacked. That is, the correspondent node can "turn off" route optimization; communication with mobile nodes will still take place through the home agent.

One of the most important advantages of the return routability procedure is that it does not require any manual configuration or infrastructure support. This feature assists with the quick deployment of Mobile IPv6 and encourages vendors to support route optimization, which would have been much harder if route optimization came with the burden of infrastructure support or the unrealistic assumption of manual configuration. However, it is important to note that this comes at the cost of having weak authentication compared to the more traditional applications of public key cryptography.

#### 5.3.4.7 HOW SECURE IS RETURN ROUTABILITY?

When assessing the security level of the return routability procedure, we need to recall the most important requirement that we had to deal with: Do no harm to the current Internet. The current Internet is predominantly based on IPv4 but moving toward IPv6 in the near future. Hence, we need to understand the types of attacks that can be launched today in IPv4, attacks that can be launched in IPv6, and finally compare that situation with the one created by introducing Mobile IPv6.

In the current IPv4 Internet, Bad Guy can launch several attacks while located on the path between two communicating nodes (on-path includes the case where Bad Guy is sharing a link with one or both of the nodes). Bad Guy can modify data between the two nodes (MITM attack), impersonate one of the nodes (masquerading), or perform DoS, including connection termination (e.g., sending a TCP RST message; more on TCP in the next chapter). The same attacks can be launched in an IPv6 Internet, but we currently have mechanisms that can protect against these attacks. Authentication and message integrity can protect against masquerading. Finally, encryption can provide protection against eavesdropping. Off-path attacks can be classified into three categories:

- *DoS attacks:* Bad Guys can launch several types of DoS attacks. These include memory/processing capability exhaustion attacks. Bad Guy can also terminate ongoing connections, as discussed above.
- *Reflection attacks:* Reflection attacks can be performed if the attacker spoofs a victim's IP address; the receiver of a message with a spoofed IP address will respond to that address, causing the message to be sent to the victim. Ingress filtering can minimize the range of addresses that can be spoofed (i.e., to the addresses that can be derived from the prefixes on one link) and expose the location (link) of the attacker. However, it cannot stop this attack.
- *MITM attacks:* These attacks require the attacker to be able to remotely compromise one of the routers on-path to allow him to be **present** on-path. This is extremely difficult and practically impossible today. Alternatively, the attacker may corrupt the routing system to be able to divert traffic to his location. For instance, an attacker may inject routes to a particular destination, hence fooling other routers to believe that traffic to that location should be delivered to him. However, routing protocols are typically protected against such attacks by setting up security associations between routers to allow for authentication and message integrity.

Thus far, it seems that both IPv4 and IPv6 suffer the same types of vulnerabilities when considering on-path, on-link, and off-path attackers. The important question is, What does return routability add to these vulnerabilities? To answer this question, let's consider the attacker's possible locations (Figure 5–9).

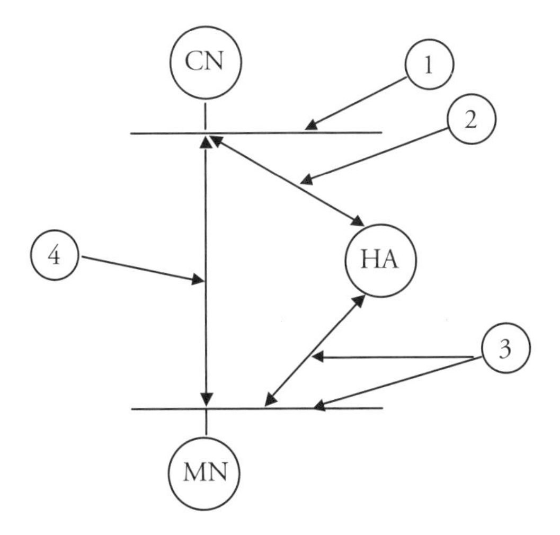


Figure 5–9 Possible locations for Bad Guy.

- *Location 1:* Here, Bad Guy is sharing a link either with the correspondent node or somewhere else where path 2 intersects path 4. We've already shown that when Bad Guy is sharing a link with the correspondent node, several attacks can be launched without using Mobile IPv6 or return routability. However, things can be made worse with return routability. Bad Guy can pick one or more home addresses for one or more mobile nodes. He can send both HOTI and COTI messages to the correspondent node. Since he can see both HOT and COT messages (assuming a shared link, e.g., Wireless LAN), he can construct $K_{bm}$ and send a binding update for those home addresses. This would stop the legitimate mobile nodes that own those addresses from communicating with the correspondent node for the duration of the binding. The same attack can be launched against a mobile node that is already communicating with the correspondent node. The difference from using Neighbor Discovery is that Bad Guy can move to another location (off-path), and continue to deny service between the mobile and correspondent node by refreshing the existing binding. If Bad Guy were somewhere else where paths 2 and 4 intersect, he can essentially launch off-path attacks, provided that ingress filtering allows him to generate a packet containing the mobile node's home address (for HOTI). In addition, Bad Guy would need to be able to snoop the HOT message containing the home keygen token.

- *Location 2:* Bad Guy is somewhere on the path between the correspondent node and the home agent. He can see the HOT message, but cannot see the COT message, and therefore cannot claim to own the mobile node's care-of address. However, he can claim to own the mobile node's home address. He can initiate a return routability procedure, intercept the HOT message (or simply snoop it), and use his current address as a care-of address; that is, the COT message will be sent to his current address. Similar to attacks from location 1, the attacker can then move to another location, keep refreshing the binding, and consequently deny service between the mobile and correspondent nodes. The fact that the attacker can move to another location, off-path, and continue to deny communication between the mobile and correspondent nodes makes this attack different from a typical MITM attack that can take place today.
- *Location 3:* We have combined the two locations (on the mobile node's link and on-path between the mobile node and home agent) for simplicity, since they have the same implications, with one exception described below. Basically, this location is different from 1 and 2 because we have a secure channel between the mobile node and home agent, which is used to encrypt both the HOTI and HOT messages. Hence, if Bad Guy is in this location, there is not much he can do that cannot be done without Mobile IPv6. However, Bad Guy can in fact be a malicious mobile node, which has a real home address. In this case, he can use his valid home address and a care-of address that belongs to another node on the visited link. He would be able to receive the HOT message securely and snoop the COT message (on a shared link). Hence, $K_{bm}$ can be generated, and Bad Guy can divert his traffic to another node on the link. That is, Bad Guy can bomb a neighbor.
- *Location 4:* In this location, Bad Guy can launch the same attacks as can be launched without Mobile IPv6 with MITM cases. He cannot see the HOTI/HOT messages and therefore cannot construct $K_{bm}$. Hence, no new vulnerabilities are added by return routability.

An interesting observation is that attacks from locations 1 and 2 can be launched even on encrypted/authenticated traffic (e.g., using ESP or AH). Recall that the identifiers used to authenticate a binding update are completely different from those used by a client to authenticate traffic for her Internet banking or to run an IPsec tunnel to her corporate network. Return routability and the resulting binding update redirect traffic from one location to another. Whether that traffic is encrypted or not is irrelevant to the binding update. This is a result of the requirements placed on identifiers used to authorize the binding update (home and care-of addresses ownership), which have nothing to do with the real identity of a user or a machine that might have been used to set up a security association between the mobile and correspondent nodes.

From the above analysis we can see that the vulnerabilities of return routability are associated with an attacker being located on the correspondent node's link, somewhere where he can see both HOT and COT, or on the path between the correspondent node and the home agent. In addition, there are two main differences between these attacks and today's attacks on the Internet:

- The attacker can disallow future connections (i.e., ones that do not exist yet) between a correspondent node and any other mobile node by picking a home address and performing a return routability procedure.
- The attacker can move to another location and refresh the binding in the correspondent node to continue to deny service to the mobile node, even while located off-path.

So what can we do to stop these attacks or minimize their impact? Let's consider these vulnerabilities and how to minimize their impact. Vulnerabilities related to the first scenario are addressed in sections 5.3.4.8 and 5.4.

Vulnerabilities related to the second scenario can be addressed using two different approaches:

1. Conservative approach: Every time a mobile node moves, it must run the entire return routability procedure. This will ensure that Bad Guy cannot move somewhere off-path and continue to refresh the correspondent node's binding cache based on the previous home keygen token.
2. Moderate approach: Set a lifetime for the home and care-of keygen tokens. A mobile node will only be able to refresh a binding while these tokens are still valid. After that, the mobile node must run the entire return routability test. This approach will limit the time during which the attacker is refreshing bindings while located off-path. That is, if Bad Guy was in location 1 (or 2), then moved off-path, he would be able to refresh the binding (and hence deny future communication between the mobile and correspondent nodes) for the lifetime of the home keygen token (he would only need to send a COTI message from his new location).

The conservative approach has the advantage of eliminating the vulnerability in question, but requires sending HOTI/COTI and receiving HOT and COT every time the mobile node moves. During this period, the mobile node cannot communicate with the correspondent node directly; outbound packets (sent from the mobile node) would need to be routed via the home agent, but packets sent from the correspondent node are routed to the mobile node's previous location and effectively lost. Therefore, we can assume that communication is impossible for the duration of the return routability procedure. The moderate approach has the advantage of sending a single message (COTI) while the home keygen token is still valid. While the overall disruption time may not vary significantly (since HOTI and COTI are sent at the same time), the moderate approach causes fewer packets to be sent by the mobile node. The Mobile IPv6 specification chose the moderate approach over the conservative one. This allows Bad Guy to continue to refresh the binding for the lifetime of the home keygen token (240 seconds). Moreover, Mobile IPv6 restricts the overall lifetime of a binding cache entry created using the return routability procedure to 7 minutes. After 7 minutes, the mobile node must perform the entire procedure to be able to send a new binding update. This must be done regardless of the mobile node's location—even if it never moved since the last binding update.

A new mobility protocol is bound to have new threats associated with it; hence, it is pertinent to understand these threats and decide which ones must be avoided and which ones can be "lived with." It is certainly likely that new threats will also emerge when the protocol is deployed. Therefore, stronger security mechanisms might be required in the future to eliminate some of these threats. Some of the efforts to develop more secure mechanisms are discussed in the next section and in section 5.4.

#### 5.3.4.8 SECURING BINDING UPDATES USING PUBLIC KEYS AND CERTIFICATES

In Chapter 4 we saw how Alice and Bob used public keys and certificates to authenticate each other and encrypt traffic. These mechanisms can also be used to authenticate binding updates between mobile and correspondent nodes. In fact, earlier revisions of Mobile IPv6 assumed that public key cryptography would be used between mobile and correspondent nodes to authenticate binding updates. However, the earlier approach was not clear on two aspects:

1. How does the mobile node prove ownership of the home address?
2. How does the mobile node prove ownership of the care-of address?

Public key cryptography typically requires a key exchange mechanism (e.g., IKE) and a method of binding the public key of the correspondent to the identity that it claims. The identity needed to answer the first question above is the home address. The home address can be included in a certificate. Two possible options exist:

1. Include the home address in the *subject* field, or
2. Include the home address in the *subjectUniqueIdentifier* field.

Including the home address in the latter field is probably a better option, as it allows the subject field to be used for other cases where a different identifier may be needed (i.e., where current applications normally expect the identifier). Hence, a correspondent node receiving a mobile node's certificate would be able to prove (given the CA's public key) that the mobile node owns the home address.

Answering the second question would involve a reachability test for the care-of address. That is, if the correspondent node can send a message to the care-of address and receive a correct reply (authenticated by the same node that claims to own the home address), then the correspondent can authorize the mobile node to direct traffic from its home address to its current care-of address. Once a security association is established, the mobile node could reuse that security association to send future binding updates. However, the mobile node still needs to prove its ownership of the care-of address every time it moves. This is needed to prevent the mobile node from using a fake care-of address or another node's care-of address (flooding attack). An example of a protocol that can be used to solve this problem is shown in Figure 5–10.

When using this protocol, the mobile node uses its care-of address as a source address in IP packets containing the IKE messages. In phase 2, the mobile node negotiates the security association that will be used to protect the binding update. IKE allows the mobile node to specify an address other than the source address to be used to identify the security association (recall Chapter 4, Figure 4–10). The mobile node can use its home address for this purpose. However, the correspondent node would need to ensure that the mobile node is authorized to use such address. This can be done by reading the subjectUniqueIdentifier field in the mobile node's certificate and making sure

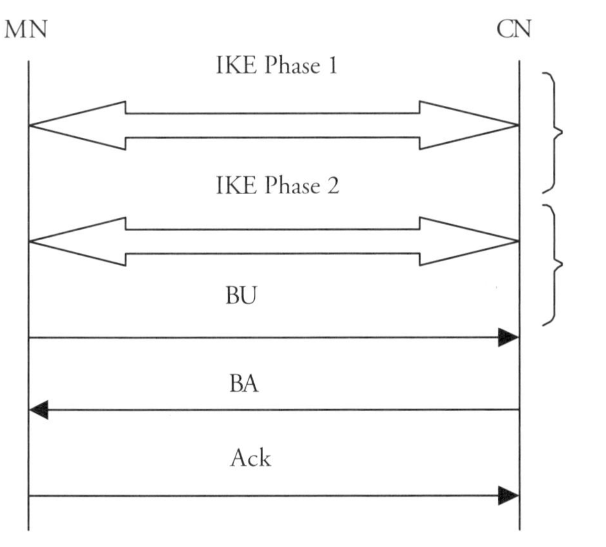


**Figure 5–10** *Authenticating binding updates using IKE and IPsec headers (ESP/AH).*

that it includes the mobile node home address. If it does, the security association is established. At this stage, the mobile node can send a binding update to the correspondent node. The correspondent node acknowledges the binding update by sending an encrypted (with ESP) binding acknowledgment to the mobile node's care-of address. The binding acknowledgment should contain a sequence number that matches the one included in the binding update. In addition, the binding acknowledgment should contain a nonce (e.g., 64 bits). So far, the mobile node has proven that it owns the home address, but not necessarily the care-of address. For example, the mobile node could have used its real care-of address during the IKE exchange but sent the binding update with a fake one. If the mobile node replies with a message containing the same nonce sent by the correspondent node (the new ACK message), then the binding can be accepted.

We must note that this scheme has the same vulnerability associated with location 3 in the return routability case (or, more accurately, a subset of location 3; that is, if Bad Guy is sharing a link with the victim). In this case, Bad Guy can actually be the mobile node itself, which is trying to bomb a victim, sharing a link with it. The mobile node would be able to successfully set up a security association with the correspondent node. It can also use a care-of address that belongs to another node on the same link (on a shared link). The mobile node would be able to see the binding acknowledgment, decrypt it, and read the nonce (since it knows the key). The mobile node can then send the correct reply in the ACK message.

### 5.3.5 Preventing Attacks Using Home Address Options and Routing Headers

In section 5.1.2, we saw how Bad Guy can launch reflection attacks using the home address option and the routing header. In Chapter 2, we saw how the routing header can be abused by Bad Guy to gain access to nodes that he is not authorized to gain access to. In Mobile IPv6, restrictions were placed on the home address option and the routing header to prevent these types of attacks. Both the home address option and the routing header (type 2) can be used only **after** the correspondent node has accepted a binding update from the mobile node. When receiving a packet containing a home address option, the correspondent node checks its binding cache to see if a binding exists between the home address (included in the option) and the care-of address included in the source address of the packet. If such entry does not exist, the mobile node sends a *binding error* message to the source address in the packet. This stops Bad Guys from using the home address option without proving their ownership of both the home address and the care-of address.

To prevent attacks against the routing header, Mobile IPv6 requires that two rules be satisfied in the mobile node in order to accept a packet containing a routing header:

1. The routing header must contain one address only.
2. The address in the routing header must be one of the mobile node's home addresses.

These rules can certainly be satisfied when using the existing routing header (with type 0). However, other concerns were also brought up during the design of Mobile IPv6. Network administrators might get concerned about attacks like the one shown in Chapter 2. After all, a firewall has no way of knowing if the address contained in the routing header belongs to the same node receiving the packet or to another node. Hence, network administrators might configure their firewalls to drop any packet containing a routing header. This would effectively prohibit the use of Mobile IPv6 (recall that binding acknowledgments from the home agent contain a routing header). To avoid the routing header (type 0) overload, which might force network administrators to throw away the baby with the bath water, Mobile IPv6 uses a new routing header (type 2). This allows firewalls to distinguish between a routing header type 0 and Mobile IPv6's routing headers, which are restricted to particular use and policed by mobile nodes. That is, a node would process routing header type 2 only if it were a mobile node. As a consequence, it must know the rules of processing such header. Therefore, it will not relay the packet to any other node.

# 5.4 Future Mechanisms for Authenticating Binding Updates

In Chapter 4, we saw how a node can generate a cryptographically generated address by generating a public/private key pair, then using the public key to produce *h*(PK, *something*). The hash function will produce 62 bits that can be used (in addition to the u- and g-bits) to generate the interface identifier. If *something* is a number that includes the prefix on a particular link, then the CGA produced can prove that a node owns an address on a particular link. For instance, the mobile node might have a cryptographically generated home address as follows:

```
64-bit Interface_identifier = First(64, (H(home_prefix|PK|
context) & 0xfcffffffffffffff
```

where `context` is a number associated with the context in which CGAs are used (e.g., set to 1 for Mobile IPv6). The `home_prefix` is the mobile node's home prefix. The `& 0xfcffffffffffffff` indicates a bitwise AND operation used to set the u- and g-bits to their correct values (zero, because the interface identifier cannot be guaranteed to be globally unique and is not part of a group).

The mobile node can prove that it owns such an address and authenticate messages sent from this address by signing the message with its private key and including its public key inside the message. The signature is calculated as follows:

```
SIG = SIGALG (H(m), SK)
```

where `SIGALG` is a signature algorithm, `H(m)` is a hash of the message being authenticated, and `SK` is the mobile node's private key (we use **S** for private key, since **P** was already used for the public key). Since the message includes the mobile node's public key, the receiver of such message is able to verify two things:

1. The sender knows the public/private key pair.
2. The sender owns the address used in the source address.

The first can be verified by verifying the signature sent in the message. The second can be verified because a change in address would show that the packet has been modified (i.e., assuming that the receiver knows **context** and can recalculate the interface identifier).

Before we consider how this mechanism can be used to secure binding updates, we need to understand an important limitation: CGAs prove that a node owns a particular address, but do not prove that this node is currently located at that address. A node may have formed this address based on a fictitious prefix or while being on that link, but it is no longer there. Hence, there is no guarantee that the node owning a CGA is actually located on that link at a particular moment in time. To prove this, we need to test that address to see if it is "live." In other words, we still need a return routability test.

## 5.4.1 Alternative 1: Using a Cryptographically Generated Home Address

Consider the case where the mobile node has a cryptographically generated home address. Such home address can be used when performing the return routability procedure to provide stronger authentication. Note that we still need to perform return routability to make sure that the mobile node is actually located where it claims to be. Figure 5–11 shows the return routability procedure.

In this protocol, the mobile node sends the HOTI and COTI messages as described earlier. After receiving the HOT and COT messages, the mobile node

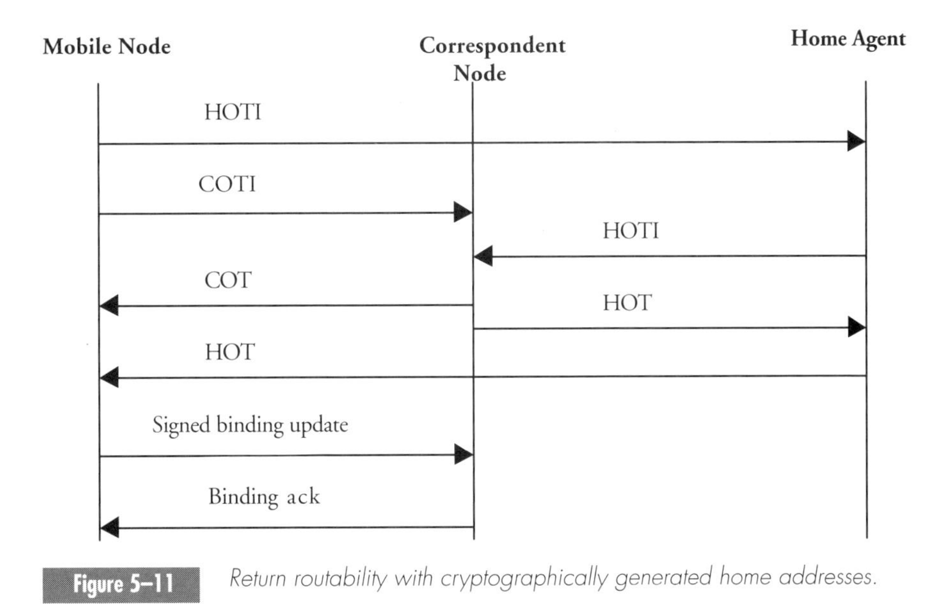


**Figure 5–11** *Return routability with cryptographically generated home addresses.*

can form $K_{bm}$ and authenticate the binding update. In addition, the mobile node can sign the binding update with its private key, as shown earlier. This assures the correspondent node that the mobile node owns that home address.

The advantage of this addition to return routability is that it protects from Bad Guys on the path between the correspondent node and the home agent (locations 1 and 2 in Figure 5–9). Bad Guy can no longer claim to own the mobile node's address without generating the corresponding key pair, which is extremely difficult. Thus, Bad Guy cannot steal the mobile node's traffic or prevent future communication between the correspondent and mobile nodes. Bad Guy cannot hijack a connection either, since he cannot sign the binding update message with the right private key. Furthermore, due to the strong cryptographic nature of this scheme, the mobile node could be allowed to generate a security association that exempts it from running the home address test. This would allow for a much longer lifetime for the home keygen token.

A disadvantage of this mechanism is that it is computationally intensive (to generate the public/private key pair). However, this might not be a major obstacle to deploying CGAs, since mobile nodes will not need to generate these keys very often. The strong cryptographic property of CGAs allows the mobile node to use the same key pair for a long period of time. A more significant disadvantage of this approach is that verifying the mobile node's signature could be computationally intensive for the correspondent node (depending on the signature algorithm used). If the correspondent node is a battery-powered wireless device, this could be a concern. Finally, public keys are very large; on a bandwidth-challenged link, sending these packets can block the link for some time and cause some jitter for ongoing sessions. In Chapter 7, we show how binding updates can be reduced for fast moving mobile nodes.

### 5.4.2 Alternative 2: Using Cryptographically Generated Home and Care-of Addresses

CGAs are useful for any application in which a node is required to prove that it owns a particular address. One of these applications (discussed below) is securing Neighbor Discovery messages. A mobile node might wish to ensure that no other node on a visited link will steal its address and might consequently produce a cryptographically generated care-of address. This is useful in providing more security to the binding update. In this case, the messages used are identical to the procedure shown in Figure 5–11, with the addition that the mobile node would use a cryptographically generated care-of address. Hence, the mobile node's signature will prove that the mobile node owns both the home and care-of addresses (both generated from the same public key but with different prefixes).

This scenario is useful because it stops a malicious node from bombing another neighbor on-link, since only the mobile node will have the right key pair to generate both home and care-of addresses. When using this approach, all known vulnerabilities of the return routability procedure can be addressed.

### 5.4.3 Other Improvements Gained from CGAs

As shown earlier, CGAs are useful for securing Neighbor Discovery messages like the following:

- *Duplicate Address Detection* (DAD): In this case, a node can prove that it owns a particular address. The DAD message can be signed by the node generating the address. If another node claims the same address (very unlikely), it must sign its Neighbor Advertisements (NAs) to prove that it generated the same key pair. This technique prevents DoS attacks that can be launched by Bad Guy to prevent a node from configuring an address by claiming that it owns that address (i.e., replying to all DAD messages with corresponding NAs).
- *Neighbor Solicitation* (NS): The Neighbor Discovery specification allows nodes to include their own link-layer addresses in an NS. This optimization saves the responding node from sending another NS to the soliciting node. If a node signs the NS with its private key and uses a CGA, the receiver would be sure that such node owns the address that it claims to own.
- *Neighbor Advertisements* (NAs): NAs can also be protected if the sending node signs the message with its private key and uses a

CGA as a source address. Since these messages bind a node's IP address to its link-layer address (not covered by IP layer security), somehow the receiver of this message will need to check (possibly from its own link layer) that the message was in fact received from the right link-layer address.

- *Router Advertisements* (RAs): This message is a bit different from the others. A router can certainly use a CGA for its source address to allow nodes to verify that the same router is sending the same advertisement. However, this would not allow other nodes to know that this address (the router's) is in fact authorized to send a router advertisement. For instance, Bad Guy could generate a CGA and a prefix (may or may not be valid, depending on the intention of the attack) and start sending RAs on the link. This could fool other nodes into believing that Bad Guy is actually a router and start sending their traffic to him. This scenario illustrates, once more, the value of authorization. Bad Guy can be authenticated using CGAs, but such authentication does not authorize Bad Guy to be a router.
- *Redirect Messages:* A redirect message instructs the sender of a particular message to send future messages to a different node; such node may be the ultimate destination of those messages (a neighbor on a link) or a router, which has a better route to the ultimate destination. If not authenticated, these messages can be used to bomb other neighbors. Bad Guy could instruct node A to use node C as a default route to send packets addressed to node B. However, node C is not a router; it is another host on the link. In this case, node A needs to ensure that a redirect message actually came from a router authorized to send this message. If the RA authorization problem is solved (e.g., using certificates that are signed by the authority delegating the prefix to the router), a node would only need to authenticate a redirect message provided that it knows that the source address (a CGA) of this message is authorized to be a router.

The SEcure Neighbor Discovery (SEND) Working Group in IETF was chartered in October 2002 to solve the problem of securing Neighbor Discovery messages. In [2] and [5], the different threats associated with Neighbor Discovery messages and ways of addressing them using CGAs are discussed.

## 5.5 Summary

This chapter started by justifying the need for Mobile IPv6 security. Mobile IPv6 messages were analyzed, and a brief threat analysis was presented to show the impact of an insecure mobility management protocol like Mobile IPv6. Next, we discussed the main requirements used to design the security solution for Mobile IPv6.

We presented Mobile IPv6 security in the context of two different security relationships:

1. mobile node–home agent relation
2. mobile node–correspondent node relation

Mobile IPv6 signals to the home agent are protected with IPsec (ESP), and an example showing manual configuration of security associations was shown. Route optimization signaling to the correspondent node is done using the return routability procedure. This procedure was discussed in detail, emphasizing its strengths and vulnerabilities.

Finally, we showed different alternatives for securing Mobile IPv6 with public keys and certificates, or with CGAs. The use of CGAs for other applications was also discussed.

This chapter concludes our look at Mobile IPv6's operation. The next part of this book looks into the performance of Mobile IPv6 handovers and different optimizations that can be used to enhance it. This be shown in the context of wireless IP networks.

## Further Reading

[1] Arkko, J., P. Nikander, G. Montenegro, E. Nordmark, and T. Aura, "Residual threats," note from the design team to the Mobile IP mailing list, 2002.

[2] Arkko, J., P. Nikander, and V. Mantyla, "Securing Neighbor Discovery Using Cryptographically Generated Addresses (CGAs)," draft-arkko-send-cga-00, work in progress, June 2002.

[3] Arkko, J., V. Devarapalli, and F. Dupont, "Using IPsec to Protect Mobile IPv6 Signaling between Mobile Nodes and Home Agents," draft-ietf-mobileip-mipv6-ha-ipsec-03, work in progress, February 2003.

[4] Aura, T., and J. Arkko, "MIPv6 BU Attacks and Defenses," draft-aura-mipv6-bu-attacks-01, work in progress, February 2002.

[5] Kempf, J., and E. Nordmark, "Threat Analysis for IPv6 Public Multi-Access Links," draft-kempf-ipng-netaccess-threats-00, work in progress, October 2001.

[6] Montenegro, G., and C. Castelluccia, "SUCV Identifiers and Addresses," draft-montenegro-sucv-02, work in progress, November 2001.

[7] Mankin, A., et al., "Threat Models Introduced by Mobile IPv6 and Requirements for Security in Mobile IPv6," draft-ietf-mobileip-mipv6-scrty-reqts-02, work in progress, May 2001.

[8] Nikander, P., E. Nordmark, G. Montenegro, and J. Arkko, "Mobile IPv6 Security Design Rationale," draft-nikander-mipv6-design-rationale-00, work in progress, February 2003.

[9] Nikander, P., "Address Ownership Problem in IPv6," draft-nikander-ipng-address-ownership-00, work in progress, February 2001.

[10] Nordmark, E., "Securing MIPv6 BUs Using Return Routability (BU3WAY)," draft-nordmark-mobileip-bu3way-00, work in progress, November 2001.

[11] O'Shea, G., and M. Roe, "Child-proof Authentication for MIPv6 (CAM)," *Computer Communications Review*, April 2001.

[12] Roe, M., T. Aura, G. O'Shea, and J. Arkko, "Authentication of Mobile IPv6 Binding Updates and Acknowledgments," draft-roe-mobileip-updateauth-02, work in progress, March 2002.

[13] Savola, P., "Security for IPv6 Routing Header and Home Address Option," draft-savola-ipv6-rh-ha-security-03, work in progress, December 2002.

[14] Thomas, M., "Binding Updates Security," draft-thomas-mobileip-bu-sec-00, work in progress, November 2001.

PART THREE

# HANDOVER OPTIMIZATIONS FOR WIRELESS NETWORKS


- Chapter 6 Evaluating Mobile IPv6 Handovers
- Chapter 7 Mobile IPv6: Handover Optimizations and Extensions
- Chapter 8 Current and Future Work on IPv6 Mobility

SIX

# Evaluating Mobile IPv6 Handovers

Mobile IPv6 had some important requirements to satisfy and was designed to meet these requirements. However, the rapid adoption of new applications (e.g., real-time applications), the desire to run traditional circuit switching applications like Voice over IP, and the merger of the cellular world with the Internet are all factors that motivated the addition of new requirements on IP mobility. In 1999, an initial attempt was made in [3] to reduce the delays caused by IPv4 mobility. Following this proposal, several other proposals were made, aiming to do the same thing for both Mobile IPv4 and Mobile IPv6. Due to the large interest in this topic, the Mobile IP Working Group formed two design teams to develop solutions that can minimize or eliminate the negative impacts of IP mobility. At the time, the requirements were not very clear; our mandate was essentially to "reduce the losses of IP packets due to Mobile IP handovers." It was not clear how much reduction was needed. Should we aim for zero losses? Do we consider all applications, or do we focus on real time only? Do we produce one solution that works for all link layers or one for each category of link layers? What categories of link layers are there? Different proposals had completely different assumptions and different answers for almost each one of these questions. It is no surprise that it took us almost two agonizing years to agree on the major issues!

In this chapter, we answer these questions first (the beauty of hindsight!). In Chapter 7, we present different optimizations needed to reduce the negative implications of IP handovers. At the end of this chapter, you will have a clear understanding of the problem space. In Chapter 7, we present the efforts made to solve these problems.

## 6.1 Layer 2 Versus Layer 3 Handovers

In the context of mobility management in wireless networks, a handover describes the movement of a mobile computer between different points of attachment to a network. This process may not have any relation with the IP layer. For instance, the Global System for Mobile communication (GSM) was designed to handover a mobile phone from one base station to another. This process is typically a result of a degrading quality of the wireless link. The link quality might be degrading because the device is moving away from a base station, so the received power is degrading due to changes in radio conditions or due to *shadowing* (i.e., when an object blocks the path of the strongest signal between the mobile phone and the base station). In any case, GSM handovers are handled by the radio link protocols.

If IP were used on top of a radio link, the handover may cause a change of the device's location within the Internet topology. This is illustrated in Figure 6–1.

In Figure 6–1, a handover from cell_1.1 to cell_1.2 can be handled by the radio link protocols. This would cause the mobile device to break its attachment from the base station in cell_1.1 and attach to the new base station in cell_1.2. In this handover, there is no change in the device's location within the Internet topology, since both cells are associated with the same router. Hence, the mobile node is still reachable through Router_1. The mobile node's addresses, based on the prefixes advertised by Router_1, are still valid. Therefore, when a mobile node is handed over from cell_1.1 to cell_1.2, the IP layer does not see any changes except for a slight disruption, which happens on most radio links.

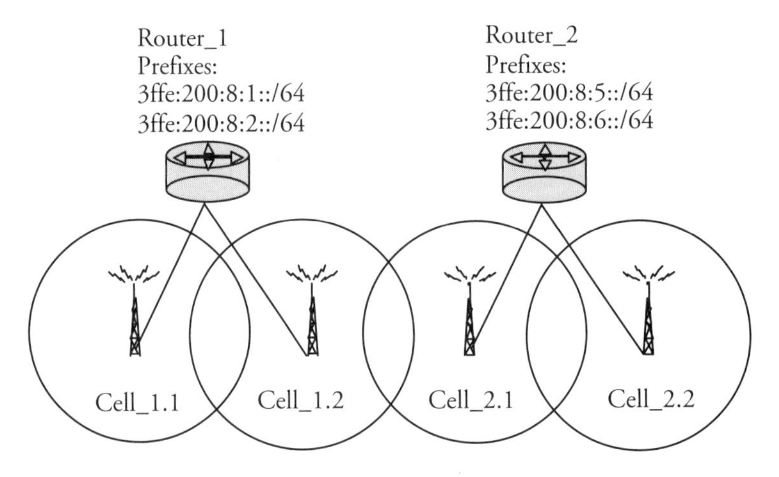


**Figure 6–1** *Layer 2 versus layer 3 handovers.*

A cell is a coverage area associated with a particular base station. Cells can vary in size depending on the radio technology being used and the amount of power being transmitted. For instance, GSM systems have an upper limit of 35 km for a single cell. GSM cells can also be as small as 100 meters. Other technologies, like Code Division Multiple Access (CDMA), have cells that vary in size depending on the number of users in a cell and the transmitted power. IEEE802.11b (typically called WLAN) cells are typically limited to 100 to 300 meters.

We observe that while a single cell might be sufficient for a small office, a house, or a suburb (depending on the radio technology), it will not be sufficient for a very large area with a large number of users. Furthermore, as the area gets larger and the number of users increases, there is typically a limit on the number of cells that can be supported by a single router. For instance, if a hotel wishes to deploy WLANs to allow its guests to access the Internet, it may be satisfied with 50 cells that can serve a few hundred guests. These cells can all be attached to a single router. However, if a large operator wished to provide nationwide radio coverage, it would be unwise to attach all cells to a single router for a number of reasons:

1. The router will most likely be unable to support the millions of users in that country.
2. This router is a single point of failure; if it fails, no one in the entire country will be able to communicate.
3. The network will be hard to manage and debug.

Regarding network management and debugging, we can draw an analogy with a large multinational corporation that has one manager. You can easily see that with a single manager it is difficult to manage everyone and locate problems within the organization. If this organization were distributed into different departments, each with its own management and responsibilities, it would be easier to monitor the organization and locate problems (although some might argue that increasing management is always the problem!). The same "divide and conquer" approach can be used when designing IP networks by splitting the topology into different links, each supported by one or more routers. However, when this division is done, we must manage nodes' mobility between the different subnets.

In Figure 6–1 we see that a handover between cell_1.2 and cell_2.1 results in a change in the mobile node's location within the topology. Hence, both IP layer handover and a link-layer handover take place. This is an example of an IP handover taking place due to a link-layer handover.

An IP layer handover does not have to be triggered by a link-layer handover. Consider a multihomed mobile node that has two different physical interfaces, one connected to a cellular link and the second to an Ethernet link (e.g., IEEE802.3). In this case, the mobile node might prefer one link to the other and decide to use its care-of address on that link. As a result, the mobile node updates its home agent and correspondent nodes. In this case, the IP handover is not taking place due to link degradation or link-layer handover, but because the mobile node prefers one interface to another. This effectively changes its location within the Internet topology.

The efforts to enhance the performance of IP handovers are primarily motivated by the former type of handover (IP layer handover triggered by link-layer handover). In order to understand the design tradeoffs of these proposals, we need to gain a high-level understanding of how link-layer handovers work in different radio technologies. We also need to understand the limitations associated with different radio link layers and consequently the assumptions that can be made by a particular solution.

## 6.1.1 Where Is Layer 2 Terminated?

When a host connects to a network, it communicates with an Access Router (AR) using Neighbor Discovery messages. IP packets are sent from the host and later received and processed by the router. Similarly, when sending packets to other off-link destinations, the default router is the first node that "sees" the IP packets after they are sent by the host. The same happens to link-layer messages. However, in this case, the node receiving and processing the host's layer 2 messages is called the Access Point (AP), or a base station. Hence, in a wireless link, the AP is the link-layer termination point. In some cases, an AP and an AR are collocated, (i.e., implemented in the same device). But in many cases, the AP and AR are physically separated in two different boxes. Figure 6–1 shows such configuration. Most currently deployed wireless systems follow the AP/AR separation shown above. In the following sections and chapters, we use *AP* and *base station* as synonyms.

## 6.1.2 Two Different Categories of Wireless Links

There are several ways of categorizing wireless links (e.g., based on whether they have encryption or how much bandwidth they offer). However, in this section, we categorize different radio links by the nature of the connection they allow between mobile nodes and ARs. The reason is that we view these links from a mobility perspective. This becomes clear when you study the Fast Handover mechanisms in Chapter 7.

Generally speaking, there are two types of radio links: shared links and point-to-point links. We refer to the former as *connectionless* links and the latter as *connection-oriented* links. Cellular links fall into the connection-oriented links category. Connection-oriented means that each host is connected to the access router over a unique logical link (or channel). That is, the host's only neighbor on the link is the AR. The link is logical because all hosts are connected to the AR using the same physical space. The uniqueness of each logical link is guaranteed by the network and depends on the radio technology being used. For instance, radio links based on Frequency Division Multiplexing (FDM) allocate a separate frequency to each host. Hence, hosts cannot communicate directly to each other. Time Division Multiplexing (TDM) systems use a single frequency but allocate different timeslots for different hosts. Each host can transmit only during the allocated timeslot.

The logical links used in these schemes are usually encrypted using secret key cryptography. Each cellular device is configured with a unique key tied to its identity. Hence, an operator supporting *n* devices maintains *n* security associations, one for each device. Connection-oriented radio links utilize the logical links established between hosts and APs to receive link state reports sent in link-layer messages. Hosts regularly scan other channels and compare their quality to the one they are currently connected to. This is done to discover whether or not a better AP can be used. If a better signal is received through one or more APs, the host informs the network. The network then selects an AP that the host can move to. Finally, the host is handed over to the new AP.

Connectionless radio links use a single channel between the AR and all hosts on the link. In some cases, the link is secure. However, all hosts use the same key to encrypt traffic. More recent work, documented in IEEE802.1X, allows devices to have unique keys for encryption. Unlike connection-oriented links, hosts do not send link state reports to the network. This is the main point of distinction where IP mobility is concerned. In other words, the network is not informed about a host's mobility in advance. When the current link is degrading, a host starts scanning other possible channels for a better link. For example, if each link were allocated a particular frequency, a host would scan all possible frequencies defined for the radio link being used and see if a better link can be used. As a result of this design, hosts on connectionless radio links make the entire decision about their mobility. This is quite different from connection-oriented links where the network is always informed and chooses the next AP that a host should move to. Figure 6–2 shows the different logical channels in the two types of links.

You might wonder why some link-layer designers make these very different design choices. The answer becomes clear when we look at the different uses of various types of wireless links. To simplify this discussion, we

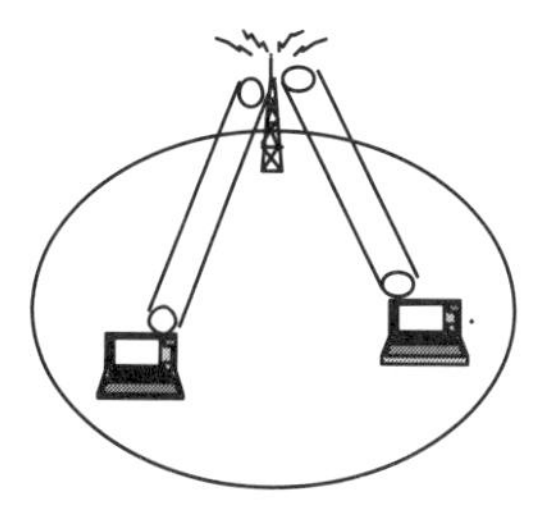

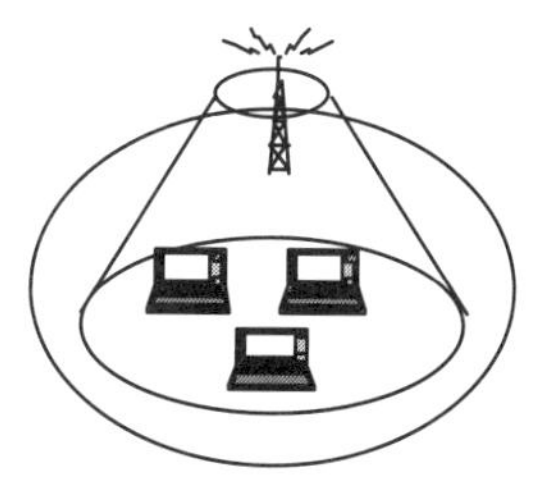

**Figure 6–2** *Connection-oriented versus connectionless radio links.*

consider one example of each type: cellular links as an example of a connection-oriented link layer and IEEE802.11b as an example of a connectionless link layer. All cellular links were designed for use in wide area networks (WANs). Therefore, they need to allow an operator to support tens of millions of devices (in a large country). The coverage area of the antenna in a base station is inversely proportional to the frequency being transmitted. A signal sent on a higher frequency experiences higher attenuation, and the coverage area of a base station is consequently smaller. Higher frequencies are more susceptible to noise. Hence, a cellular link cannot use extremely high frequencies (as opposed to a link designed to cover a small area). If higher frequencies are used, an operator needs to use more APs in the same coverage area, which results in further cost (the radio network costs the most compared to the rest of the network). All these factors are among the main reasons for not choosing very high frequencies for cellular networks. We are therefore limited to relatively low frequencies (700 MHz to 2.5 GHz, depending on the technology). For this reason, cellular link designers are extremely careful with the bandwidth utilization on links. Having a point-to-point connection with each device allows the network to get link state reports from devices. In addition, the network has some intelligence to decide where the device should move next. This is one of the components of Radio Resource Management (RRM). The decision is based on the network's knowledge of bandwidth utilization in different cells and the user's needs.

In conclusion, the point-to-point design allows the network to have more control over its RRM by receiving the radio link state reports from the mobile devices. Control over RRM is essential for a network operator to use its bandwidth in the most efficient manner. In addition to RRM issues, Quality of Service (QoS) is easier in connection-oriented links. In this context, QoS means reserving the right amount of bandwidth requested by the user and making sure that user_a does not use user_b's bandwidth. It's much simpler to enforce this when each user is allocated a separate secure channel. Achieving the same goal in a shared link involves an assumption that users will "behave" correctly. However, it is difficult to make this assumption in a nationwide public network where anyone can connect and misbehave. Therefore, preventing these actions is pertinent (as opposed to detecting them when they take place).

IEEE802.11b was designed for an entirely different reason. The goal was to provide a wireless link for local area networks (LANs). QoS was not a design goal. The intention was to provide a simple wireless equivalent to wired Ethernet links. Hence, a radio technology with small coverage areas and plenty of bandwidth was designed. The simplicity of this design made WLANs very attractive for use in offices, shopping centers, hotels, and airports.

From this discussion, we conclude that networks utilizing connection-oriented link layers are, by design, aware of the mobile nodes' mobility; a base station is always aware of the mobile node's connectivity and its most likely new base station. On the other hand, connectionless link layers are generally unaware of a mobile node's connection status (whether a mobile node is still connected or not). In addition, connectionless link layers are not aware of a mobile node's next likely AP. Such knowledge is determined by the mobile node and is not communicated to the AP before the handover.

### 6.1.3 Make-Before-Break Versus Break-Before-Make Handovers

From the previous section, we know that different link layers have their knowledge of the mobile node's "future AP" (due to the mobile node's movement) in different entities: the mobile node (connectionless links) or both the mobile node and the network (connection-oriented links). Another important aspect to consider is whether or not a mobile node is always connected during the handover.

We consider two types of handover:

- Break-before-make handovers
- Make-before-break handovers

A break-before-make handover means that the mobile node disconnects from its current link, then reconnects to the new link. What does "disconnect" mean in a wireless link? It means that the mobile node loses the synchronization between its own transceiver and the current AP's transceiver and is therefore unable to send or receive data. The mobile node does that to be able to use its radio resources to synchronize with the new AP. Hence, there is a definite period of time in which the mobile node is unreachable because of the layer 2 handover (i.e., resynch with the new AP). That period of time can vary depending on the technology. Some measurements of IEEE802.11b show that this period can vary between 60 and 200 ms depending on the vendor equipment. In Wideband Code Division Multiple Access (WCDMA), this period can be as long as 100 ms. Note that these figures are used to give some indication of how long this period can be: they should not be assumed absolute figures. Measurements usually depend on the implementation of the radio interface and the environment in which they were taken. Break-before-make handovers are by far the most common handovers for both types of link layers discussed in the previous section.

Make-before-break handovers involve a mobile node "making" the new link while maintaining the old link; when the new link is ready, the old link can be broken. Clearly, this scheme would minimize, if not eliminate, the break times discussed earlier. On most cellular links, this scheme is currently not supported in existing products. However, we can see an example of a make-before-break handover when we consider a multihomed host moving traffic from one of its interfaces to the other. In this case, the host would ensure that the new interface is "up," move its traffic to the new interface, then break the old link. If radio coverage is available for this period of time, the handover can be seamless, (i.e., does not cause any packet losses).

## 6.2 How Long Does a Mobile IPv6 Handover Take?

An IP layer handover is a result of a change in a mobile node's location within the topology. Regardless of the reason for such change (layer 2 handover or interface switching), the mobile node needs to inform the correspondent node(s) and the home agent of such change. Until they are informed, all packets continue to be sent to the mobile node's old address. This results in packet losses for a period of time. We refer to this time as the *handover time*. Enhancing the performance of a Mobile IPv6 handover essentially means that the handover time must be reduced or, ideally, eliminated. Note that this does not necessarily mean that the break introduced due to a link-layer handover will be eliminated. This can be illustrated by the following equality:

$$d_t = d_i + d_l$$

where $d_t$ is the time during which the mobile node is unable to send or receive packets, $d_i$ is the time during which the mobile node is unable to send or receive packets due to Mobile IPv6 actions, and $d_l$ is the time during which the mobile node is unable to send or receive packets due to link-layer handover. Ideally, we would like to eliminate $d_i$. To do that, we need to understand what contributes to $d_i$. Consider a mobile node moving between router_1 and router_2, as shown in Figure 6–3.

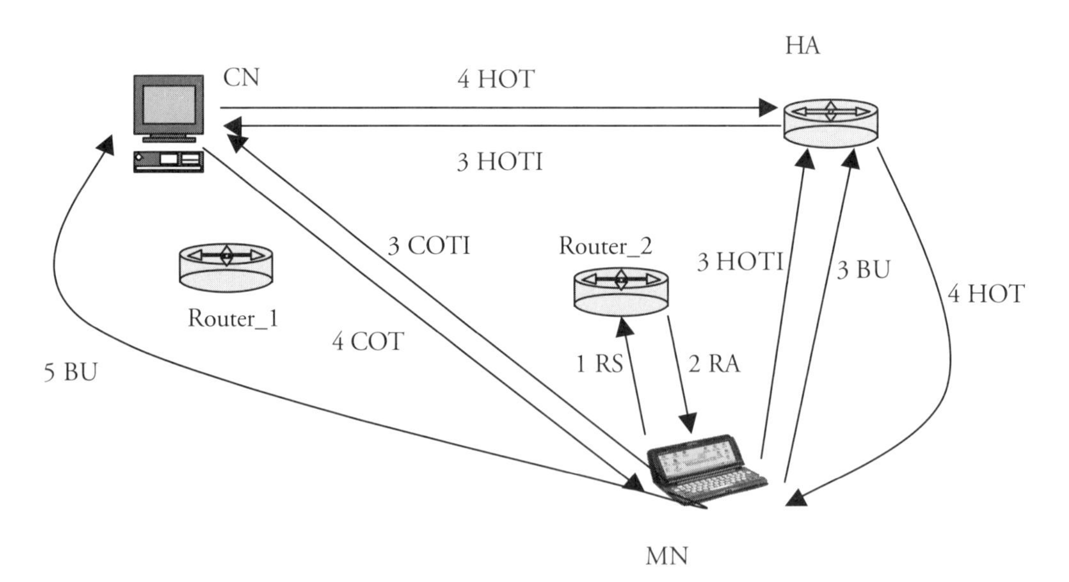


Figure 6–3 Mobile IPv6 handover time.

In Figure 6–3, messages that have the same numbers are ones that can be sent simultaneously (or almost simultaneously). For instance, the Router Advertisement (RA, message 2) needs to be sent before the binding update (BU, message 3) can be sent, since the mobile node has to discover that it moved, and it must configure a new care-of address.

The first task is for the mobile node to discover that it has moved to a new router. This is done by sending a Router Solicitation (RS) message to the new router. But how does the mobile node know that it should send a router solicitation? There are several possibilities:

- The RA interval has lapsed, and the mobile node has not received a router advertisement.
- The IP stack in the mobile node received a "hint" from the link layer, informing it that a link-layer handover has just taken place. We refer to the hint as *layer 2 trigger*. Clearly this trigger does not necessarily mean that a handover between two routers has taken place (which should trigger a Mobile IPv6 handover). Link-layer protocols do not necessarily know whether the handover is between two APs attached to the same router or to different ones. However, a mobile node's implementation could use this hint to solicit the router anyway, and depending on the answer, it would decide if a Mobile IPv6 handover is needed.

Obviously, if the new router is sending advertisements frequently (e.g., every 50 ms), the mobile node might receive an unsolicited router advertisement, which would save it from sending the RS message. After receiving the RA message, the mobile node will know if it has moved. An *eager* movement detection algorithm will assume that a new prefix means that the mobile node has moved. This is acceptable provided that the mobile node does not oscillate back and forth between prefixes. This can be avoided by comparing a newly advertised prefix to those included in the prefix list. After receiving the router advertisement, the mobile node will configure a new care-of address. Now consider the movement detection delays encountered so far:

$$T_{mv} = T_{rs} + T_{ra} + T_{coa}$$

where $T_{mv}$ is the time required for a mobile node to detect that it has moved and to form a new care-of address, and $T_{rs}$ is the time it takes router_2 to receive the RS message. This includes the delays over the wireless link (mobile node–router_2 delay) as well as the time taken to receive lower-layer hints or lapses of the inter-router advertisement interval that cause the mobile node to send an RS. In addition to these times, $T_{rs}$ also includes a random delay period imposed by the Neighbor Discovery specification. This random delay is between 0 and 1 second. The reason for this delay is to avoid collisions on a shared link when several devices start at the same time and all start sending router solicitations. $T_{ra}$ is the time period that starts after $T_{rs,}$ and ends when the mobile node receives the router advertisement. Finally, $T_{coa}$ is the time it takes the mobile node to form a new care-of address and test it for duplication (DAD). According to RFC 2462, this could take 1 second or more depending on how quickly a response is received (in case of duplication) or whether the node retransmits the DAD message (the timeout for this message is 1 second). Figure 6–4 illustrates these times.

After detecting movement, the mobile node must update its home agent and correspondent nodes. In this example, we consider only one correspondent node for simplicity. If more than one correspondent node exists, the mobile node must perform the same actions simultaneously; hence, no additional delays will be incurred. We define $T_{bu}$ as the time it takes the mobile node to update its home agent and correspondent nodes. Defining this time depends on the delay incurred by packets sent between the mobile node and the home agent (i.e., HOTI/HOT and a binding update to the home agent), the delay between the home agent and the correspondent node (for HOTI/HOT messages), and the delay between the mobile and correspondent node (i.e., for COTI/COT and the binding update). However, the HOTI/HOT messages always take more time than the COTI/COT messages. In other words,

$$T_{mn\text{-}ha\text{-}cn} \geq T_{mn\text{-}cn}$$

where $T_{mn\text{-}ha\text{-}cn}$ is the time it takes a message to travel between the mobile and correspondent nodes via the home agent (HOTI/HOT messages), and $T_{mn\text{-}cn}$ is the time it takes the same message to travel directly between the mobile and

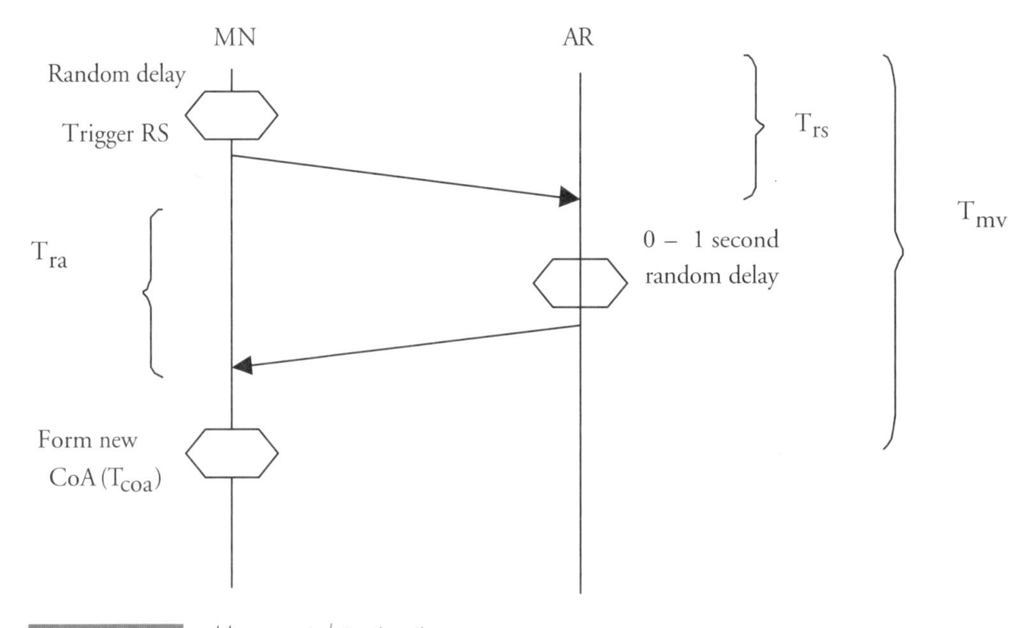


Figure 6–4 Movement detection time.

correspondent nodes (COTI/COT/BU). These two times will be equal only if the correspondent node is located on the mobile node's home link. Hence, we will assume that the return routability delay is bound by the time it takes to complete the HOTI/HOT exchange (note that this assumption is based on a best-case scenario; i.e., none of the packets is lost). Furthermore, we assume that the mobile node can send the HOTI message immediately after sending the binding update message to the home agent; that is, it does not wait for a binding acknowledgement. In a best case scenario, where the binding update arrives at the home agent before HOTI, this will allow the return routability test to work. Hence,

$$T_{bu} = T_{mn\text{-}ha\text{-}cn} + T_{mn\text{-}cn}$$

From this, we can calculate the total handover time, $T_{ho}$, as

$$d_i = T_{ho} = T_{mv} + T_{bu} = T_{rs} + T_{ra} + T_{coa} + T_{mn\text{-}ha\text{-}cn} + T_{mn\text{-}cn}$$

Incidentally, the major delay components in this equation are not related to the return routability procedure or the binding updates; they are caused by DAD delays ($T_{coa}$) and the random delay that is used in the mobile node and router_2 before an RS/RA are sent respectively (i.e., $T_{rs}$ and $T_{ra}$). Before we consider how disruptive the handover time is, let's consider some proposed optimizations that can reduce these major delay components.

## 6.2.1 Reducing Neighbor Discovery and DAD Delays

In Figure 6–4 we observed that there are three major delay components introduced by a random delay period before sending a router solicitation, router advertisements (0–1 second), and a third due to the waiting period during which a host is performing DAD.

The Neighbor Discovery specification introduces a random delay before sending router advertisements for two reasons: first, to avoid collisions between different router advertisements sent at the same time by routers on a shared link. A random delay would allow routers to send the advertisements at different times. Second, this delay protects routers against DoS attacks where Bad Guy sends many solicitations to try to keep a router too busy to forward packets. A possible solution for the collision problems is to ensure that only one router responds immediately to solicitations; other routers on-link could still use the random delay before responding. This "designated" router can be manually configured to respond immediately. We refer to advertisements sent without delay as *fast RAs*. Fast RAs are sent directly to the mobile node's address (i.e., not multicast to the entire link). To avoid the DoS attacks, we can set a limit on the number of fast RAs sent between the normal multicast RAs—let's say 10. Hence, a router will only send 10 fast RAs between the normal multicast RAs. If the period between the normal multicast RAs is not too long, this mechanism minimizes the delay period associated with sending a router advertisement. For example, suppose that router advertisements are normally sent once a second. If many mobile nodes are moving to the link, each soliciting for a fast RA, the result would be 10 fast RAs in one second instead of just one multicast RA. However, since fast RAs are unicast to the soliciting node, if more than 10 mobile nodes move to the new link in 1 second, only the first 10 nodes will receive the router advertisement; the rest will suffer normal delays.

DAD delays can be eliminated by assuming that a mobile node can use its new address while performing DAD. Before we consider the impacts of this mechanism, let's observe that the probability of collision between two randomly generated interface identifiers is approximately $10^{-12}$ [1]. Considering such low probability and comparing it with the probability of the neighbor solicitation message (used for DAD) being lost or corrupted, we can see that the value of DAD for randomly generated interface identifiers is questionable (assuming an unreliable link layer that has no retransmissions).

We refer to a node that uses an address before DAD completion as an *optimistic node*. An optimistic node can send neighbor advertisements on the new link to check whether duplication exists. However, the neighbor advertisements must not have the *O* flag (override) set. This ensures that another node's entry will not be overridden. When these steps are performed, we can minimize any negative impacts resulting from address duplication. However, there will still be some failure cases when address duplication exists. For instance, suppose that node_1 and node_2 have the same address and node_2 is the optimistic node. Suppose that node_3 is off-link and had an ongoing session with node_2 before it moved to the new link. When packets are directed to the new link (because node_2 sent a binding update containing its new care-of address to node_3), the router will already have a mapping between the IP address and either node_1 or node_2's link layer address; if not, it will send a neighbor solicitation to resolve that IP address. If it already has node_1's link-layer address, the traffic will be directed to node_1, and optimistic node (node_2) will not receive it. Clearly node_1 will simply drop the packets. If the router has no entry for node_1's address, it will send a neighbor solicitation; as a result, it will get responses from node_1 and node_2, since they both have the same address. Since node_1 will have the *O* flag set, its link-layer address will be used, resulting in the same effect (i.e., node_2's packets going to node_1). When considering these failure modes, it is important to note the rarity of address duplication for randomly generated addresses. When duplication is detected by the optimistic node, it immediately configures a new address and avoids the conflict. Hence, such failures may be acceptable, considering the very low probability of address duplication when interface identifiers are randomly generated. The method described above is called *optimistic DAD*.

Note that the problems discussed in this section are applicable to shared links. In connection-oriented links, the only neighbor is a router. Hence, the link can be designed in a way that avoids the possibility of address collision between a mobile node and an access router (we see an example of this design in Chapter 10).

## 6.3 Handover Impacts on TCP and UDP Traffic

So far, we have discussed the delay associated with Mobile IPv6 handovers, or handover time ($d_i$). Let's represent $d_i$ below:

$$d_i = T_{bo} = T_{mv} + T_{bu} = T_{rs} + T_{ra} + T_{coa} + T_{mn\text{-}ha\text{-}cn} + T_{mn\text{-}cn}$$

If we eliminate $T_{coa}$ (using optimistic DAD), we end up with

$$d_i = T_{rs} + T_{ra} + T_{mn\text{-}ha\text{-}cn} + T_{mn\text{-}cn}$$

Note that $T_{ra}$ is still included even if the random delay were eliminated; in this case $T_{ra}$ is equal to the delay over the wireless link, or the time it takes to send a router advertisement from a router to the mobile node. Also note that $T_{rs}$ might be zero in the case where the mobile node receives a router advertisement immediately after attaching to the new link (i.e., **if** the mobile node arrived at the new link when an RA is being sent).

The handover time is known only to the Mobile IPv6 implementation in the mobile node. All knowledge about mobility and service disruption is (by design) limited to this part of the mobile node's software. Hence, applications and transport layers will view this interruption as an indication of a change in the network's availability. This will cause applications and transport layers to behave differently. We will use TCP and UDP as examples of two different transport layers and see how they react to this disruption.

### 6.3.1 How Does TCP Work?

TCP is a connection-oriented reliable transport protocol. It achieves reliability by retransmitting lost segments and discovering errors in received packets by utilizing a checksum, calculated over each transmitted segment. In addition, TCP passes information to applications in the same order they were sent. In this section, we examine the operation of TCP in order to understand how Mobile IPv6 handovers impact connections using TCP. It is not this book's intention to give a detailed description of TCP. The main goal of this chapter is to give enough background of TCP to allow you to understand the impacts of Mobile IPv6 handovers. If you are interested in knowing TCP in detail you should read [12] or [14].

In addition to reliable transmission, TCP offers end-to-end flow-control mechanisms controlled by the sender and receiver. These mechanisms were designed to allow TCP connections to adapt to network congestion by modifying the transmission rate according to changing network conditions. These features take the burden of flow control from applications and assure a predictable behavior from TCP, as opposed to having each application devise its own flow-control mechanisms.

A TCP connection is established using the *three-way handshake* shown in Figure 6–5. Node_A, initiating the connection, sends a SYN (synchronize both ends for this connection) with an Initial Sequence Number (ISN). Node_B responds with an acknowledgment for the SYN message that also includes its ISN, then Node_A acknowledges the reception of the Node_B's acknowledgment.

Every TCP segment contains a sequence number. The sequence number is incremented by the number of bytes sent in each segment. When a node sends its ISN, the receiver acknowledges the segment and includes the sender's ISN, incremented by 1. This means that the receiver has received all segments containing the sequence numbers—less than ISN + 1. TCP connections are full duplex (i.e., they allow the sender and receiver to transmit and receive simultaneously). Hence, the receiver needs to include its ISN in the *ACK* message to synchronize with the sender. The sender then acknowledges the receiver's ISN by sending an ACK message containing ISN + 1. From this point on, both ends can start sending segments. For the duration of a TCP connection, each TCP segment sent by one end is acknowledged by the other end. The acknowledgment contains the sequence number of the last byte received in order, plus 1. For instance, suppose that a node received four segments, each containing 100 bytes and sequence numbers 100, 200, 300, and 500 (segment containing 400 is missing); it will send an ACK containing sequence numbers 101, 201, 301 respectively (the sequence number is incremented by one in the ACK message). However, it will not acknowledge the reception of the segment with sequence number 500. Instead, when the segment containing 500 is received, the receiver resends the acknowledgement containing 301. As a result, the sender will retransmit that segment (containing 400). When this segment is received at the other end, an ACK will be generated, containing the sequence number 501 (i.e., indicating that all segments up to sequence number 500 have been received correctly). Hence, a receiver will only acknowledge the last segment received in order. Later in this section, we see how further developments in TCP allow the receiver to selectively acknowledge certain TCP sequence numbers.

When Node_A finishes transmitting data, it sends a message with the FIN flag (a flag in the TCP header indicating that it has finished transmitting) set with a sequence number $k$, as shown in Figure 6–6. Node_B acknowledges the reception of the FIN with an ACK message containing the sequence number $k$ + 1. Following this, Node_B sends a FIN with sequence number $n + 1$, which results in an ACK for $n + 2$ from Node_A that would terminate the connection in both directions. Note that a FIN from Node_A need not trigger a FIN from Node_B. TCP allows connections to be *half open*. Hence, one end may close one direction of the connection, but the other end will still be allowed to send packets to the sender.

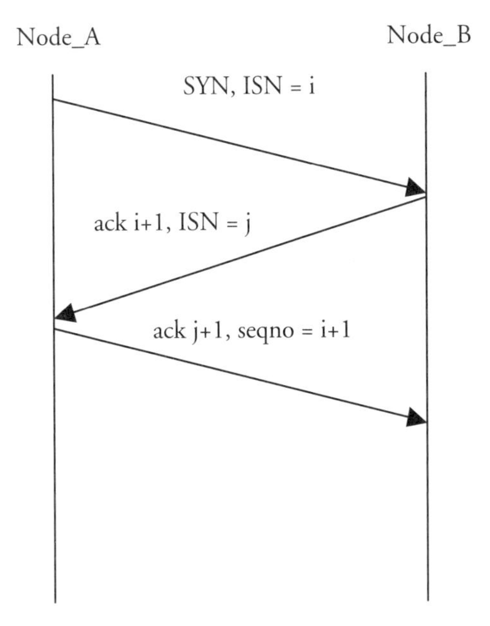


**Figure 6–5** *TCP connection setup.*

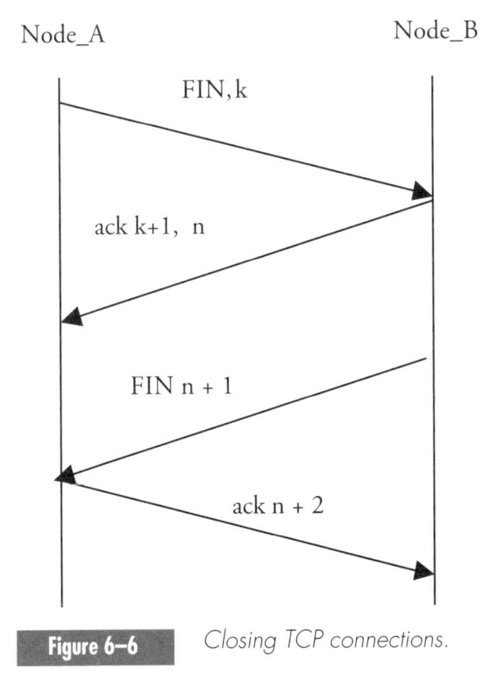


**Figure 6–6** *Closing TCP connections.*

TCP implementations contain four different timers:

- *Retransmission timer*: This timer is used to detect when a segment should be retransmitted because it has not been acknowledged by the receiver.
- *Persist timer*: This timer keeps the flow control of information between the two ends.
- *Keepalive timer*: This timer detects when the other end is unreachable.
- *2MSL timer*: This timer measures how long an open connection has been idle.

The most important timer for our purpose is the retransmission timer. This timer detects when a TCP segment has not been acknowledged for a particular period of time and retransmits that segment if necessary. The method used to calculate this timer and its use is shown in the following two sections.

#### 6.3.1.1 HOW DOES TCP FLOW CONTROL WORK?

A TCP implementation must make sure that the sender does not flood the receiver with more information than it can handle. Every receiver has a buffer associated with each connection (e.g., 16K bytes). If the sender exceeds this limit, the buffer overflows and some TCP segments are dropped. For this reason, every receiver advertises a *window* to the sender. The window size (in bytes) tells the sender how many bytes it can send. When the segments are received, the window size is reduced by the number of bytes received, and the new window size is included in the ACK message sent to the sender. When the receiver processes a TCP segment, its window gets larger. This is conveyed to the sender in the next ACK message. This mechanism ensures that the sender does not send more segments than the receiver's window size at any given time.

The receiver's window alone cannot avoid network congestion if many senders are transmitting TCP segments corresponding to their receivers' window size. Consider Figure 6–7. In this example, a router is connected to two links: A and B. Link A has more than twice the bandwidth of Link B. If all computers on Link A start sending packets at high rates, they will eventually fill the outgoing queue of the router (Link B). If this continues, the incoming queue eventually gets full as well, causing the router to start dropping packets. TCP avoids this situation by using flow control between the sender and the receiver.

We discussed how the receiver limits the number of packets sent by advertising its window in every acknowledgment. The sender has another window called the *usable window* that is calculated based on the receiver's window and local knowledge within a TCP implementation. The sender limits the sending rate to the minimum of the receiver's window and usable window. When the sender starts sending segments, the usable window is set to 1. Every time the sender receives an ACK for any segment, the sender increments its window by 1. Hence, when the receiver acknowledges the first segment (containing *n* bytes), the

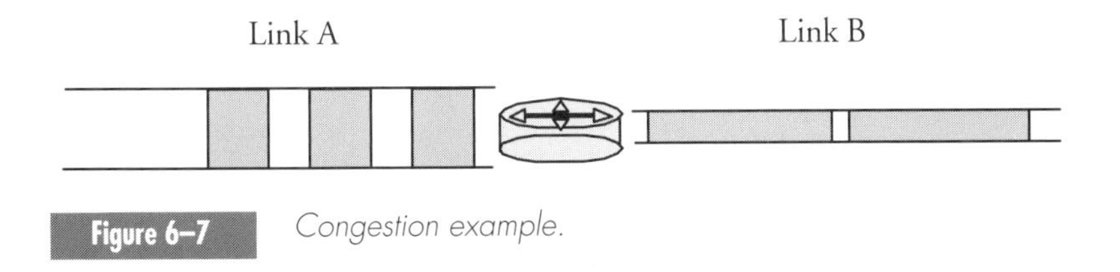


**Figure 6–7** *Congestion example.*

sender can send $2n$ bytes without waiting for an ACK. When a second ACK is received, the sender can send $3n$ bytes immediately, without waiting for an ACK, and so on until the usable window size matches the receiver's window. This behavior is called TCP's *slow start.* If we plot the number of bytes sent against time (given no congestion), we see that this number increases exponentially. While this behavior alone does not prevent the congestion situation shown in Figure 6–7, we will see in the following section how this knowledge can be utilized to allow TCP to avoid and recover from network congestion.

#### 6.3.1.2 TCP TIMEOUTS AND CONGESTION AVOIDANCE

When a TCP segment is not acknowledged for some time, it causes the retransmission timer to expire; TCP will resend that unacknowledged segment. The retransmission timer is calculated dynamically based on the mean variance of the roundtrip time (RTT) between the sender and receiver. When the first segment is sent, the RTT calculation starts. When the ACK for that segment is received, TCP calculates the RTT for the first segment. The RTT timer is started again when sending a new segment after the last calculation was done (i.e., at any given time, only one RTT calculation takes place for a connection). TCP uses the mean variance between the two RTTs to calculate a new RTT value. This process continues for the duration of a connection. The mean variance is used to calculate the new RTT (as opposed to the average) because it gives better indication of the real RTT and avoids cases where the RTT varies drastically from one segment to another.

We saw how TCP utilizes the receiver's window and the usable window to control the rate of sending segments. However, a network might become overloaded with large amounts of traffic. This will result in packets being dropped by routers. When this happens, TCP's exponential increase of the sending rate effectively adds fuel to the fire. Hence, TCP needs to adapt its sending rate according to the network's condition.

TCP assumes that packet losses are a result of network congestion and not due to transmission errors (the validity of this assumption for wireless links is discussed later). Based on this assumption, there are two different ways that allow TCP to detect congestion: timeouts and duplicate acknowledgments. As described above, timeouts are detected when the retransmission timer expires. On the other hand, duplicate acknowledgments take place when one or more

TCP segments are lost. For instance, suppose that a sender sent segments 1, 2, and 3 (segment numbers instead of bytes are used for simplicity) and that the receiver acknowledged these segments. The sender's usable window allows it to send segments 4, 5, and 6. If segment 4 were lost, the receiver would have no way of indicating that; so when it receives segment 5, it will send an ACK for the last segment received in order (i.e., 3). The same will happen when the receiver gets segment 6 because TCP has no other option but to send ACKs in response to a received segment. Hence, the sender will receive 3 ACK messages indicating that the last segment received in order was 3. Three duplicate acknowledgments are a strong indication to the sender that at least segment 4 was lost. Therefore, TCP uses this as an indication that network congestion has occurred. When congestion is detected, TCP needs to reduce the sending rate to help the network recover.

To avoid increasing network congestion, TCP maintains a *congestion window* (sometimes abbreviated *cwnd*; this is typically implemented with slow start's usable window) and a *slow start threshold* (sometimes abbreviated *ssth*) value. When the usable window exceeds the slow start threshold value, TCP gets into *congestion avoidance*. In this mode of operation, the congestion window size starts to increase linearly. When a TCP retransmission timer expires, TCP reduces its congestion window back to one; that is, it will go back to slow start. In addition, the slow start threshold value is reduced to half the value of the usable window size when congestion occurred. When the next packet is sent and acknowledged by the receiver, the usable window size is increased. The increase depends on whether TCP is in slow start (exponential increase) or congestion avoidance (linear increase).

#### 6.3.1.3 FAST TRANSMISSIONS

The TCP retransmission timer is used to indicate whether a segment should be retransmitted. A fast retransmission algorithm allows the sender to retransmit segments before the retransmission timer has expired. This is based on the assumption that a reception of three duplicate acknowledgments provides a strong indication that a segment has been lost. Since a receiver will only send an acknowledgment if a new segment has been received, duplicate acknowledgments do not indicate severe congestion. But they show that there is a "hole" in the order of segments received at the other end. In other words, the receiver is still receiving the sender's segments. Hence, the sender's TCP will retransmit the missing segment(s) without waiting for the retransmission timer to expire.

#### 6.3.1.4 SELECTIVE ACKNOWLEDGMENTS

Suppose a sender has sent segments 1, 2, 3, 4, 5, 6, and 7 to a receiver, but only segments 1, 2, 3, 6, and 7 were received at the other end. Normally, the receiver's TCP will send an ACK for segment 3 when it is received and repeat the ACK when segments 6 and 7 are received. A sender's TCP will resend segment 4. The receiver will acknowledge segment 4, causing the sender to resend segment 5. At this stage, the receiver will acknowledge segment 7, indicating that segments 1 through 7 were received. In this case, resending segments 4 and 5 will take two RTTs. Selective Acknowledgments (SACK) allow the receiver to indicate exactly which segments were lost. To do this, two options are defined and included in the TCP header: *SACK-permitted* and *SACK*. When a connection is established, the receiver sends the SACK-permitted option to see whether the sender will accept this mechanism (i.e., for backward compatibility). The SACK option includes the number of segments that were actually received correctly.

Now let's reconsider the same example where segments 4 and 5 were lost and segments 6 and 7 were received. In this case, the receiver can use the SACK option (assuming that the sender understands this option) to indicate that it received segments 6 and 7. This explicit acknowledgment will allow the sender to know that segments 4 and 5 were lost and need to be retransmitted immediately. Obviously, the advantage of this mechanism is that it uses one RTT to inform the sender that more than one segment are lost.

#### 6.3.1.5 TCP ASSUMPTIONS IN WIRELESS NETWORKS

Interestingly enough, TCP's assumptions about network congestion are not always applicable to wireless links. These links are typically slow and have high Bit Error Rates (BER). The main reason for losing packets on a wireless link is not necessarily network congestion, but errors. These wireless links usually utilize flow-control mechanisms similar to those implemented by TCP. For instance, cellular links (e.g., WCDMA) allow for link-layer retransmissions. However, due to the significant delays experienced by traffic sent on these links, it is very likely that TCP's retransmission timer will expire before an ACK is received (possibly due to link-layer retransmissions). For instance, suppose that a node sent segment 2 on an error-prone cellular link. Let's also assume that this segment was corrupted the first time it was sent and needed to be retransmitted by the link layer. Due to the large RTT between the cellular host and the network, retransmissions take a significant amount of time. If a particular segment's retransmission is done more than once, it is possible that the TCP retransmission timer will expire before it receives an ACK for that segment. This will cause TCP to resend the unacknowledged segment. Both the originally sent segment and the retransmitted segment will be received at the other end, causing the receiver to send two acknowledgments for the same segment. As we saw earlier, three duplicate acknowledgments will result in fast retransmits and a reduction in the usable window size. This artificially perceived network congestion would cause a reduction in the sending rate by TCP, which unnecessarily reduces the throughput on a wireless link. We now see how the Eifel algorithm can be used to solve this problem.

#### 6.3.1.6 EIFEL ALGORITHM

Wide area coverage wireless networks (i.e., cellular networks) all share two distinct properties: they have large delays on the radio link and high BER (typically $10^{-3}$). To accommodate reliable connections based on TCP, cellular links usually implement reliable links that utilize forms of flow control similar to those implemented by TCP. However, the combination of delay and high BER can cause a number of problems that will result in TCP timeout, and consequently slow start, or in TCP fast retransmit and recovery. A timeout of the TCP retransmission timer can be very costly as it forces TCP to go to slow start. On the other hand, a fast retransmit is also costly because it reduces the usable window size and, as a result, prevents the sender from fully utilizing the radio link. Hence, it is crucial that timeouts or fast retransmits are not done unnecessarily due to an inaccurate perception of the network congestion.

Current TCP implementations have no way of distinguishing between an ACK message sent in response to the original segment or an ACK sent due to a retransmission. In other words, if a timeout occurs and a segment is retransmitted when a TCP implementation receives an ACK for that segment, it does not know if the ACK is for the originally transmitted segment (i.e., it was never lost) or for the retransmitted segment. If the ACK were sent for the original segment, then the segment was not lost and slow start was prematurely invoked. The same applies for the reception of three duplicate acknowledgments; if they were sent in response to receiving the same segment twice, then there is no need to initiate fast retransmit and resend the same segment for the third time. That is, the reception of a late ACK in this case merely indicates that the segment in question was delayed by a period of time larger than the previously calculated RTT, but it was not lost.

The Eifel algorithm allows a TCP sender to distinguish between an ACK sent in response to the original TCP segment and an ACK sent in response to the retransmitted segment. This is done by including a *Timestamp* option in the TCP header for all segments sent to the receiver. The receiver is required to echo this timestamp when sending an ACK for that segment. The Eifel detection algorithm is triggered when a TCP connection goes to slow start or fast retransmit. The detection algorithm is concerned with distinguishing ACKs received for original transmission from those received due to retransmission. Before a segment is retransmitted, the Eifel algorithm stores the timestamp associated with that segment. When an ACK is received, the timestamp option in the ACK is compared with the timestamp used in the retransmitted segment; if it is less than the timestamp of the retransmitted segment, then the ACK is for the original segment, and going to slow start or fast retransmit was premature. As a result, the Eifel response algorithm is initiated, which restores the original window size for the sender and recalculates the RTT.

As illustrated, this algorithm addresses cases where the ACK simply arrived too late or the retransmission timer expired too early. In these cases, slow start or fast retransmit is a premature action and would result in poor link utilization.

## 6.3.2 Mobility Impacts on TCP

Let's consider TCP's behavior shown in Figure 6–8. Initially, we observe the exponential increase in the allowed sending rate (congestion window). When the congestion window size reaches the slow start threshold (we picked 16 in this example), we observe the linear increase in the congestion window. The sender will continue to send segments at the rate allowed by the congestion window until it reaches the limit specified by the receiver's advertised window.

However, during a Mobile IPv6 handover, we will experience a handover period during which the mobile node is unable to receive new TCP segments containing data or acknowledgments from the receiver. Depending on the duration of the handover time and the way TCP tracks its retransmission timer, TCP is likely to experience timeout. In [14], it is noted that TCP implementations track the retransmission timer every 500 ms. That is, the timer granularity is 500 ms. Suppose the mobile node set the retransmission timer (based on

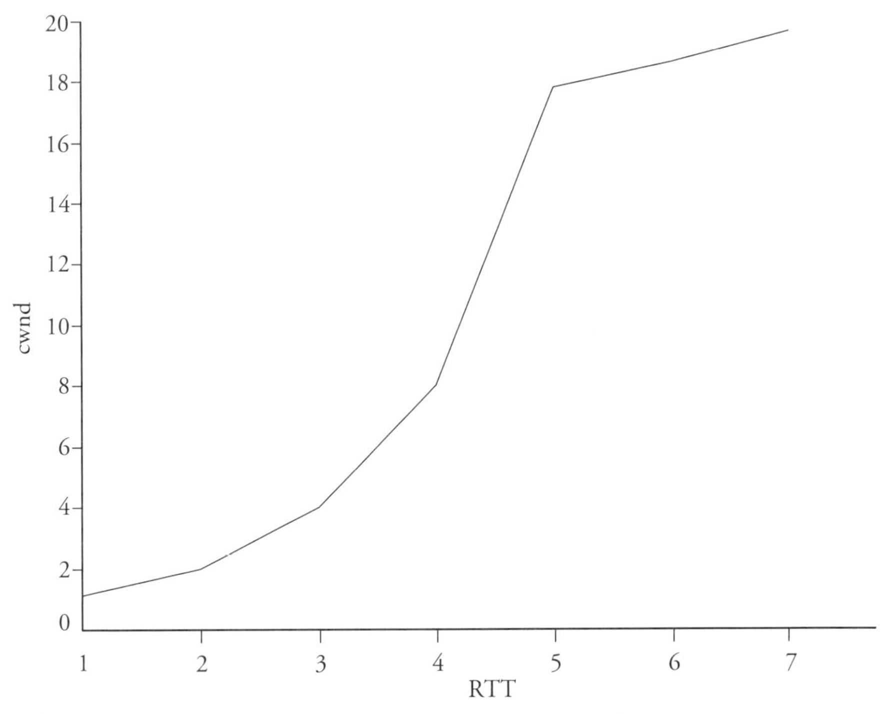


**Figure 6–8** *TCP slow start and congestion avoidance phases.*

calculated RTT) for its TCP connection with a correspondent node to be 1 second. At the 900 ms mark, the mobile node performs a handover. Suppose that the handover time was 200 ms. The retransmission timer will be invoked after 100 ms from the start of the handover. TCP will resend the unacknowledged segment and go to slow start. Clearly slow start was unnecessarily invoked in this scenario. There is no network congestion, but TCP is unaware of this fact. If the mobile node is unlucky, as shown in [4], or is attached to an error-prone wireless link, the first transmitted packet might be lost, causing the TCP timeout period to be doubled. That is, the first timeout is 1 second, followed by a wait of 2 seconds for the retransmitted segment. If the retransmitted segment were lost, it would get retransmitted again, with a timeout period of 4 seconds, and so on. We use Figure 6–9 to show an example of what could happen to TCP during a handover.

In Figure 6–9 we can see that the interruption during the handover period causes TCP to time out. The segment causing the timeout will be retransmitted during the handover period. However, one of two things can happen: the mobile node might be able to send the segment, but the ACK will not be received, since the mobile node had not updated its correspondent node or home agent yet; or the mobile node will send the segment, but the segment will

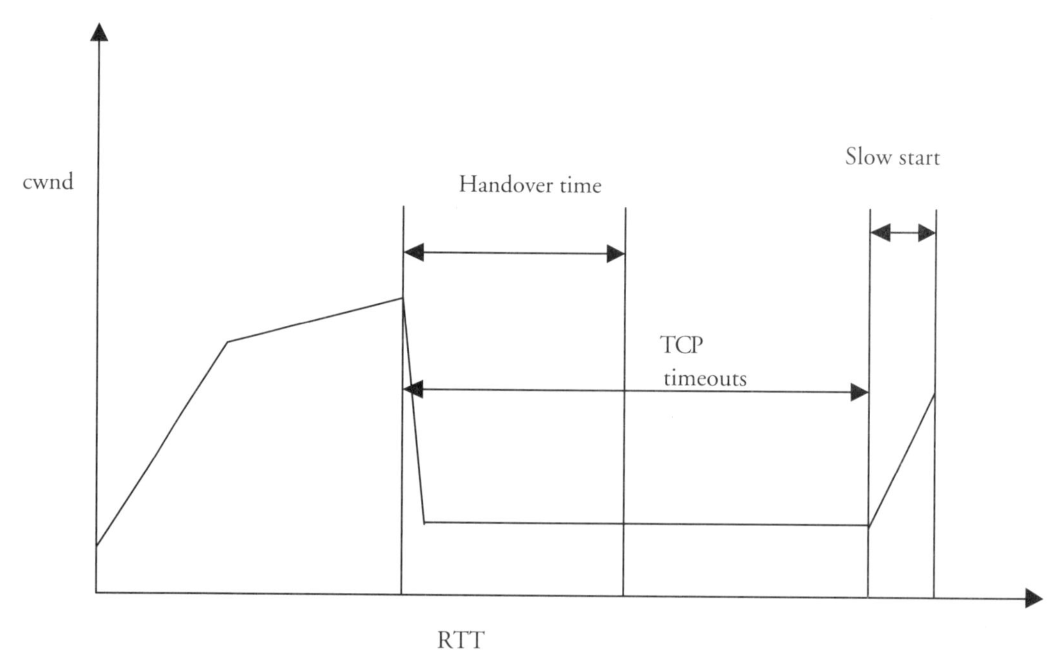


**Figure 6–9** *TCP behavior during Mobile IPv6 handovers.*

be dropped by the new AR because the mobile node hasn't detected that it moved and therefore used its old care-of address, which is not valid on the new subnet. This will cause another timeout and retransmission (the new timeout is twice the previous one). If the handover time is too long, multiple timeouts can occur, followed by slow start when a segment is finally acknowledged. Note that the TCP timeouts can in fact be larger than the handover time. Moreover, TCP may take several RTTs to be able to restore its original sending rate. That is, depending on the number of timeouts that are caused by the handover (and potential packet losses after that), it will take a significant amount of time for TCP to restore its original sending rate (after going through slow start again). Also note that SACKs may not help during the Mobile IPv6 handover period that caused the timeouts, because they will not be delivered to the mobile node until it has updated its home agent and/or correspondent node (i.e., after the handover has occurred).

The Mobile IPv6 implementation can certainly notify TCP that a handover is taking place. However, TCP flow control and congestion avoidance is maintained on an end-to-end basis; notifying one end will not help the other end's perception of network congestion. Another approach is to try to minimize or eliminate the Mobile IPv6 handover time in order to reduce the chance of timeouts. We see how this can be done inChapter 7.

### 6.3.3 What About UDP?

UDP is an unreliable transport protocol. It does not include any flow control or retransmission. Hence, it does not provide any guarantees to upper layers about the order in which the data is received. When applications use UDP, they need to exercise their own flow-control mechanisms.

Some applications use UDP because it's unreliable. For example, applications encapsulating real-time voice or video over IP would not want a transport layer to buffer packets or retransmit them if they're lost. It is more important for these applications to continue to receive most of the traffic and show it to users than to receive all of the traffic. For instance, it is more important for most football fans to watch the game live, despite some interruption, than to wait for some video frames to be retransmitted. Such real-time applications usually number their payload sequentially; when a packet experiences delays and arrives late (the decision of how late a packet is depends on the delay sensitivity of the application), it is simply dropped by the receiver because it will only confuse the user. Therefore, applications using UDP will lose packets during a Mobile IPv6 handover. If the handover time is too long, users will notice a significant disruption in service. Hence, frequent handovers will degrade the service provided to users.

## 6.4 Summary

The aim of this chapter was to set the scene for optimizations that are currently proposed for Mobile IPv6. We explained the causes for handovers and distinguishing between layer 2 and layer 3 handover. We also saw that different types of link layers handle handovers differently, depending on the link layer's requirements.

The definition of handover time was discussed. We also analyzed the major contributing factors to handover delay. Optimizations for eliminating DAD delays and the random delays imposed by Neighbor Discovery were presented. Next, we showed that the remaining factors contributing to handover delays were the time it takes to update the home agent and correspondent node.

Finally, this chapter introduced TCP in order to analyze the impacts of handover on its operations. We saw that TCP is significantly impacted by handovers, as they reduce its throughput due to packet losses that cause TCP to assume that network congestion has taken place. As a result, TCP goes to slow start. UDP applications are also affected by the packet losses during handover, as they can reduce the quality of the service provided to the user.

The next chapter builds on this knowledge by showing different optimizations for various aspects of the problems shown in this chapter. We also show how those optimizations can be combined to achieve maximum gain.

## Further Reading

[1] Bagnulo, M., I. Soto, A. Garcia-Martinez, and A. Azcorra, "Random Generation of Interface Identifiers," draft-soto-mobileip-random-iids-00, January 2002.

[2] Caceres, R., and L. Iftode, "Improving the Performance of Reliable Transport Protocols in Mobile Computing Environments," *IEEE Journal on Selected Areas of Communications (JSAC)*. June 1995.

[3] ElMalki, K., N. A. Fikouras, and S. R. Cvetkovic, "Fast Handoffs Method for Real-Time Traffic over Scalable Mobile IP Networks," draft-elmalki-mobileip-fast-handoffs-01, June 1999.

[4] Fikouras, N. A., K. ElMalki, and S. R. Cvetkovic, "Performance of TCP and UDP During Mobile IP Handoffs in Single-Agent Subnetworks," Proceedings of IEEE WCNC, 1999.

[5] IEEE std, 802.11-1997, "Wireless LAN Medium Access Control (MAC) and Physical Layer (PH) Specifications".

[6] Kempf, K., M. Khalil, and B. Pentland, "IPv6 fast router advertisement," draft-mkhalil-ipv6-fastra-03, work in progress, March 2003.

[7] Ludwig, R., and M. Meyer, "The Eifel detection algorithm for TCP," draft-ietf-tsvwg-tcp-eifel-alg-05, work in progress, October 2002.

[8] Ludwig, R., and A. Gurtov, "The Eifel response algorithm for TCP," draft-ietf-tsvwg-tcp-eifel-response-00, work in progress, July 2002.

[9] Ludwig, R., "Eliminating Inefficient Cross-Layer Interactions in Wireless Networking," Thesis Dissertation, April 2000. Online: http://iceberg.cs.berkeley.edu/papers/Ludwig-Diss00/.

[10] Mathis, M., J. Mahdavi, S. Floyd, and A. Romanow, "TCP Selective Acknowledgment Options," RFC 2018, October 1996.

[11] Moore, N., "Optimistic Duplicate Address Detection," draft-moore-ipv6-optimistic-dad-02, work in progress, February 2003.

[12] Postel, J., "Transmission Control Protocol," RFC 793, September 1981.

[13] Stevens, W., "TCP Slow Start, Congestion Avoidance, Fast Retransmit, and Fast Recovery Algorithms," RFC 2001, January 1997.

[14] Stevens, W. *TCP/IP Illustrated*, Vol. I. Addison-Wesley, 1994.

SEVEN

# Mobile IPv6: Handover Optimizations and Extensions

In light of the discussion presented in the Chapter 6, it is clear that there is a need to reduce the packet losses due to Mobile IPv6 handovers. We saw that these delays were due to two main reasons: first, the delays between the mobile node and its home agent, and second, the delays associated with movement detection. In addition, there is the delay of layer 2 handovers. In this chapter, we present different mechanisms that can be used independently to address certain aspects of these problems or combined to improve the performance of Mobile IPv6 handovers. In addition, we show an optimization that can be added to Mobile IPv6 to allow finer granularity of movement—that is, allowing mobile nodes to associate some connections with certain interfaces (care-of addresses) as opposed to the Mobile IPv6 signaling, which moves all connections with a correspondent to one care-of address. This is particularly useful for multihomed mobile nodes.

It is important to note that most of the proposals discussed in this chapter are currently documented in Internet drafts. Some of them are working group drafts in the Mobile IP Working Group, and some are personal drafts. Therefore, the content of these drafts may change in the near future. Nevertheless, learning these techniques will give you some good insights about improving the Mobile IPv6 handover performance.

## 7.1 Fast Handovers for Mobile IPv6

Fast handovers (sometimes called fast handoffs or just FMIPv6) [4] are concerned with minimizing the delay associated with movement detection and removing Mobile IPv6 signaling to the home agent/correspondent nodes from the critical handover time. To achieve this goal, fast handovers are designed to allow mobile nodes to anticipate their IP layer mobility. In other words, mobile nodes can discover the new router and its prefix before being disconnected from the current router. A radio link that allows make-before-break handovers automatically provides this service because the mobile node is able to listen to advertisements from both routers simultaneously (provided that there is enough overlap between cells). However, the most commonly deployed link layers today provide a break-before-make handover. Therefore, some tricks are needed on the IP layer to anticipate movement to a new router. In the following sections, we show how anticipation is done and, just as importantly, how we can avoid the negative implications of anticipation.

### 7.1.1 Anticipation and Handover Initiation

To understand how anticipation works, let's consider a mobile node on a connectionless link like IEEE802.11b. Let's also assume that the mobile node detects that the current link is degrading and therefore starts searching for a new Access Point (AP) to attach to. To do this, the mobile node scans all possible frequencies (specified by IEE802.11b specification) and compares the received signal with the one currently received. If the mobile node finds a better signal, it can switch to the new AP. However, the mobile node's link-layer implementation does not know whether or not such AP is attached to a new Access Router (AR). The link layer only knows of link-layer addresses and AP names. However, if the AP name/link-layer address (or Basic Service Set identity, BSS ID, in IEEE802.11, which identifies an AP) is known, the mobile node's IP layer's implementation can request that the current AR provide the prefix/router address, which the new AP is attached to. This idea assumes that an AR is configured with a table (see Table 7–1) containing its own and the neighboring AP's link-layer addresses and the corresponding AR. This is a realistic assumption if the mobile node is moving within a single operator's domain. In Table 7–1,

**Table 7–1** *Mapping Between AP MAC Addresses and their Corresponding Routers*

| Router | AP MAC Address |
|---|---|
| Router_1 | 00601DF02CB2 |
| | 00901DF02CB2 |
| | 00901DF02AB2 |
| Router_2 | 00601DA02CB2 |
| | 00601DA02DB2 |

the first three APs are connected to Router_1 and the following two are connected to Router_2.

If the mobile node is connected to Router_1 and requests that it provide it with information on AP 00601DF02CB2, Router_1 informs it that this AP is connected to the same router. Hence, no IP mobility is invoked. However, if the mobile node requests information on AP 00601DA02CB2, Router_1 informs it that it is connected to Router_2. This way, the mobile node knows that a Mobile IPv6 handover is about to take place when it moves to this AP. The message sent by the mobile node to its AR to request this information is called a *router solicitation for proxy* because it is requesting another router's information. This message contains the link-layer address of one or more APs that may be selected by the mobile node in its next handover. The router's response is called a *proxy router advertisement*. This contains the neighboring router's advertisement (in this example, it's Router_2's advertisement), which includes at least one prefix option valid for the anticipated link. It may also suggest a new care-of address for the mobile node to use on the new link. Figure 7–1 shows the anticipation process.

Note that in our IEEE802.11b example for a connectionless link, the mobile node initiates the handover by sending a proxy router solicitation. This action is triggered by the underlying link layer in the mobile node, which is aware that a handover is about to take place. This is natural because in these link layers, the mobile node is the only entity aware that it is about to attach to a new AP. However, in a connection-oriented link, the current AP is aware that the

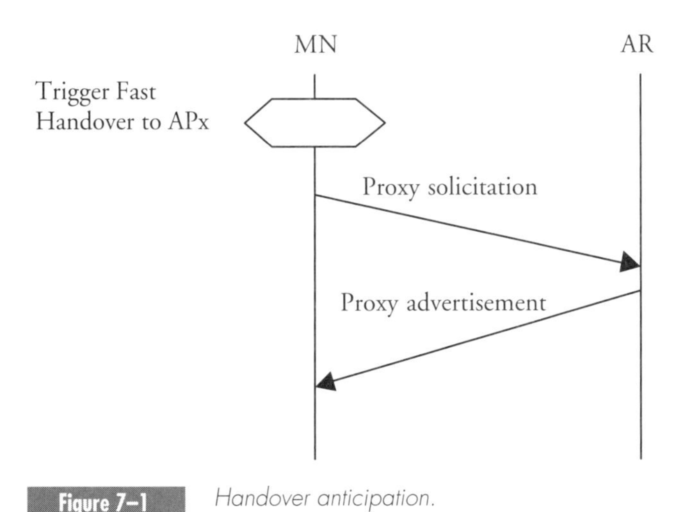


**Figure 7–1** *Handover anticipation.*

mobile node is about to move to the new AP.[1] In this scenario, the link-layer triggers can take place in the network. Hence, a router can send an unsolicited proxy advertisement to the mobile node, which saves the time required to send the solicitation message. Therefore, the solicitation message is not needed for all handover cases. However, it is worth noting that having the two messages allows the protocol to work independently of the underlying link layer. When the mobile node is aware of its mobility, it solicits for the new AP's information; otherwise, the information is received unsolicited. This works for both types of link layers discussed in Chapter 6

Both the proxy router solicitation and advertisement messages are identical to normal router solicitation and advertisement messages shown earlier in Chapter 2, with the exception that the ICMP *type* field value is different (to be allocated when this specification becomes a standard in IETF). The ICMP *code* values in the proxy router advertisements will also have different implications; these are discussed in detail in the next section. In addition, the fast handover specification defines a *new care-of address* option that can be included in the proxy advertisement. This option may be included to suggest a new care-of address that the mobile node should use on the new link. The details of this option are discussed in the following section.

### 7.1.2 Updating the Current Access Router

After anticipating movement, the mobile node will know the prefix for its new link and can form a new care-of address while still connected to the current AR. The mobile node also needs to check whether this address is valid on the new link (i.e., DAD). However, this cannot be done until it moves to the new AR.

We already saw that Mobile IPv6 provides a home agent function that redirects traffic addressed to a mobile node's home address to its care-of address. The current AR can implement the home agent function. Hence, the mobile node can send a binding update to the current access router to tunnel packets addressed to its current care-of address. That is, the current AR can act as a temporary home agent for the mobile node's current address (which is treated like a temporary home address). However, to avoid address duplication causing the packets to be delivered to another node on the new link, the current AR checks the address's validity on the new link by sending a Handover Initiate (HI) message to the new AR. This message serves two purposes: first, it checks if the new AR is aware of any duplication for the mobile node's new care-of address, and second, it sets up a tunnel between the current AR and the

[1] We use the term AP loosely. In most cellular networks, there is some intelligence in the network that detects (based on power measurements, among other factors) where the mobile node is moving. However, since this book is not concerned with the architectures of such networks, we simplify matters by using *AP* to refer to the layer 2 termination and the intelligence in the network responsible for detecting the mobile node's location and possible movements.

new one. After the tunnel is set up, the new AR replies with a Handover Acknowledgment (Hack) message to the current AR. Figure 7–2 shows the full sequence of messages exchanged during fast handover.

We discussed the first two messages in the last section. The fast-binding update (F-BU) is used to inform AR1 that traffic addressed to the mobile node should be forwarded to AR2. At the same time, AR1 sends an HI message to AR2. In Figure 7–2, it is assumed that the F-BU message will arrive at AR1 after Hack because the mobile node is probably connected to AR1 through a wireless link, which typically has more delays than those imposed by a wired link (between AR1 and AR2). AR1 acknowledges the F-BU message by sending a binding acknowledgment. The acknowledgment is sent to either the mobile node's current location or its new location (AR2's link), depending on where the mobile node sent the F-BU. That is, if the mobile node sent the F-BU while connected to AR1, the acknowledgment will be sent on that link. Otherwise, the acknowledgment is sent to the mobile node's new location. The details of this operation are discussed in the following two sections.

After receiving both the F-BU and Hack messages, AR1 starts forwarding the mobile node's packets to its new location. However, unlike the normal home agent operation, the tunnel is terminated in AR2, not in the mobile node. This allows the mobile node to use its old care-of address while verifying the

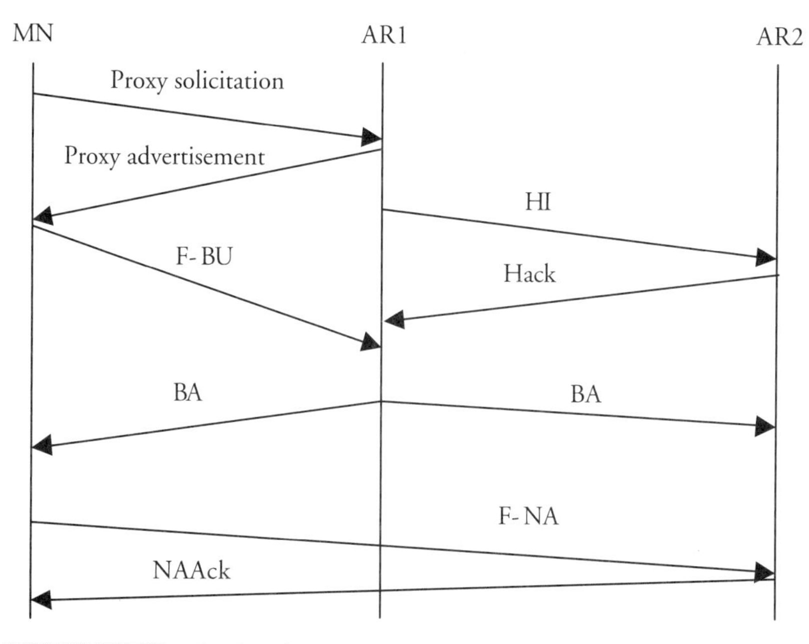


**Figure 7–2** *Fast handovers.*

new one. This is especially pertinent if AR2 could not verify that the new care-of address was unique on its link. However, this also means that the mobile node will be using a topologically incorrect address on the new link. Hence, AR2 will be required to "open" its ingress filter for this mobile node. In this case, the mobile node's outgoing traffic will be tunneled back to AR1, which will forward it as usual. This process will continue until the mobile node validates its new care-of address and updates all correspondent nodes and the home agent. When all these nodes are updated, the mobile node will receive traffic directly at its new care-of address.

Hence, the validity of the new care-of address can be achieved in two different ways: first, if the new AR knows that the address is valid, it informs the current AR (using the Hack message), which in turns informs the mobile node (using the binding acknowledgment); and second, if the new AR cannot validate the address, the mobile node must perform DAD on the new link and use its previous care-of address in the mean time.

The fast handovers specification allows AR1 to establish a tunnel to AR2 for a particular mobile node without receiving an F-BU or a solicitation message. This "optimization" assumes that AR1 has information about the mobile node's movement and its next link. Hence, it does not need to wait for the mobile node to inform it of such movement. However, this is not a generic solution for all different types of radio links. In particular, on a connectionless link, the AR does not have this information a priori and therefore cannot establish the tunnel without being informed by the mobile node. Furthermore, this approach assumes that the redirection process does not require an authenticated binding update (i.e., it trusts layer 2 triggers to be authentic). This assumption is also questionable in the general case.

### 7.1.3 Moving to a New Link

Two different events need to be considered when the mobile node moves to a new link. First, the mobile node needs to know that the layer 2 handover has taken place. This can be known if the mobile node's IP layer receives explicit or implicit information from the lower layer. Explicit information indicates that a handover has taken place. Alternatively, implicit information indicates that the mobile node is now attached to an AP whose MAC address is different from the previous one. Second, the mobile node must know whether the old AR has accepted the binding and must check the validity of the new care-of address. If a binding acknowledgment (indicating a successful operation) were received before the mobile node moved, then the mobile node assumes that data will be forwarded to its new location. However, if the mobile node detached from the old link before it had a chance to receive the acknowledgment, it will not have this information. If the mobile node advertises its presence on the new link, the new AR can forward all buffered messages to it (including the binding acknowledgment).

The Fast Neighbor Advertisement (F-NA) message is used to inform the new AR that the mobile node is now attached to its link and essentially requests two things: first, the new AR forwards buffered packets (including the acknowledgment), and second, the new AR tells the mobile node if the new care-of address is invalid. The latter is delivered to the mobile node in a new option called the *Neighbor Advertisement Acknowledgment* (NAACK), which is included in a router advertisement unicast to the mobile node's previous care-of address. The new option tells the mobile node whether it should use its new care-of address or keep using the old one (on the new link) while testing another care-of address.

### 7.1.4 Failure Cases

Fast handovers are designed to eliminate time delays associated with movement detection and care-of address testing. Furthermore, fast handovers eliminate updates sent to the home agent and correspondent nodes (including return routability testing) from the critical path (i.e., from the handover time). However, fast handovers depend on handover anticipation in the mobile node or the AR. Recall that layer 2 handovers are typically triggered by a degrading link. Therefore, there is no guarantee that the mobile node can be connected to its current AR long enough to be able to send and receive all the fast handover messages. When anticipation is used, there is a chance that the mobile node will lose its connectivity to the current AR before it has time to update it with the F-BU message. Let's reconsider the message flow shown in Figure 7–3 and see where the anticipation could fail.

Figure 7–3 shows the time period during which the mobile node could lose its connectivity to the AR due to a sudden degradation in the link, which causes the link-layer implementation to switch to the new AR before the anticipated time. We consider four different failure cases, as follows:

- Router solicitation for proxy not sent: If the link to the AR is lost before a solicitation is sent, anticipation fails. Unless the AR knows where the mobile node is moving (e.g., a connection-oriented link), the mobile node reverts to normal Mobile IPv6 movement detection and fast handover fails.
- Proxy advertisement not received: In this case, the mobile node has no information about the new link before it moves. This leads to anticipation failure, and consequently the mobile node reverts to normal Mobile IPv6 movement detection.
- F-BU not sent: In this scenario, the mobile node anticipated its movement but did not have time to send an F-BU. That is, anticipation was successful, but there was no time to request that the AR forward packets to the new link. When the mobile node attaches to the new link, it can send the F-BU message to the old AR. Since the new AR already received the HI message, a tunnel is set up to the

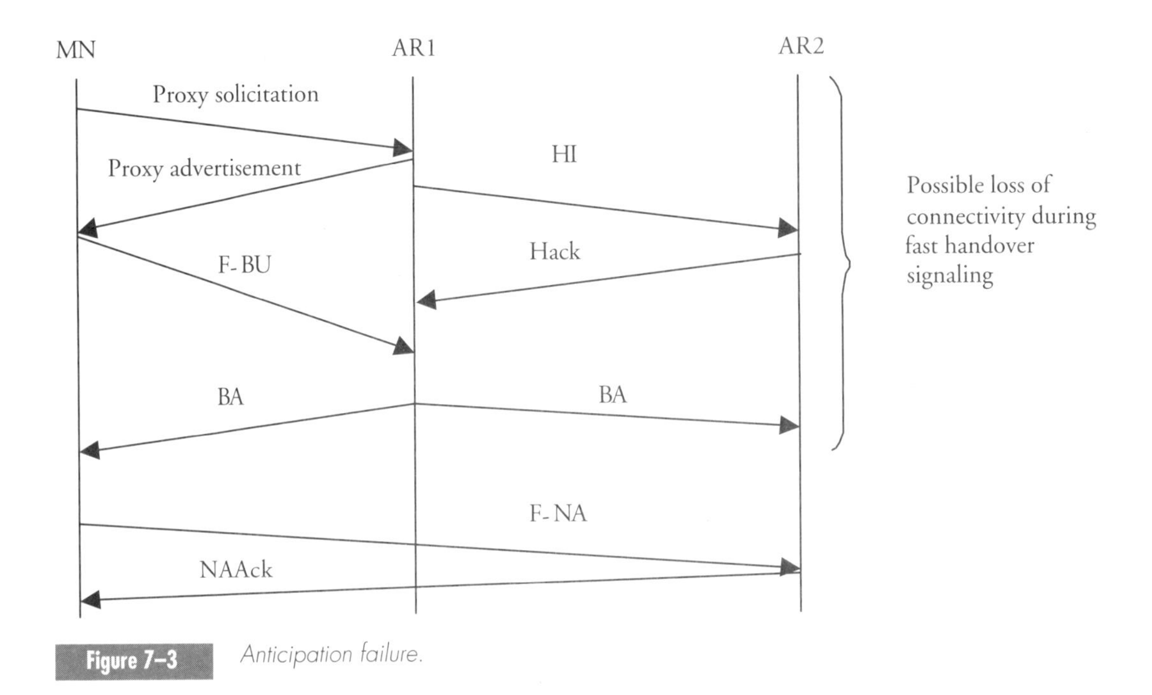


Figure 7–3 *Anticipation failure.*

old AR. Hence, the F-BU is forwarded through the tunnel and packets are rerouted to the mobile node's new location. In this case, some packets may be lost; however, the movement detection time is eliminated and the overall performance is better than that experienced with normal Mobile IPv6 handovers.

- Binding acknowledgment not received: In this case, the mobile node sent the F-BU but did not receive the acknowledgment before moving. The mobile node uses the F-NA message to find out whether the binding was accepted (see section 7.1.3). In other words, if the binding acknowledgment were sent to the new location and buffered by the new AR (because it sent neighbor solicitations but got no response), the F-NA message allows the new AR to learn the mobile node's MAC address and forward the buffered packets.

## 7.1.5 The Cost of Anticipation

Fast handover's performance hinges on anticipation. If anticipation is successful, the movement detection time in Mobile IPv6 can be eliminated. In addition, all Mobile IPv6 signaling to the home agent and correspondent nodes can be removed from the handover time, since the traffic is rerouted to the mobile node's new location while it is updating those nodes. However, anticipation does not come without a price; it is important to understand the price associated with anticipation when applied to current cellular technologies.

Suppose a node on a connection-oriented link is moving from Cell_1 to Cell_2, and let's ignore IP mobility for the moment. Normally, cellular networks are designed such that the cell overlap is sufficiently long to provide enough certainty that the mobile node is in fact moving in a certain direction. If the cell overlap is too small, the mobile node does not have enough time to ensure that it is in fact moving to Cell_2 and to inform the network. The mobile node could lose connectivity very quickly (e.g., when moving at high speeds). In addition, the network does not have enough time to prepare for the handover (in case of connection-oriented links). On the other hand, if the overlap is too large, there is too much interference between the two cells, which causes the cell's coverage area to be reduced, resulting in a reduced capacity (i.e., fewer nodes supported) and consequently more cost to support a certain number of nodes (due to the need for more cells). Hence, radio network designers need to pick a cell overlap that is not too small to provide enough time for the mobile node to predict its movement between cells (without a sudden loss of radio connectivity), but not too large that it significantly reduces the cell's capacity. This is especially true for cellular networks where the network is optimized for maximum bandwidth utilization.

We illustrate the impact of anticipation by a considering a typical case for a connection-oriented radio link (Figure 7–4) used today in current cellular systems and examining how layer 2 handovers are done. Using this example, we can observe the challenges involved in introducing fast handovers to these links. In Figure 7–4, a node is moving from Cell_1 to Cell_2, where both cells are part of a cellular network. We assume that the layer 2 handover procedure starts at

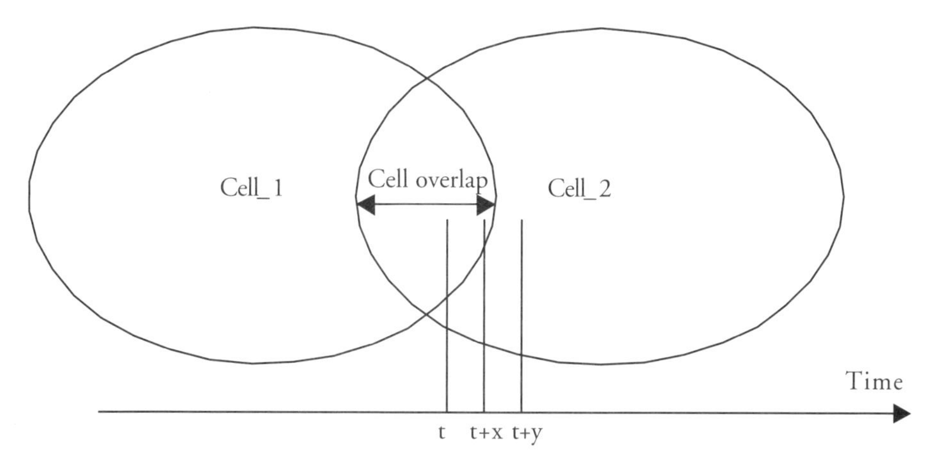


**Figure 7–4** *Anticipation time during handover.*

time $t$; the mobile node will send information to the network indicating that it has a better signal from the AP in Cell_2. The network will decide that the mobile node can move to Cell_2 (for example, the network would verify that Cell_2 has enough resources for this mobile node). The current AP will also send the mobile node information to prepare it for the handover and allow it to be able synchronize to the new AP. For instance, the current AP would provide the frequency of operation in the new cell. Other parameters may also be provided, depending on the technology being used by the cellular link. This process will take about $x$ milliseconds. At time $t + x$ the mobile node will lose its connection to the current AP. Following this, the mobile node will attempt to synch with the new AP. At time $t + y$ (where $y > x$), the mobile node will be attached to the new AP. It is crucial to understand that even in today's cellular networks, time $x$ is not always predictable. That is, a mobile node cannot assume that it will detach from its current link in exactly $x$ milliseconds. The cell overlap is planned (depending on the cellular technology used) such that period $x$ is at least as long as needed for the link-layer message exchanges between the host and the network during the link-layer handover. That is, $x$ will be large enough to allow for the link-layer signaling to be successful in most cases.

Currently deployed cellular networks assume that the chosen overlap will be long enough to allow a fast-moving node to exchange the necessary handover messages between the moving node and its current AP; that is, it will be large enough to allow for receiving reliable measurements and sending the necessary information to the mobile node ($x$ milliseconds in our example). If we add fast handovers signaling to this process, we must add some time $a$ to the handover signaling, where $a$ is the time required to successfully anticipate movement. In other words, we need $t + x + a$ to end at the same time that we originally had (i.e., where $t + x$ is located in Figure 7–4) without fast handovers. We can achieve this by starting the handover at time $t - a$. However, depending on $a$, this will add more uncertainty about when and whether the handover will take place. If $a$ is too large, the mobile node's mobility direction might change due to changing radio conditions. As a result, the mobile node might not move to the originally anticipated AP. Radio conditions can change rapidly for various reasons; if the handover starts too early, by the end of the handover signaling, the mobile node might not move after all or it might move somewhere else. Hence, it is important to understand that a large anticipation time (caused by $a$) is likely to reduce the certainty of the handover.

The value of $a$ is the anticipation time required in order to allow fast handovers' anticipation to work. Note that $a$ need not be added to the link-layer handover time. That is, one could possibly send link-layer and fast handover signals simultaneously. However, in any case, it is useful to determine the value for $a$ independently of the use of fast handovers in a particular system. We can calculate $a$ as follows:

$$a = 3T_{mn\text{-}AR}$$

where $T_{mn\text{-}AR}$ is the time it takes to send a message from the mobile node to the AR. Since we have three messages (solicitation, advertisement, and binding update), we need to multiply it by three (see Figure 7–5). Obviously, we are assuming that all messages are the same length for simplicity; we also did not include the binding acknowledgment because it is not required to be received on the current link (more on this in Section 7.1.5.2). On a connection-oriented link, we can reduce the anticipation time to $2T_{mn\text{-}AR}$, since the mobile node will not need to solicit for a proxy router advertisement. This is beneficial because it reduces the number of messages exchanged and therefore adds more certainty to the exact timing of the mobile node's movement.

Based on the above discussion, we can conclude that the additional anticipation period ($3T_{mn\text{-}AR}$) will add uncertainty about the mobile node's mobility during a handover. Hence, premature forwarding of data by the AR (upon reception of an F-BU) could be harmful because the mobile node might not move to the new link, so the data would be unnecessarily lost! If the mobile node does not eventually move to the new AR, it will need to inform the current AR (using a binding update) that its binding cache entry should be deleted. Hence, another message would need to be sent to the current AR. In the meantime, all packets sent to the mobile node will be lost. This is one part of the price that mobile nodes pay for the anticipation process introduced in fast handovers. The other part is presented in the following section.

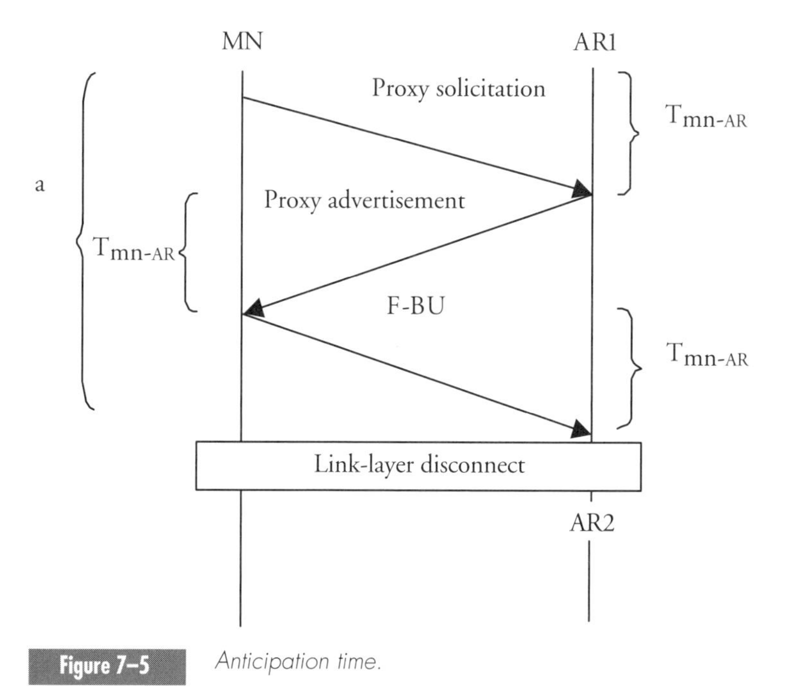


**Figure 7–5** *Anticipation time.*

#### 7.1.5.1 PING-PONG MOVEMENT

Another important factor involved in deciding on the overlap between cells is to ensure that the overlap is long enough so that the signals received from the new AP are not deceptively large.

In reality, the borders between radio cells are not as clear as shown earlier, but are more like the one in Figure 7–6. This squiggling border is a natural consequence of having objects that block the signal between the AP and the mobile node. These objects could be fixed (buildings, trees, etc.) or moving (e.g., a bus or a car blocking the signal). Hence, the border may change dynamically. A consequence of this is that a mobile node may move back and forth, or *ping-pong*, between cells. Cell planners and radio engineers attempt to avoid this problem by having a large enough overlap area between cells. For instance, a handover heuristic might be based on comparing the power measurement reports from a mobile node for a particular AP; when the received signal is, for example, 3 dbm larger than the first measurement, then there is enough certainty that the mobile node can be served by that AP. While these rules do not guarantee that ping-pong will be eliminated, they do significantly reduce the chance of ping-pong between two cells.

Radio engineers consider these types of movement when designing and deploying radio links. If ping-pong occurs between two APs connected to the same AR, the problem is handled by the link layer. However, if the mobile node moves back and forth between two APs connected to different ARs, the mobile node invokes Mobile IPv6, resulting in sending binding updates to its AR, home agent, and possibly correspondent nodes every time it moves. This problem cannot be addressed on the link layer, which is not aware of the impact of such movement on Mobile IPv6. Furthermore, anticipating the handover on the IP layer increases the probability of ping-pong movement. This is due to the additional anticipation time imposed by fast handovers, which causes the handover to start earlier than originally planned by the link layer, hence

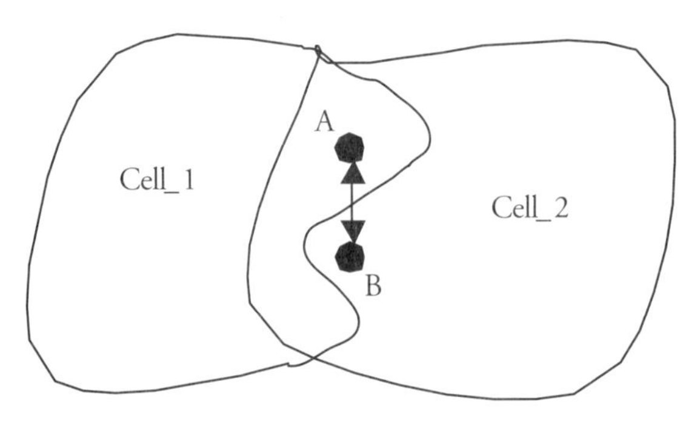


Figure 7–6 *Ping pong between locations A and B.*

reducing the certainty about the mobile node's movement and increasing the likelihood of ping pong movement.

When ping-pong happens, the mobile node suffers the following:

- Every time the mobile node moves, it will attempt to send a new F-BU to the old AR or a normal binding update to the home agent and correspondent nodes. This results in additional signaling. Furthermore, by the time the binding updates are received, the mobile node could move again.
- The mobile node will not be able to receive any packets while moving back and forth. In our earlier example of a mobile node moving from AR1 to AR2, if it moves back, AR1 would have already forwarded the packets to AR2, so it will not receive packets.

From the discussion in this and the previous sections, we can conclude that anticipation, while useful, introduces new problems that need to be addressed. In general, anticipation, in most links, means that the handover must start earlier than originally planned by a radio link that does not use fast handovers. Starting the handover sequence earlier reduces the certainty about the mobile node's movement. That is, a mobile node may move to a new cell and then move back (ping-pong). Alternatively, the mobile node may not move at all (false alarm). Either scenario will have undesirable effects on the mobile node's traffic. To solve these problems, we must try to minimize the anticipation time imposed by fast handovers. In addition, we must address the ping-pong case described above. These issues are discussed in more detail in the next section.

#### 7.1.5.2 HOW CAN WE MINIMIZE PROBLEMS RESULTING FROM ANTICIPATION?

We can reduce one of the impacts of anticipation by reducing the anticipation time. This can be done by reducing the number of messages sent as well as reducing the sizes of these messages. For instance, in an IEEE802.11b link, the number of neighboring APs is not large and movement is typically slow (walking speed). Hence, the mobile node could send the solicitation for proxy message some time before the handover. This allows the mobile node to gather information about the neighboring default routers before it detects that it needs to move. On a connection-oriented cellular link, we already saw that the solicitation for proxy message is redundant, since the network already knows where the mobile node is going. A further optimization would be to send Mobile IPv6 signaling at the same time the mobile node sends the lower-layer handover signaling. That is, it might be possible to overlap the link-layer handover with the IP layer handover and consequently minimize the additional anticipation time required for fast handovers. However, the link-layer handover sequences may not match Mobile IPv6 sequences in all link layers. That is, a complete overlap between the link-layer handover and the IP layer handover may not always be possible. But even if a complete overlap were possible, adding fast handovers signaling increases the sizes of the messages exchanged between the mobile node and the network (when compared to link-layer messages without fast handovers), which results in longer transmission periods. Hence, a "seamless" application of fast handovers (i.e., without increasing the anticipation time) is difficult to assume for all link layers.

One solution to the impacts of anticipation and solving the ping-pong problem is to use *bicasting*. A mobile node could request (in F-BU) that the AR send packets to its current location in addition to the new location for a short period of time (e.g., 2 seconds). After 2 seconds, the mobile node's packets will be forwarded to its new location only. This ensures that the mobile node continues to receive packets while being connected to the current AR, even after the handover signaling is complete. In addition, the mobile node continues to receive packets during the ping-pong period. The advantage of this mechanism is that it is completely decoupled from lower layers. The cost is clearly that traffic will be duplicated on the wired network (between the two ARs). However, if this period is short enough, the cost may not be significant. Figure 7–7 shows the bicasting of the mobile node's packets from AR1 to the mobile node's current and anticipated care-of addresses.

Note that when bicasting is used, packets need not be sent on both wireless links (i.e., the current link and the new link). When packets arrive at the AP, the AP will not be able to find the right MAC address (on a shared link) or the right channel (on a connection-oriented link) to send them to if the mobile node were not connected to that AP. Hence, whenever the mobile node is not connected to the AP, its packets are dropped instead of being transmitted on the wireless link. Clearly, packets will be duplicated on the wired links (i.e., between old and new ARs and between an AR and an AP).

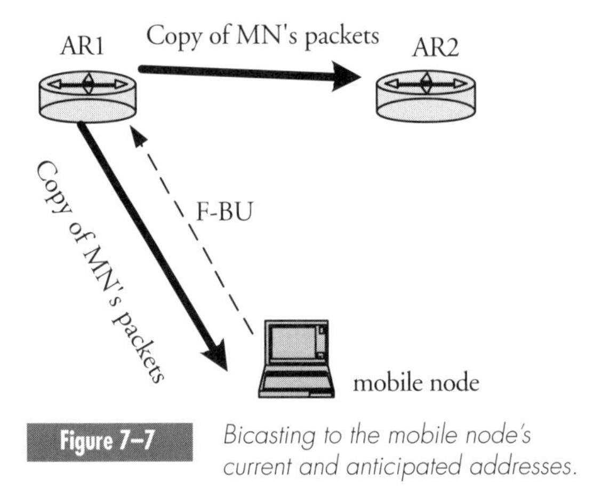


**Figure 7–7** *Bicasting to the mobile node's current and anticipated addresses.*

Bicasting can also be useful for ensuring that the mobile node receives the binding acknowledgment regardless of its location. After accepting an F-BU that requests bicasting, the AR sends the binding acknowledgment (like all other packets) to the mobile node's new location as well as to the current one. Hence, the binding acknowledgment is received even if the mobile node moves from the old link.

The concept of sending two different copies of a packet to two different locations is not new to wireless systems. In fact, this concept is currently used by Wideband Code Division Multiple Access (WCDMA) systems, as will be shown in Chapter 10.

## 7.1.6 Security Issues

There are a number of open questions that need to be resolved in terms of securing fast handovers. In this section we present some of the attacks that can be launched using this mechanism.

- *DoS attacks*: Bad Guy could send a large number of router solicitations for proxy messages to keep an AR too busy to process legitimate ones. This attack can be more significant if the AR uses the proxy solicitations to trigger the HI/Hack message exchange. This would mean that every time the AR receives a solicitation, it creates state in its memory and sends the HI message to a new AR. The problem is that the AR is using an unauthenticated message to trigger a costly message exchange (HI/Hack) with all its associated state. This problem can be solved if the AR waits for an authenticated F-BU message to start the HI/Hack exchange with the new AR or by eliminating the HI/Hack exchange, as will be shown in section 7.1.8.
- *Attacks on F-BU*: The same attacks discussed in Chapter 5 are applicable to the F-BU. When the mobile node sends an F-BU to an AR, it uses its current care-of address as a home address and its address in the new link as care-of address inside the binding update message. This essentially makes the current AR act as a temporary home agent. Hence, the relationship between the mobile node and the AR is similar to the one between the mobile node and its home agent. However, a generic identity like bob@example.com will not suffice to prove that the mobile node owns its current address (home address as viewed by the current AR). A return routability test is not useful in this case, since the movement is anticipated; the mobile node is not at the new location yet and therefore cannot send or receive the COTI/COT messages respectively.
- *Proxy advertisements*: There is no need to secure proxy advertisement for fast handovers. However, like all Neighbor Discovery messages, this message can also be spoofed by Bad Guy. Fast handovers do not introduce this vulnerability, though; if Neighbor

Discovery messages were secure, this problem would have been solved automatically.

- *HI/Hack messages*: The HI/Hack message exchange establishes a tunnel between the mobile node's current and future ARs. In addition, the new AR opens its ingress filter to allow the mobile node to send and receive packets using its old (topologically incorrect) address. Therefore, these messages must be authenticated and protected against replay attacks. A network administrator could configure two ARs with a secret key to allow them to authenticate these messages using ESP or AH. However, since replay protection is bound to a finite number of sequence numbers, the key will need to be changed eventually. The use of IKE with public or secret key cryptography could solve this problem, since new session keys would be negotiated when the security association expires.

One of the main problems with securing fast handover signaling is that the mobile node needs to negotiate a new security association with its AR every time it moves. If IKE is used, it means that the mobile node will initiate an IKE negotiation with each new AR. On bandwidth-challenged (e.g., cellular) radio links, this method is far from ideal. Furthermore, as we saw in Chapter 5, it is difficult to use IKE to solve the address ownership problem, which needs to be solved in order to secure F-BU and Neighbor Discovery messages. In the following section, we see how fast handovers can be secured by using CGAs.

### 7.1.7 Can We Use CGAs to Secure Fast Handover Signaling?

We showed in Chapters 4 and 5 that CGAs can be useful for securing Neighbor Discovery messages and binding updates to correspondent nodes. CGAs can also be useful for fast handovers.

If the mobile node has a cryptographically generated care-of address (in the F-BU, this is treated as the home address), it can use that to sign the F-BU message. The AR can verify the signature because the mobile node will include its public key in the binding update. In this case, the AR (acting like a temporary home agent) does not need prior relationship with the mobile node. Unlike a "real" home agent, this AR acts as a temporary home agent for only a limited period of time. Similarly, when the mobile node sends an F-NA message on the new link, it can sign it. The new AR could verify this signature without any prior negotiation with the mobile node. Note that CGAs will not help the AR in securing the proxy advertisements.

Cryptographically generated care-of addresses also help reduce the likelihood of address collisions. Recall that the probability of finding a pair of nodes that cause a hash collision in this case is $2^{-31}$, which is quite low (half the size of the IP version 4 address space). We can be confident that this number of nodes will not be located on a single link—at least not in our lifetime!

## 7.1.8 An Alternate Approach to Fast Handovers

As shown earlier, the purpose of the HI/Hack exchange is to set up a tunnel between the current and new ARs and check if the mobile node's predicted address is valid on the new link. However, if the interface identifier in the mobile node's care-of address is generated using a good random number generator (or cryptographically generated), the probability of address duplication becomes very small and practically insignificant. In fact, on a wireless link (with no link-layer retransmissions), it is more likely that one of the messages will be lost, causing the handover to fail due to reasons other than address duplication. Hence, we can eliminate the HI/Hack exchange without causing any significant problems. If we eliminate the HI/Hack messages, the tunnel from the current AR must be terminated in the mobile node (as opposed to the new AR in the current approach).

The F-NA message is needed to ensure that the mobile node knows whether its care-of address is valid on the new link in the absence of a binding acknowledgment (i.e., if the message were lost or the mobile node moved to the new link before receiving it). However, the binding acknowledgment could be sent to the mobile node's current location and the new link simultaneously (bicast). If the mobile node receives the acknowledgment message on the old and new links, the second acknowledgment message received by the mobile node can be ignored. We now present a different fast handover scheme in Figure 7–8, based on the above assumptions.

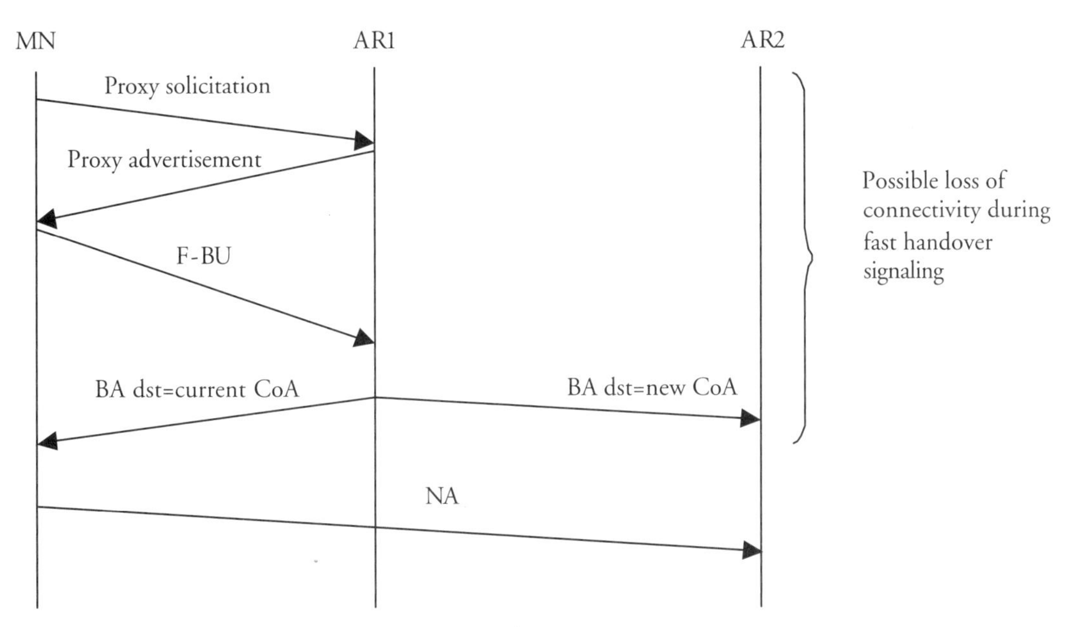


**Figure 7–8** *A handover scheme with no inter-AR communication.*

The handover scheme shown in Figure 7–8 relies on the same anticipation mechanisms discussed earlier. However, in this scheme, the mobile node sends an F-BU to AR1 requesting it to bicast packets to its current location as well as to the anticipated location. We note that the mobile node's movement will be interrupted by a period of time due to the layer 2 handover (this period is equal to $y - x$, as shown in Figure 7–4). Hence, there is still a good chance that some packets sent to the new AR will be lost because the mobile node is unreachable during this period. Therefore, we suggest that AR1 initially delay the packets destined to the new location by time $D$, where $D \geq (y - x)$. The period $D$ is link layer-dependent and must be configured in AR1 (if AR1 is not aware of the type of radio link it is connected to—e.g., if the AP and AR are physically separated). This allows the mobile node to receive packets that would normally have been lost while it was disconnected from both links.

When the mobile node arrives at the new link, it can immediately use its new care-of address while testing it for duplication, using the *optimistic DAD* algorithm discussed in Chapter 6. Optimistic DAD allows the mobile node to use its address immediately without waiting for DAD. This eliminates the address autoconfiguration time.

When the mobile node arrives at the new link, it must advertise its presence to the new AR. This is necessary to allow the AR to forward new packets to the mobile node. During the period when the mobile node was disconnected from both links, the new AR might have received packets destined to the mobile node; it might have sent a neighbor solicitation to the mobile node's address, but since the mobile node was not present, no neighbor advertisements were received. Hence, upon arrival at the new link, the mobile node should send a neighbor advertisement to the new AR. If the new AR has already sent a neighbor solicitation to the mobile node, it will assume that this is a late reply (even if the *S* flag is not set) and create a new entry for the mobile in its neighbor cache. However, if there are no packets waiting for the mobile node (i.e., no ongoing connections), the new AR ignores the neighbor advertisement received.

The advantage of the mechanism shown in this section is that it does not require any prior security relationship between AR1 and AR2, hence less manual configuration. This is based on the assumption that the probability of address duplication is too small to justify having the HI/Hack messages and the associated overhead (security and implementation). Reducing the dependency between AR1 and AR2 is particularly important if the two ARs are not administered by the same entities (i.e., belong to different ISPs). An ISP may not wish to establish a security association between routers in its domain and other routers in another ISP's domain. For instance, an ISP may not trust the security practices of another ISP. In addition, establishing the security association between routers adds additional burdens to administrators and they may not wish to have them. In some cases, any of these reasons can cause the HI/Hack exchange to fail, and consequently, fast handovers will fail in some cases. This mechanism also contains fewer messages, which simplifies implementations and therefore reduces the complexities of the state machines in the mobile node and ARs.

Another advantage of this mechanism is that it significantly reduces the problems associated with anticipation and ping-pong. If the mobile node moves back and forth between the two ARs, it will continue to receive packets without sending any more signaling. After the bicasting period, AR1 will forward packets to the mobile node's new location only. There is certainly a possibility that the handover will fail and the mobile node never moves to the new location. In this case, the mobile node can send another binding update to AR1 to remove the existing binding cache entry.

The disadvantage of this mechanism is that it is not as deterministic as the previously discussed fast handover scheme. This scheme trades off simplicity for reliable delivery of the binding acknowledgment. The main problem is that the period $(y - x)$ or $D$ is not an exact science; it greatly depends on the radio environment and the link-layer handover implementation. It may vary from one mobile node to another based on the device vendor. Therefore, there is a chance that the binding acknowledgment message is never received. In this case, the mobile node will resend the F-BU message to AR1. However, if the first binding update was accepted by AR1, the mobile node will be receiving packets at the new location. Hence, while the mobile node will unnecessarily send another binding update, the traffic will not be interrupted.

Another disadvantage of this mechanism is that since $D$ cannot be predicted in an exact manner, there is a chance that the mobile node will receive the same packet twice. For instance, an AR might assume $D$ to be 100 ms, but during a particular handover, it was 70 ms. This results in a 30-ms overlap between packets delivered to the old and the new links. Packet duplication can happen today due to network congestion, but bicasting increases the likelihood of duplication. This is not a problem for UDP traffic because the application running on top of UDP usually accounts for duplicate packets and drops them. However, receiving duplicate TCP segments will cause a TCP implementation to send duplicate acknowledgments, which could cause a correspondent node to go to fast retransmit and possibly congestion avoidance (see Chapter 6). This problem can be avoided when the Eifel detection and response algorithm is used . When the Eifel algorithm is used, the correspondent node will detect that the duplicate acknowledgment was due to receiving the same segment twice and not due to a "hole" in the received TCP stream. Upon detecting this, the Eifel algorithm will restore the original congestion window size and normal transmission rate will resume.

## 7.2 Hierarchical Mobile IPv6 (HMIPv6)

In the previous sections, we saw how fast handovers are used to eliminate movement detection time and care-of address configuration, but without reducing the number of messages sent by a mobile node over a wireless link. A bandwidth-challenged wireless link will suffer from sending many binding updates each time the mobile node forms a new care-of address in addition to the return routability signaling to correspondent nodes. Consider a mobile node communicating with two correspondent nodes. Whenever the mobile node moves, it must send a binding update to its home agent and perform a return routability test with both correspondent nodes before sending them two binding updates. If the *home keygen tokens* have not expired, the mobile node sends and receives a total of six messages to and from the correspondent nodes (a COTI/COT and binding update with each one). However, eventually the home keygen tokens will expire, adding another two messages (HOTI/HOT) between the mobile and correspondent nodes and increasing the total number of messages to 10; this number will clearly increase if messages are lost.

Fast handovers rely on "anchoring" the mobile node's traffic in the old AR while it updates its correspondent nodes and the home agent. New packets (route-optimized based on the mobile node's new location) will arrive at the mobile node's new address. This could cause a problem for TCP connections because some segments could arrive out of order. We illustrate this in Figure 7–9.

Figure 7–9 shows an example of a network where AR1 and AR2 are not sharing a physical link. When a node connected to AR1 communicates with another connected to AR2, traffic is routed upstream to an aggregate router that can route them to AR2. This is not an unusual network design. In fact, this is a very common way of designing IP networks, especially if links B and C are not high-bandwidth links or if AR1 and AR2 are geographically far away from each other. This type of network design saves operators money when compared to connecting all ARs in a fully meshed network and therefore requiring $N(N-1)$ physical links, where $N$ is the number of ARs.

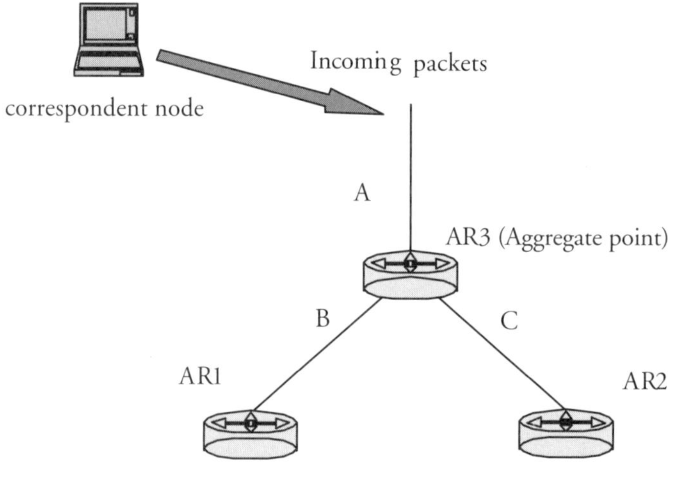


**Figure 7–9** *Routing incoming packets to the mobile node.*

Let's consider what happens to a mobile node's traffic during a handover from AR1 to AR2 while communicating with a correspondent node that sends traffic through link A (i.e., a correspondent node that is not connected to AR1 or AR2). When the mobile node updates AR1 with an F-BU, traffic destined to the mobile node travels from link A to link B, where it is intercepted by AR1 and tunneled to AR2. Hence, it is routed back onto link B, then onto link C. After a correspondent node has successfully processed a binding update from the mobile node, it sends packets directly to its new care-of address, which belongs to AR2's link. Therefore, packets are routed from link A to link C. Now consider a case where the mobile and correspondent nodes have a TCP connection and the correspondent node's window size is 8 (we assume 8 segments for simplicity instead of counting bytes, which is how the window size is represented). Suppose that TCP segments 20 to 24 were sent from the correspondent node just before processing the binding update, and segments 25, 26, and 27 were sent immediately after processing the binding update. The time difference between sending segment 24 and segment 25 is the time it takes the correspondent node to process a binding update; let's call it $t_1$. Let's assume that the delay between the aggregate point (AR3) and AR1 is $t_{ag\text{-}ar1}$. Assuming that all segments will experience the same forwarding delays until they arrive at link A, segments 20 to 24 will experience an additional delay of $2t_{ag\text{-}ar1}$. This is because they are routed twice on link B—that is, from AR3 to AR1, then back to AR3—before they are forwarded to link C.

If $2t_{ag\text{-}ar1} > t_1$, segments 25, 26, and 27 will be received by the mobile node out of order (i.e., before the preceding segments 20–24). In other words, if the forwarding delay on link B is larger than the time it takes to process a binding update, segments 20 to 24 will be sufficiently delayed between AR1 and AR3 to allow segments 25, 26, and 27 to arrive at the mobile node first.

When the mobile node's TCP implementation receives segment 25, it sends a second acknowledgment for segment 19 (the last one received in order). Hence, a duplicate acknowledgment arrives at the correspondent node. Furthermore, if $2t_{ag\text{-}ar1} >> t_1$ (this is likely, since processing time in a fast computer is usually smaller than delays on loaded links), segment 26 and 27 will also arrive at the mobile node before segments 20 to 24, causing three duplicate acknowledgments, which triggers TCP fast retransmit in the correspondent node.

Selective acknowledgments are widely deployed in all major operating systems. If the mobile node, upon receiving segment 25, sends a SACK option indicating that it did not receive segments 20 to 24, the correspondent node retransmits these segments unnecessarily.

We can create variations of this scenario based on the delays between the aggregate point and AR1. However, at this point, we can conclude that anchoring traffic in the old AR adds delays to packets, which can have negative impacts on TCP connections. The impact will vary depending on the delays between the aggregate point and AR1. In addition to these impacts, link B may experience bursts of overload if many mobile nodes are moving at the same time. Every mobile node's packets will consume twice the amount of bandwidth needed on that link, since every packet will be routed twice between the aggregate point and AR1.

Note that the same argument can be made for the first time the mobile node sends a binding update to the correspondent node to inform it of its care-of address (i.e., with standard Mobile IPv6 and no fast handovers); the path will change for some TCP segments (from the home address to a care-of address), and they may arrive ahead of the others. However, handovers can be frequent, while this case will only happen once, when the initial binding update is sent. Hence, it is more important to solve this problem for the case where the mobile node is moving between two foreign links.

In the following sections, we see how Hierarchical Mobile IPv6 (HMIPv6) solves these problems and introduces other new features.

## 7.2.1 HMIPv6 Overview

HMIPv6 was designed to allow mobile nodes to move within a particular domain without having to update their home agents or correspondent nodes every time they move. Instead, they send one binding update to a Mobility Anchor Point (MAP) located in the visited network. But is it not trivial to do this with Mobile IPv6 if the mobile node updates only the home agent? Certainly; this would minimize the number of binding updates sent to the home agent. However, this comes at the price of eliminating route optimization. That is, all correspondent nodes will send packets to the mobile node's home address, which will be tunneled to the mobile node's care-of address. The aim of HMIPv6 is to reduce the number of messages to one while maintaining an optimal route to the mobile node (more on route optimality later).

The solution for the above problem is trivial; if we can have a home agent close to the mobile node, we can optimize the routing while reducing the number of signals sent by the mobile node when it moves. A MAP is essentially a *local* home agent located in the visited network. When a mobile node visits a network, it will discover the IPv6 address of the MAP. The mobile node will then request to be allocated a *temporary* home address on the MAP's link and request that the MAP defend that home address in the same way that a home agent defends the mobile node's home address on its home link. The temporary home address is called the *regional care-of address* (RCoA). It is certainly a care-of address because it is not the mobile node's real home address, and it is *regional* because, as will be shown, it is valid on more than one link, which is different from the normal care-of addresses explained earlier. To distinguish the mobile node's address on the MAP's link from its care-of address configured based on a foreign link's prefix, we use the terms *regional care-of address* (address on the MAP's link) and *on-link care-of address* (LCoA; address derived from the foreign link's prefix). Figure 7–10 points out where each address belongs in the topology. Clearly the mobile node would be configured with all of these addresses, independently of their topological locations.

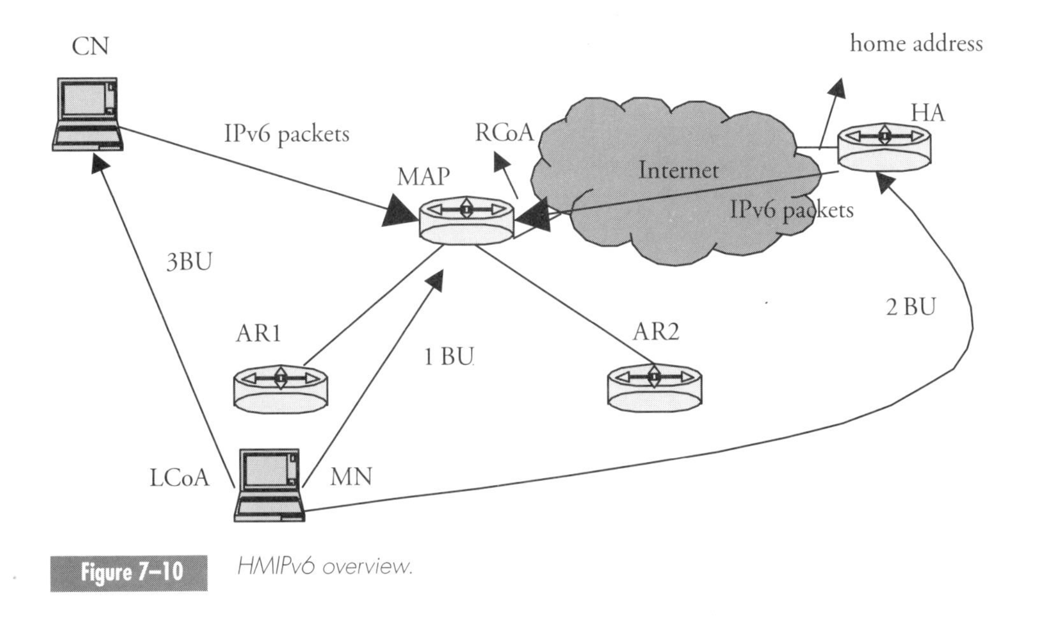


**Figure 7–10** *HMIPv6 overview.*

When a mobile node attaches to a new link, it receives a router advertisement, including a MAP option. The MAP option informs the mobile node of the MAP's IP address. The mobile node generates a regional care-of address based on the MAP's prefix and an interface identifier generated by the mobile node. The mobile node then sends a binding update to the MAP, using its regional care-of address as a home address (i.e., in the *home address* option) and its on-link care-of address as the care-of address (i.e., included in the source address of the IP header). Note the similarity between this binding update and the one sent to the home agent. The regional care-of address essentially replaces the home address as far as the mobile node's communication with the MAP is concerned. The MAP knows nothing about the mobile node's real home address. A new flag, *M*, is included in the binding update message to distinguish it from binding updates sent to the home agent. When sending the binding update, the *M* and *A* flags are set to indicate that this is a *local binding update* (i.e., sent to a MAP), and it requires the MAP to check for address duplication and acknowledge the reception of this message. The only difference between this binding update and the one sent to the home agent is the *M* flag, which is set instead of the *H* flag for binding updates sent to the home agent. The reason for adding this flag is that we need to distinguish between local binding updates sent to a MAP and those sent to a home agent. For instance, a MAP may also be a home agent for other mobile nodes. If it gets overloaded, it is likely to prioritize mobile nodes for which it acts as a home agent to other mobile nodes sending local binding updates. This distinction could have been made based on the security association used by the mobile node to protect the binding update.

However, having the *M* flag makes it easier to decouple the mobility implementation from the security implementation in the MAP, especially if new mechanisms are developed in the future for securing binding updates.

If the MAP accepts the local binding update, it sends a binding acknowledgment to the mobile node; with an appropriate status value; no new values are required for this use of the binding update. At this point, the mobile node is ready to update its home agent with its current location. When the mobile node sends a binding update to the home agent, it binds its regional care-of address to its home address. Hence, whenever the home agent receives packets destined to the mobile node's home address, it will intercept them and forward them to the regional care-of address. Since the MAP is also acting as a local home agent for the mobile node's regional care-of address, it will intercept those packets and forward them to the mobile node's current location (the on-link care-of address stored in the MAP's binding cache). This process illustrates the reason for calling this mechanism *Hierarchical* Mobile IPv6, as it introduces a hierarchy of home agents for the mobile node. The hierarchy is limited to one additional logical hop. In other words, the MAP will not tunnel packets to another MAP lower in the hierarchy; the tunnel exit point is the mobile node's on-link care-of address.

When the mobile node starts communicating with a correspondent node, it can send a binding update indicating that its care-of address is the regional care-of address. Hence, all traffic addressed to the mobile node will be intercepted by the MAP and forwarded to the mobile node's on-link care-of address. The mobile node's outgoing traffic is tunneled to the MAP in a manner identical to the tunneling of outgoing packets to the home agent.

In Figure 7–10, AR1 and AR2 belong to the MAP domain; that is, they both advertise the same MAP option (described in section 7.2.2) to the mobile node. Hence, when a mobile node moves from AR1 to AR2, it must send only one binding update to the MAP, informing it of its new on-link care-of address. This results in replacing the mobile node's care-of address in the MAP's binding cache, and consequently the tunnel exit point becomes the mobile node's new on-link care-of address derived from the prefix advertised on AR2. We can see how this mechanism reduces the number of binding updates sent (due to movement) to one binding update, regardless of the number of correspondent nodes the mobile node might be communicating with. At the same time, we haven't lost route optimization. Since all correspondent nodes (and the home agent) include the mobile node's regional care-of address in their binding caches, there is no need for the mobile node to update any other nodes unless its regional care-of address changes, which happens only when the MAP changes. Hence, while moving within the same MAP domain, the mobile node would need to send only one binding update—to the MAP—to continue to receive traffic from correspondent nodes.

A deeper look at route optimization will tell us that if the MAP is located at such a point that the distance between the MAP and all ARs in its domain is approximately the same, packets will travel an optimal route between the mobile node and a correspondent node outside the MAP domain. However, a correspondent node might be located on the same link as the mobile node. If packets are sent to the mobile node's regional care-of address, they will experience more delays than if sent directly to the mobile node's on-link care-of address (i.e., without HMIPv6), because they will be routed to the MAP first, then to the mobile node. A mobile node knows whether a correspondent node is on the same link (if the first 64 bits match the on-link prefix in the router advertisement). If the correspondent node were on the same link, the mobile node can choose whether it should send it its on-link care-of address or its regional care-of address in the binding update. This is clearly a tradeoff between the number of messages sent every time the mobile node moves and the required degree of route optimality. Generally, it is acceptable to simply inform every correspondent and the home agent about the regional care-of address. If the MAP is only one or two hops away from the AR, the additional delay is insignificant. This is particularly true on wireless links where the largest delay is experienced over the wireless link (i.e., the first hop).

In many large networks, the ARs are geographically and topologically sparse; as the mobile node moves within the same operator's domain, it is bound to move further away from the MAP. If the operator's domain is very large, this distance can add significant delays to packets and reduce some of the benefits of route optimization. Hence, a large network is likely to need more than one MAP domain. The handover between MAP domains involves a change in the regional care-of address of the mobile node. Since this is the care-of address seen by the home agent and correspondent nodes, the mobile node must inform them about the change by sending a new binding update that includes its new care-of address (the new regional care-of address acquired from the new MAP). Section 7.2.3 discusses this case in more detail.

## 7.2.2 MAP Discovery

Mobile nodes discover the presence of a MAP within a network through router advertisements. If an AR is part of a MAP domain, its router advertisement will include the MAP option shown in Figure 7–11.

> The *dist* field is an unsigned 4-bit number that indicates how far the MAP is from the mobile node (topologically). This number does not need to indicate the exact number of hops; it should be used on a scale of 0 to 15, where 15 is the largest possible distance.
>
> The *pref* field shows the preference of the MAP on a scale of 0 to 15, where 15 indicates the highest availability and 0 indicates that this MAP should not be used for new local binding updates (i.e., binding updates that will create new binding cache entries, but not those refreshing an existing entry).

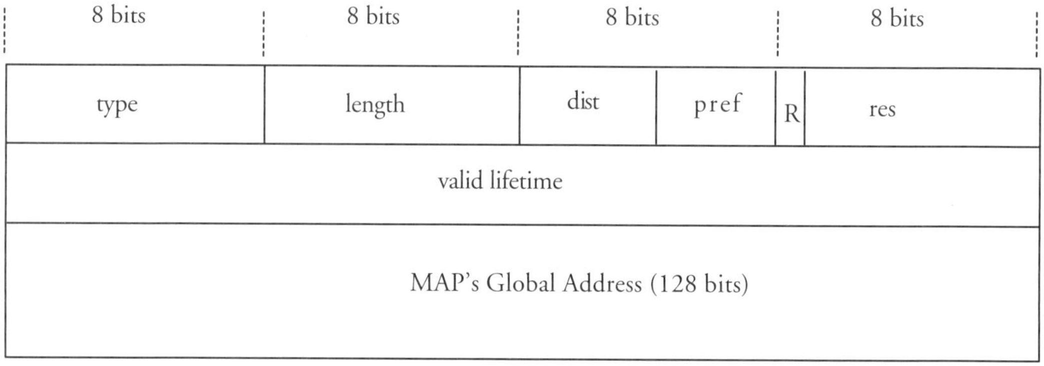

| 8 bits | 8 bits | 8 bits | | | 8 bits |
|---|---|---|---|---|---|
| type | length | dist | pref | R | res |
| valid lifetime | | | | | |
| MAP's Global Address (128 bits) | | | | | |

**Figure 7–11** *MAP option.*

The *R* flag (when set) indicates that the mobile node should use the first 64 bits in the MAP's global address as a prefix for its regional care-of address. This flag was included to allow for future modes of HMIPv6, where the regional care-of address can be derived in a different manner.

The *res* field contains 7 bits that are reserved for future use and are set to zero in current implementations.

The *valid lifetime* field indicates (in seconds) the validity of the MAP's prefix included in this option. It is the minimum value of the valid lifetime (in the MAP's router advertisement) and *preferred lifetime* associated with the MAP's prefix. Hence, this value tells us how long the MAP and the regional care-of address can be used. When this field is set to zero, mobile nodes will assume that the MAP has failed and will attempt to discover another MAP.

The last 128 bits in this option include the MAP's global IPv6 address, which is needed for two reasons: first, the mobile node must know where to send the local binding update, and second, the mobile node needs to know the prefix used to derive its regional care-of address. HMIPv6 requires the prefix length to be 64 bits.

#### 7.2.2.1 HOW DO ROUTERS KNOW ABOUT THE MAP OPTION?

Routers are configured to send certain router advertisements on specific interfaces. Options (e.g., the prefix option) are manually configured in the router. The simplest way an AR can know which MAP option it should send (and the default in the HMIPv6 specification) is through manual configuration, yet this adds slightly more work for the administrator.

Ideally, a router should be configured in a dynamic manner. This allows MAPs to change their preferences based on their load. Furthermore, it allows a router to discover whether the MAP has failed and consequently to send a MAP option with a valid lifetime value set to zero. HMIPv6 defines *dynamic MAP discovery* to allow ARs to advertise different values in the MAP option, depending on the MAP's conditions.

Dynamic MAP discovery allows the MAP option to be propagated between the MAP and ARs within its domain through a network of routers. This can be described as follows:

- The MAP is configured with the default values for each field in the option. The default value for the *dist* field must be larger than one. The default value for the pref field is 10. Other flags are set based on the administrator's choice.
- Each router in the network between the ARs and the MAP (the MAP may be a number of hops away from ARs) is configured to send the MAP option on certain interface(s).
- Whenever a router receives a MAP option, it increments the dist field by one and includes the option in its router advertisement sent on the interfaces configured to include a MAP option.
- If a router receives more than one MAP option for the same MAP, it sends the option with the smallest distance after incrementing it by one.
- Finally, when an AR receives the MAP option in another router's advertisement, the option is copied and included in that AR's advertisement after incrementing the dist field in the option by 1.

This mechanism will work even if routers accidentally forward the option on the wrong interface (note the check in the fourth bullet).

Clearly we cannot rely on having a message from the MAP to all ARs in its domain that says, "I have failed, please set the lifetime in my option to zero." If the MAP fails, it will not be able to send anything. To detect MAP failures, ARs can send regular ICMP *Echo request* messages. If the MAP does not reply after several attempts, it can be assumed to have failed, and a MAP option with a valid lifetime of zero can be sent. However, the interval between ICMP Echo requests should be small enough to avoid the case where a MAP reboots (losing its entire binding cache) between two messages. A contradictory requirement, to avoid overloading the MAP, is that the messages should not be sent too often.

#### 7.2.2.2 HOW DO MOBILE NODES SELECT A MAP?

Operators may choose to deploy several MAPs in one domain, each located at a different distance from the ARs within its domain. The MAP option contains enough information for the mobile node to select a MAP based on its distance and availability. Distance-based selection is not trivial; the assumption here is that a "far away" MAP will have a larger domain than a close one. Hence, when a distant MAP is chosen, the mobile node is less likely to change its regional care-of address and consequently update the home agent and correspondent nodes. Changing the regional care-of address too often results in sending several binding updates and possibly return routability messages, which would use a radio link's bandwidth and cause packet losses. A fast-moving mobile node can (based on some heuristics that can be deduced from previous lower layer measurements and mobility trends) choose a more distant MAP in the hierarchy to reduce the probability of changing MAP domains. However, this also implies that the local binding update will experience more delays. A mobile node can measure the RTT to the MAP and use that as an input to an algorithm that selects the MAP based on a tradeoff between RTT and speed of mobility. This leads to complex heuristics based on link-layer measurements and previous patterns of mobility.

The HMIPv6 specification suggests a simple default algorithm that can be used. Mobile node implementers can use other optimizations; however, it is likely that implementers will choose the simplest method. The default selection algorithm is described below:

1. Receive and parse all MAP options.
2. Arrange MAPs in descending order based on the content of the dist field: the furthest MAP is the first on the list.
3. Select the first MAP in the list.
4. If the pref or the valid lifetime fields are set to zero, select the next MAP in the list. Otherwise, pick this MAP, then exit.
5. Repeat step 4 while other MAP options are available.

This algorithm causes the mobile node to select the furthest MAP (step 4), provided that it is available (i.e., the pref and valid lifetime fields are larger than zero).

### 7.2.3 Deploying HMIPv6

A MAP's domain is defined by all ARs advertising its presence. If AR1 to AR5 advertise the same MAP, then the MAP's domain covers all coverage areas of AR1 to AR5. How do we decide on the size of a MAP domain? This is part of the network design done by an operator. Given a particular AR, an operator estimates the amount of bandwidth it can handle and, based on that estimate, decides how large the coverage area of the AR will be. If mobile nodes in a particular area require *X* amount of bandwidth, an operator would choose a router that can handle *X*, or alternatively, would use *N* routers, each with a capacity of *X/N*. The same concept applies to the MAP; if it aggregates the traffic addressed to *N* ARs, it needs to be able to handle this amount of traffic. Otherwise, more MAPs will be needed. In Section 7.2.2, we saw that HMIPv6 allows for more than one MAP to be located in the same MAP domain. However, as a mobile node moves within the topology, it is bound to get further away (topologically) from a MAP. In other words, additional delays will be experienced.

The above discussion tells us that there are two boundaries to the size of the MAP domain:

1. MAPs within a single mobility domain have to be able to handle the amount of traffic generated by mobile nodes connected to ARs in its domain.
2. It should not be so large that mobile nodes could become too far from the MAP and lose some of the benefits of route optimization.

Point 1 can be accommodated by allowing more than one MAP within a single domain. For instance, if we choose to have three MAPs per domain, all ARs within that domain will advertise three different MAP options. Mobile nodes will select a MAP based on the information received in the MAP option. Hence, as the load on one MAP increases, others will be chosen. As a result, the traffic generated by mobile nodes will be shared between the three MAPs. Point 2 requires that operators place MAPs in locations where they approximately have the same distance (in terms of hops) to all ARs in their domains. Ultimately, a large operator's network will need to be divided into more than one domain to satisfy point 2.

A fast-moving mobile node may move beyond a single MAP domain. Let's consider a case where a fast-moving mobile node moves beyond the MAP domain, as shown in Figure 7–12.

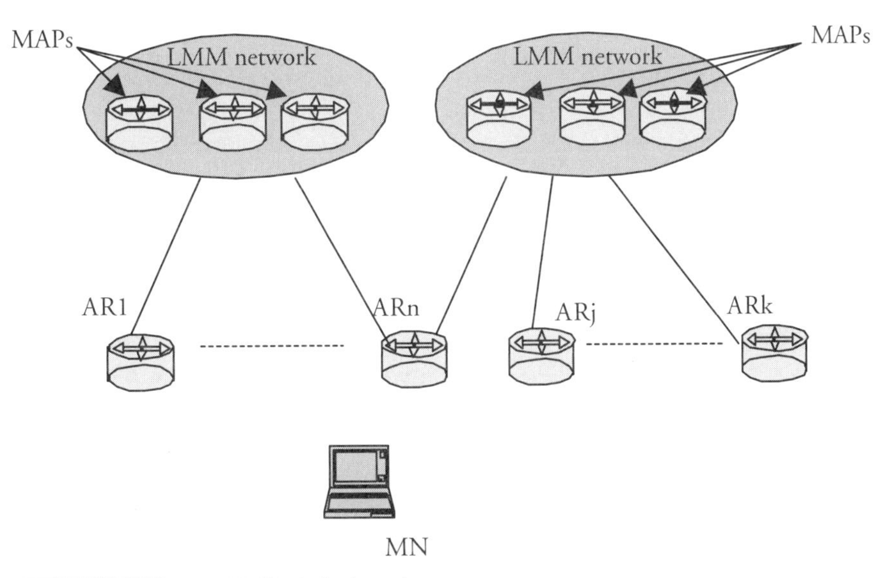


**Figure 7–12** *An HMIPv6 deployment scenario.*

In our deployment example in Figure 7–12, we introduce the term *Local Mobility Management* (LMM) network. An LMM network contains a number of MAPs covering the same domain. The first LMM domain consists of AR1 to ARn. All ARs advertise MAP options for all MAPs within their LMM network. In this example, the mobile node initially attaches to AR1. The mobile node selects a MAP (in this case, they all have the same distance, so the mobile node selects the one with the highest availability). The mobile node may also create more than one regional care-of address on different MAPs. That is, it may create a binding cache entry in each MAP. This allows the mobile node to recover from MAP failures more quickly; if MAP failure is detected, a mobile node immediately sends binding updates to its home agent and correspondent nodes, informing them of its new regional care-of address. If the mobile node has only one regional care-of address, it must create a new one (send a local binding update) before it can inform its home agent and correspondent nodes of such change.

After creating a regional care-of address, the mobile node sends binding updates to its home agent and correspondent nodes to request that they forward traffic to its regional care-of address. When the mobile node moves from AR1 to another AR within the same domain, it sends one binding update to the MAP.

When the mobile node reaches ARn, it will be located at the border between two different LMM networks. In this example, we chose to have ARn advertise MAPs in both LMM networks to allow the mobile node to send a local binding update a new MAP in the new LMM network (i.e., the mobile node is anticipating that it is about to cross the border between two LMM networks and hence reduces the number of messages that will need to be sent after moving).

When the mobile node moves from ARn to ARj, it will not see the MAP options for the previous LMM network; only the new MAP options (a subset of those advertised by ARn) will be seen. Hence, the mobile node will need to inform its home agent and correspondent nodes of the new regional care-of address. To reduce the packet loss, the mobile node should also update the previous MAP with its current on-link care-of address. Hence, packets addressed to the previous regional care-of address will be received at the new location while the mobile node is updating correspondent nodes and the home agent with its current location.

Note in our example that the mobile node sent local binding updates to more than one MAP within the same LMM network, and even to ones in different LMM networks (when connected to ARn). In other words, the mobile node was *eager* to create new bindings and *lazy* about releasing existing ones. This *eager-lazy* behavior has several benefits:

- The mobile node will spend less time creating a new local binding with a MAP when its MAP fails. This time is particularly significant because the mobile node must create a security association with the new MAP.

- When located at the border between two LMM networks, the mobile node will be able to save the time required to send a local binding to a new MAP in the new LMM when it moves to ARj, by sending a local binding update before it moves.

The *eager-lazy* behavior comes at the price of increased signaling, which is not supposed to happen when HMIPv6 is used. However, this can be mitigated by adding an explicit indication in ARn's router advertisement to tell the mobile node that it is bordering two LMM networks. This would save the mobile node from "guessing" the border and consequently save unnecessary signaling. This extension is currently not specified in HMIPv6.

### 7.2.4 Location Privacy

Mobile nodes may require a certain degree of (topological) location privacy while moving within the Internet. In some cases, geographical locations can be deduced from topological locations. For instance, if a mobile node is assigned `3ffe:200:8:9::1/128`, we can deduce that it is attached to an AR with prefix `3ffe:200:8:9::/64`. If that AR's coverage area is known (e.g., by another node on the same link), and it happens to be small (e.g., an office complex), one can deduce the mobile node's geographical location. The mobile node can hide its real location from correspondent nodes if it does not send them binding updates and it communicates through the home agent. This can be done at the cost of eliminating route optimization and showing the real location to the home agent.

The visited network always knows the mobile node's location. If it did not, the mobile node would not be able to send and receive packets. On the other hand, a mobile node might wish to hide its location from correspondent nodes and the home agent (e.g., taking a day off work without telling the boss!). HMIPv6 provides location privacy for mobile nodes in the same way that a home agent provides location privacy. Mobile nodes tunnel outgoing packets to the MAP. The outer header contains the mobile node's on-link care-of address in the source address field and the MAP's address in the destination address field. When a mobile node communicates directly with a correspondent node, the inner header contains the mobile node's regional care-of address in the source address field and the correspondent node's address in the destination address field. Therefore, the correspondent node cannot know the mobile node's on-link care-of address, which would give away the mobile node's exact topological location. This form of location privacy allows the mobile node to be traced to the MAP's subnet. Since an LMM domain's coverage area is the sum of the coverage areas of all ARs within its domain, correspondent nodes and home agents cannot track the mobile node's exact location. For instance, a correspondent node might know that a mobile node is in Sydney, Australia, but will not know which part of Sydney it is located in.

### 7.2.5 Local Mobility Without Updating Correspondent Nodes

One of the features of HMIPv6 is that it extends the concept of the home agent to cover several links; this is essentially what the MAP does. When mobile nodes are located on their home links, they can use their home addresses without sending binding updates, and they can receive packets in an optimal route. Mobile nodes can get the same effect when using HMIPv6. A mobile node can provide its regional care-of address as a source address for applications while roaming within the MAP's domain. In this case, the regional care-of address will be seen as a home address. Hence, a mobile node does not need to send binding updates to correspondent nodes when using the same regional care-of address (i.e., when moving within the same MAP domain). Correspondent nodes will send packets directly to the regional care-of address, and the MAP will intercept packets and tunnel them to the mobile node's on-link care-of address. Since a connection, in this case, will be associated with the regional care-of address (used as a source address), it will not survive when the mobile node moves across MAP domains and changes its regional care-of address. Nevertheless, since the MAP domain includes several links, this feature is useful for short-term or medium-term connections—or ones that can be easily restarted without losing a lot of information (e.g., Web browsing).

The benefit of this feature is that it allows mobile nodes to communicate, using an optimal route, with correspondent nodes that don't support any Mobile IPv6 signaling (i.e., binding updates and return routability).

### 7.2.6 Securing Binding Updates Between a Mobile Node and a MAP

We showed that a MAP is essentially a local home agent for the mobile node. Securing binding updates between the mobile node and its home agent can be done in a manual and dynamic fashion (IKE). In either case, the mobile node must be authorized to use a particular home address. However, in the case of HMIPv6, there is no need to tie certain addresses to certain mobile nodes. Unlike home addresses, regional care-of addresses are not permanent and are not used to allow mobile nodes to be reachable. Hence, any mobile node can request any regional care-of address provided that it is not used by another node. This eliminates the requirement of associating a specific address with a specific mobile node, as is the case in the home agent.

When a mobile node establishes a security association with a MAP, it can do so using IKE. In phase 1, the mobile node uses an identifier included in its certificate. However, in phase 2, the mobile node should generate a regional care-of address and use that as its phase 2 identity. This is similar to the security gateway example shown in Chapter 4, Figure 4–10, where two security gateways were acting on behalf of Alice and Bob. Essentially, the mobile node pretends that it is acting as a proxy for another address (the regional care-of address). When the MAP receives the mobile node's phase 2 identity, it checks whether a binding cache entry for this address exists (i.e., another mobile node picked the same regional care-of address). If so, the security association is rejected and the mobile node can try again with a different address.

When a security association is successfully established between the mobile node and the MAP, all local binding updates and acknowledgments can be secured (using AH or ESP) with that security association. Note that in this scheme, we assume that the MAP trusts the mobile node to not bomb another node by using a fake on-link care-of address. That is, the MAP does not test the mobile node's on-link care-of address to see if it is really owned by the mobile node. This action is based on another assumption: the relationship between the mobile node and the MAP is similar to that between the mobile node and its home agent.

## 7.3 Combining Fast Handovers and HMIPv6

In this chapter, we saw how enhancing Mobile IPv6 handover performance can be done in different ways, each addressing different aspects of the problem. Fast handovers allow the mobile node to anticipate its movements and prepare for traffic forwarding before moving to a new AR, while HMIPv6 reduces the amount of signaling due to handovers while maintaining optimal routes to the mobile node. Another feature provided by HMIPv6 is the symmetry of the MAP's location with respect to ARs in its domain. That is, a MAP can be placed in a way that allows it to have approximately the same delay to each AR in its domain. The benefit of this feature can be understood if we reconsider Figure 7–9. A network operator can place the MAP at the aggregate point of paths B and C (note that we don't assume that the MAP is one hop away from the ARs). Therefore, when the MAP receives a binding update indicating that the mobile node has moved to AR2, it can immediately change the tunnel exit point to the new on-link care-of address. Hence, packets will arrive in order at the new link. Figure 7–13 shows an example of how the combination of fast handovers and HMIPv6 can take place.

In this example, we continue to allow the anticipation using the router solicitation for proxy and proxy router advertisement messages between the mobile node and the current AR (as discussed earlier, on connection-oriented links, the solicitation message is not needed). We assume that the mobile node will generate a random interface identifier for its anticipated on-link care-of address; the interface identifier can be assumed valid on the new link (optimistic DAD). Following the reception of the proxy router advertisement, the mobile node configures its new on-link care-of address and sends a local binding update to the MAP (note that we eliminate the sending of F-BU to the current AR).

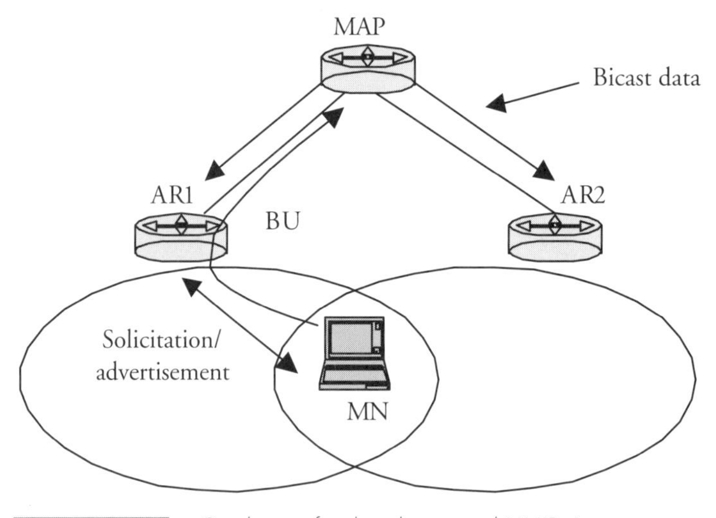


Figure 7–13 *Combining fast handovers and HMIPv6.*

The local binding update can be extended to request bicasting of packets to both the current on-link care-of address and the anticipated one for a short period of time. When the MAP accepts the binding update, it bicasts new packets addressed to the mobile node's regional care-of address to its new (anticipated) on-link care-of address as well as to the old one.

A very important question here is, what are the impacts on anticipation time? Previously, we assumed that when fast handovers are used, the anticipation time, $A$, is

$$A \geq 3T_{mn\text{-}ar}$$

or (if the link layer does not require the solicitation message),

$$A \geq 2T_{mn\text{-}ar}$$

When we combine HMIPv6 and fast handovers, the anticipation time can be represented by the following inequality:

$$A \geq 2\,T_{mn\text{-}ar} + T_{mn\text{-}map}$$

where $T_{mn\text{-}map}$ is the time it takes to send the binding update to the MAP (see Figure 7–14):

$$T_{mn\text{-}map} = T_{mn\text{-}ar} + T_{ar\text{-}map}$$

Hence,

$$A \geq 3T_{mn\text{-}ar} + T_{ar\text{-}map}$$

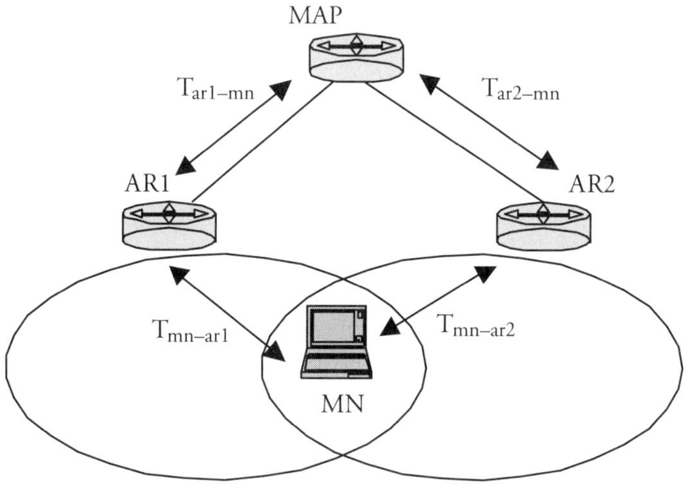


**Figure 7–14** *Delays between mobile nodes, ARs, and the MAP.*

In wireless networks, the largest delays are typically experienced over the wireless link between the mobile node and the AR (i.e., $T_{mn\text{-}ar} >> T_{ar\text{-}map}$). This is due to the latency caused by the coding required by wireless air interfaces, which is always higher than that required by wired networks where bit error rates are very small. Therefore, in current cellular networks, $T_{ar\text{-}map}$ can be ignored. This leads to the conclusion that anticipation time will not be significantly impacted when HMIPv6 and fast handovers are combined as shown above. That is, $A \geq 3T_{mn\text{-}ar}$.

The above inequality represents the amount of time required by the mobile node to be able to successfully anticipate movement and update the MAP before moving. However, it does not tell us anything about packet losses. In order to understand how to eliminate packet loss due to the mobile node's movement, we define new parameters and a new inequality, but first let us consider the events that take place during a handover, as shown in Figure 7–15.

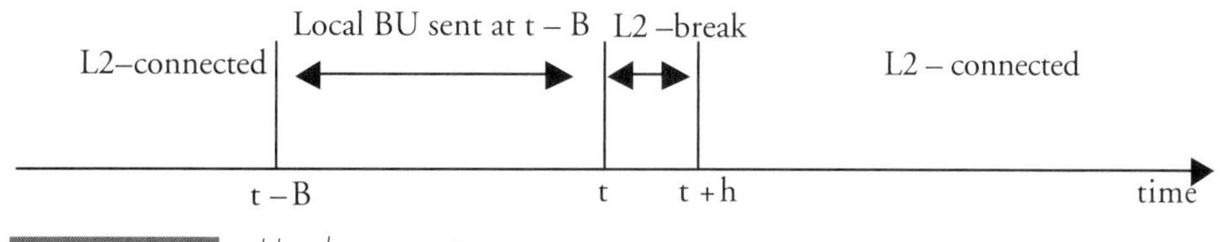


**Figure 7–15** *Handover events.*

Where

> $B$: The time period starting when the mobile node sends a local binding update and ending when the mobile node disconnects from the current link.
>
> $t$: The point in time at which the mobile node disconnects from the old link.
>
> $h$: The time it takes the mobile node to connect to the new link after disconnecting from the old one.

The handover signaling takes place during the anticipation period. Between $t$ and $(t + h)$, the mobile node is unreachable. At time $t + h$, the mobile node connects to the new link and is ready to receive packets. In addition, we define the following parameters:

> $T_{mn\text{-}map\text{-}1}$: The delay experienced by packets sent from the mobile node to the MAP while it is connected to the current AR (AR1).
>
> $T_{mn\text{-}map\text{-}2}$: The delay experienced by packets sent from the mobile node to the MAP while it is connected to the new AR (AR2).
>
> $T_{mn\text{-}ar1}$: The delay experienced by packets sent from the mobile node to the current AR (AR1).
>
> $T_{ar1\text{-}ar2\text{-}mn}$: The delay experienced by packets sent between the current AR (AR1) and the new AR (AR2), then to the mobile node at its new location (on AR2's link).
>
> $p$: The sum of the processing time in the mobile node and the MAP for the local binding update generation and verification respectively.

Delays experienced by packets over any transmission path depend on the packet size, queuing delays in routers, and the network loading. However, in a wireless network, these delays are usually much lower than those experienced over the radio link. Hence, in our calculations, we assume that the variation in delays within the routed network will not have a significant bearing on the outcome of our calculations. Note that we also assume that the delays involved in sending packets from the mobile node to the current AR or the MAP are the same as those experienced by packets traveling in the opposite direction. This condition is true in most cases. However, on some wireless links (cellular), the uplink (packets originating at the mobile node to the network) can be slower than the downlink (packets arriving at the mobile node). Since the downlink bit rate is sometimes higher than the uplink bit rate, this restriction results in conservative estimates for $B$. That is, in cases where the uplink is slower, $B$ will be larger than necessary, which will not cause any packet losses. This becomes obvious when we show how $B$ is calculated.

We can use the mobility scheme discussed earlier in this section to calculate the anticipation time required to reduce the packet losses to zero when the mobile node performs a handover between AR1 and AR2 in Figure 7–13. To achieve this, two requirements must be met:

1. Before moving, the mobile node needs to stay connected to the current AR long enough to ensure that its local binding update was received at the MAP and that all packets sent by the MAP prior to receiving the local binding update have arrived at the mobile node. The following inequality represents this requirement on $B$:

   $$B \geq 2\ T_{mn\text{-}map\text{-}1} + p$$

   Thus, the minimum value for $B$ is $(2\ T_{mn\text{-}map\text{-}1} + p)$. This time period allows the mobile node to send a local binding update to the MAP and receive the first packet that was forwarded from the MAP before the local binding update was received from the mobile node. Assuming

   $$p << T_{mn\text{-}map\text{-}1},$$

   we get

   $$B \geq 2\ T_{mn\text{-}map\text{-}1}.$$

   We show the implications of $B = 2\ T_{mn\text{-}map\text{-}1}$ below. For now we assume the following condition:

   $$B > 2\ T_{mn\text{-}map\text{-}1} \qquad (1)$$

   Now we can see that if the delays between the mobile node and the MAP are asymmetric (i.e., the uplink slower than the downlink), the estimate for $B$ will be larger than necessary. That is, the time it will take the MAP to receive the binding update ($T_{mn\text{-}map\text{-}1}$) will be larger than the time needed for the mobile node to receive the binding acknowledgment and other packets on the downlink (say $T_{map\text{-}mn\text{-}down}$). In other words, since $T_{map\text{-}mn\text{-}down} < T_{mn\text{-}map\text{-}1}$, it follows that $2\ T_{mn\text{-}map\text{-}1} > T_{mn\text{-}map\text{-}1} + T_{map\text{-}mn\text{-}down}$. This means that the estimate for $B$ is conservative and will not cause any additional packet losses. However, it may cause packet duplication when bicasting is used.

2. The mobile node must be present on the new link before any packets arriving at AR2 are lost. The following inequality represents this requirement on $T_{mn\text{-}map\text{-}2}$:

   $$T_{mn\text{-}map\text{-}2} - T_{mn\text{-}map\text{-}1} \geq h \qquad (2)$$

   This means that the delays on the new link need to be long enough to ensure that the mobile node will arrive there before the first "new" (i.e., not received on the old link) packet is lost. This statement can be said differently:

> $h$ must be less than or equal to the difference between $T_{mn\text{-}map\text{-}2}$ and $T_{mn\text{-}map\text{-}1}$. However, we cannot make this requirement on $h$, as it depends on the link layer being used and the radio conditions at the time of the handover. Hence, it is easier to control $T_{mn\text{-}map\text{-}2}$ based on some knowledge of the link layer used. In other words, packets destined to the mobile node's new on-link care-of address can be delayed slightly to meet the inequality (2). This delay can be done in two ways: The MAP can add an artificial delay on the first packet, or the new AR can delay this packet until the mobile node arrives at its link. Either the MAP or AR would then need to buffer packets only for the duration of $h$.
>
> If $B$ were equal to 2 $T_{mn\text{-}map\text{-}1}$, then the inequality in (2) would need to be modified as follows: $T_{mn\text{-}map\text{-}2} \geq h$. That is, since $B$ is exactly the same as the delay between the mobile node and the MAP through AR1, the mobile node would disconnect from AR1's link immediately after receiving the last packet sent from the MAP before processing the local binding update. Hence, the packets sent from the MAP to the new link would only be delayed by a period equal to $h$, such that the mobile node could receive the first packet sent from the MAP after processing the local binding update.

If fast handovers were used without HMIPv6, ensuring packet loss would imply the same requirements as above. However, the inequalities will change. The following inequality represents the requirements on $B$ (compare with inequality 1):

$$B > 2\ T_{mn\text{-}ar1} \tag{3}$$

Again, this is the time required to ensure that the mobile node's binding update (F-BU) was received and processed (we've already ignored the processing time $p$) and that the mobile node has received the last packet sent from the AR before the binding update it received from the mobile node was processed.

To ensure that the mobile node arrives at the new link before the first new packet is lost, the handover must meet the following requirement (compare with inequality 2):

$$T_{ar1\text{-}ar2\text{-}mn} - T_{mn\text{-}ar1} \geq h \tag{4}$$

We already know that in a wireless link $T_{mn\text{-}ar1} \approx T_{mn\text{-}map\text{-}1}$ because the delays on the wireless link represent the main part of the delays within the local network. Hence, using fast handovers alone will not provide a significant advantage in terms of reducing the anticipation time or the time needed to ensure that all old packets are received. On the other hand, when the two mechanisms are combined, we avoid the packet reordering problems described earlier in addition to reducing the number of binding updates and return routability signaling to correspondent nodes and the home agent. Finally,

HMIPv6 allows us to control the delay between the MAP and AR based on the operator's network design. This eliminates the dependency on $T_{ar1\text{-}ar2\text{-}mn}$ (see inequality 4), which is more difficult to control because the two ARs need not be sharing a physical length and may be several hops away from each other. Hence, the distance (delays) between ARs may not always be as deterministic as when using HMIPv6.

## 7.4 Flow Movement in Mobile IPv6

In this chapter, we presented handover mechanisms that are triggered by link-layer degradation, (i.e., a link-layer handover that causes a Mobile IPv6 handover). We also saw that fast handovers rely on anticipating movement and discovering the new AR before connecting to it. Anticipation is used because the majority of deployed radio links rely on a break-before-make handover; hence, the mobile node cannot expect to be connected to two different links simultaneously.

A multihomed host can have more than one interface active at the same time. Such host can attach to more than one link simultaneously and configure different care-of addresses on each interface. One interface's care-of address cannot be used as a source address when sending packets from another interface because ingress filtering will drop the packet if the two interfaces are on different links. These links are likely to vary in their characteristics. For instance, a host might have a cellular interface attached to a cellular network and another WLAN interface (e.g., IEEE802.11b) or a wired Ethernet interface (e.g., IEEE802.3). In this case, a mobile node might wish to use different interfaces for different types of applications. For instance, a multihomed mobile node may wish to have a Voice over IP (VoIP) connection with a correspondent node while downloading a large file from the same correspondent node. Since the mobile node has different interfaces connected to different technologies, each with different characteristics (e.g., bandwidth, QoS, error resilience, reliability, and cost), the mobile node should be able to choose the appropriate interface for each connection. This will work with Mobile IPv6 if each connection is made to a different correspondent node. If a mobile node has more than one connection with the same correspondent node and wishes to use a different care-of address (corresponding to a particular interface), it will not be able to do so because the binding update binds the mobile node's home address to **one** care-of address. Why should the mobile node split connections on different interfaces? Because different applications have different requirements that can be better satisfied by certain link layers. There is no point in using a 56-Kbps link to download 10 MB of data when the mobile node is also connected to a 10-Mbps link! On the other hand, the slow link may be better suited for a voice application that requires a certain level of QoS not supported by the 10-Mbps link. In order to make this distinction, the mobile node needs to achieve the following:

- Provide a flexible, source care-of address–selection algorithm based on the application's (and possibly other factors discussed below) requirements. Here we refer to the source address appearing on the wire (not the home address) because this address corresponds to the interface used to send packets.
- The mobile node must inform the correspondent node that it should use certain care-of addresses for certain connections. That is, more than one care-of address should be communicated to the correspondent node. This is not supported by Mobile IPv6 today.

Figure 7–16 shows an example of the types of policies that can be included in the mobile node's source address–selection implementation (note that the source address in this discussion is the one seen in the source address field in the outgoing packet, not the one shown to applications, which is the home address included in the home address option). In Figure 7–16, we can see that different types of information are communicated to the source address–selection algorithm. Applications can have preferences based on, for example, bandwidth and QoS requirements. The link layer can also provide information about its characteristics (e.g., bandwidth, QoS, BER, average RTT, and reliability). Other types of information may also be available (e.g., the cost associated with the use of a particular link). All this information is given to a *decision point*. The decision point receives several inputs from different sources, including the user, applications, and lower layers. Each input is weighed to prioritize some requirements over others; for instance, the user's preference can be the dominant factor in making the decision. An implementation of the decision point can consider a large number of inputs and become quite complex, but it can also be as simple as

*App_x use if*

*App_y use if1*

*Default if*

where *App* is the name of the application or the port number used, and *if* specifies the interface that should be used; *Default* specifies the interface that should be used in case a specific rule for a particular application was not found. Based on the algorithm used in the decision point, it can provide the appropriate interface for each connection.

The above discussion illustrates the need for allowing for *flow mobility* in addition to Mobile IPv6's host mobility. A flow can be defined as one or more connections that are identified by a flow identifier. A single connection

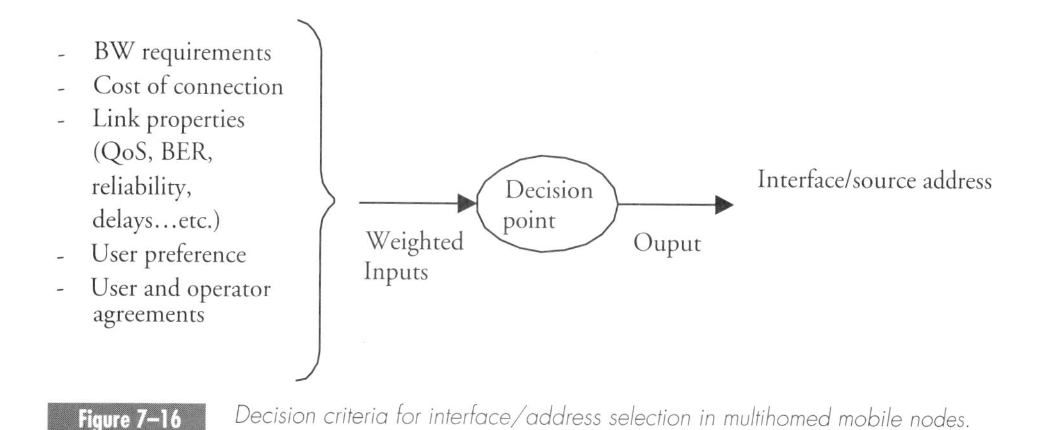


**Figure 7–16** *Decision criteria for interface/address selection in multihomed mobile nodes.*

is identified by the source and destination IP addresses, transport protocol number, and the source and destination port numbers. In IPv6, the Flow label field in the IPv6 header can also be used to aggregate more than one connection when combined with the source and destination addresses.

So far, we have discussed the factors involved in selecting the right care-of address to a multihomed mobile node to use different interfaces for different connections. How does the mobile node move these flows? In other words, how can it inform the correspondent node about the right care-of address for each flow? Flow movement can be achieved by extending the binding update to include new options that allow mobile nodes to indicate to correspondent nodes (or the home agent or MAP) to direct certain flows to corresponding care-of addresses. The receiver of the binding update must store this information in its binding cache. Whenever a packet is being sent (or tunneled, in the case of the home agent and MAP) to the mobile node, the sender checks its binding cache as usual; the flow information can be checked to see which care-of address should be used in the destination field (or the tunnel exit point in the case of the MAP or home agent). An example of this option is shown in Figure 7–17.

The *flow movement* option aims to uniquely identify a flow by specifying the flow label value and the source address used by the correspondent node. Essentially, this option requests that all traffic addressed to the mobile node, which contains the source address and the flow label included in this option, be forwarded to the mobile node's care-of address included in the source address of this IPv6 packet.

The *source address* field in the option is the correspondent node's source address that appears in the IPv6 header when it sends a packet to the mobile node. If the correspondent node were also a mobile node, the source address

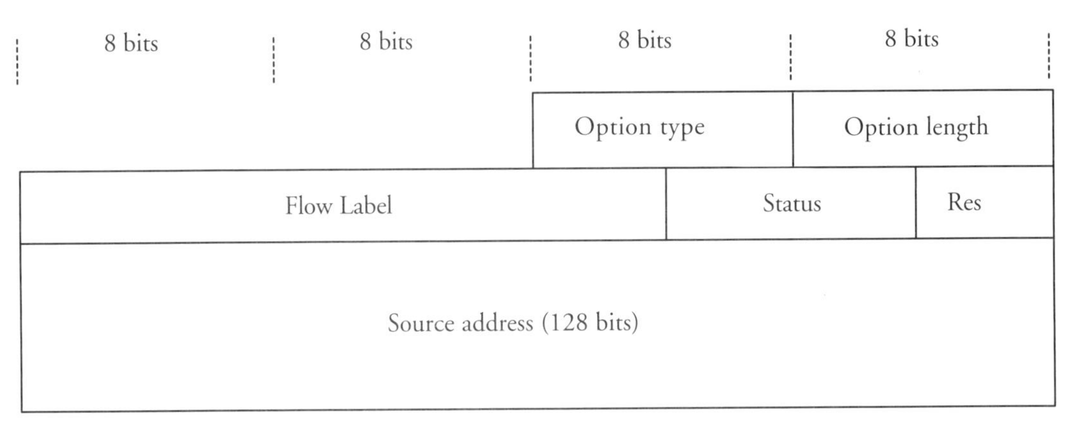


Figure 7–17 Flow movement option.

used should be its care-of address. It is important that the correspondent node's care-of address is used, because the correspondent node's home address will not appear in the header. Hence, if a MAP or a home agent were to classify packets based on the flow label field and the correspondent node's home address (included in the home address option), they would need to search through the daisy chain of IPv6 extension headers, which could impact the performance of the router.[2]

Since this option is carried in the binding update, which must be authenticated using the return routability procedure, the mobile node must ensure that the care-of address included in this binding update has been verified by the COTI/COT message exchange. This can be done by exchanging the HOTI/HOT message to obtain a home keygen token, then exchanging different sets of COTI/COT messages, one for each possible care-of address. This allows the mobile node to use different care-of addresses when communicating with a correspondent node. Clearly, the mobile node will also need to authenticate each flow movement option with the corresponding care-of keygen token to prove its ownership of the care-of address. This requires a separate authorization option for each flow movement option. The flow movement option allows the correspondent node to store more than one care-of address for the mobile node, which is not possible with a normal binding update.

[2] The use of the flow label was standardized a few years after IPv6 standards were produced. Hence, it is likely that some nodes will not support the flow label for some time. For this reason, [7] specifies another option that can use source and destination port numbers and the transport protocol number to classify flows.

## 7.5 Summary

The aim of this chapter was to build on the knowledge gained earlier to show how Mobile IPv6 handovers can be improved. Improving the performance of Mobile IPv6 handovers is important for all types of applications.

We showed how the movement detection delays can be reduced or eliminated using fast handovers. The anticipation mechanism was described in detail, followed by an analysis of the difficulties introduced by anticipation. Two issues were discussed: the uncertainty about the handover and ping-pong movement. We saw how the use of bicasting can mitigate these problems.

Another aspect of improving the handover performance is reducing the number of signals sent during the handover. This is one of the main motives behind HMIPv6. We discussed HMIPv6 in detail and saw that it reduces the number of binding updates sent (due to movement) to one message. In addition to signaling reduction, HMIPv6 allows for a certain degree of location privacy while maintaining an optimal route to the mobile node. Furthermore, it allows mobile nodes to communicate with correspondent nodes using an optimal route, without sending any binding updates. This is useful for several types of applications that a mobile node might start and terminate while moving within a MAP domain.

The combination of fast handovers and HMIPv6 provides maximal benefit to the handover performance. Such combination was discussed in this chapter. We also showed the requirements that need to be met in order to avoid packet losses.

Finally, flow movement was discussed in relation to multihomed mobile nodes. Mobile nodes can use flow movement extensions to optimize their use of the different types of links available to them. With the proliferation of various types of wireless technologies, flow movement could be a useful feature for multihomed mobile nodes.

In Chapter 8, we discuss some of the current and future work in the area of mobility.

## Further Reading

[1] ElMalki, K., and H. Soliman, "Simultaneous Bindings for Mobile IPv6 Fast Handoffs," draft-elmalki-mobileip-bicasting-v6-05, work in progress, October 2003.

[2] ElMalki, K., and H. Soliman, "Local Mobility Management and Fast Handoffs in IPv6." In proceedings of the International Workshop on Services and Applications in the Wireless Public Infrastructure, 2001.

[3] Kempf, J. (Ed.), "Requirements for Layer 2 Protocols to Support Optimized Handover for IP Mobility," draft-manyfolks-l2-mobilereq-00.txt, January 2002.

[4] Koodli, R. (Ed.), "Fast Handovers for Mobile IPv6," draft-ietf-mipshop-fast-mipv6-00, work in progress, March 2003.

[5] Moore, N., "Optimistic Duplicate Address Detection," draft-moore-ipv6-optimistic-dad-02, work in progress, February 2003.

[6] Soliman, H., C. Castelluccia, K. ElMalki, and L. Bellier, "Hierarchical Mobile IPv6 Mobility Management (HMIPv6)," draft-ietf-mipshop-hmipv6-00, work in progress, October 2003.

[7] Soliman, H., K. ElMalki, and C. Castelluccia, "Flow Movement in MIPv6," draft-soliman-mobileip-flow-move-04, work in progress, June 2003.

[8] Williams, C., (Ed.), "Localized Mobility Management Requirements," draft-ietf-mipshop-lmm-requirements-00, work in progress, March 2003.

E I G H T

# Current and Future Work on IPv6 Mobility

In the previous chapters, we saw how Mobile IPv6 can be used to manage a host's mobility as it moves within the Internet. We also saw how Mobile IPv6 can be extended to minimize the service interruption introduced by mobility. Our focus in the previous chapters was on how Mobile IPv6 manages to reroute packets from one point in the topology to another. However, when systems are deployed, other aspects related to or enabling mobility need to be considered in order for network operators to offer a complete service to the user. While these issues may not directly enhance mobility management, they enable operators and vendors to build systems that will benefit the end user.

In this chapter, we list some of the open areas that are needed in order to fill in the gap between designing a mobility management protocol and deploying a system that uses such protocol. In addition, we discuss some of the known unsolved mobility problems.

## 8.1 AAA as an Enabler for Mobility

Mobility allows devices to attach to the Internet through arbitrary access networks within the topology. These access networks are managed by operators who want to ensure that Bad Guys do not access their networks; moreover, they usually need to know that users will be able to pay for accessing the Internet. In order to achieve these two main goals, operators must control access to the Internet through their networks. To do this, they need Authentication, Authorization, and Accounting (AAA). Devices (owned by and operating on behalf of their users) must be authenticated; based on the authentication process, a decision is made whether these devices are authorized to gain access. Furthermore, some form of accounting is needed in order to charge users. That is, based on the billing model, an operator might decide to make the user pay based on the amount of information sent through the network or based on the duration of time during which the user had access to the network. Alternatively, an operator may impose a flat monthly charge on the user, which is independent of network use. All these charging models (or a mix of more than one) exist today on the Internet and in telecommunication systems.

The critical step is authentication: How does a network operator authenticate an arbitrary user? The authentication model currently being developed by the IETF is shown in Figure 8–1.

Figure 8–1 shows a functional split for AAA within an operator's network. The model consists of a *front-end AAA server* located on the same link as the mobile node, an *Enforcement Point* (EP) on the same link, a *back-end AAA server* (AAAL) somewhere inside the network, an *AAA broker* and a *home AAA server* (AAAH). We assume that the mobile node has a preconfigured security association with AAAH but no prior relationship with the front-end AAA server or the back-end one. AAAH is not required to be on the mobile node's home link or even in the same administrative domain. In fact, AAAH does not need to be part of another ISP's network (although in many cases it will be). AAAH can be owned by anybody—for example, a credit card company. The aim of AAAH is to act as a trusted third party (by the mobile node and AAAL) that can verify the user's credentials and possibly guarantee that the user will pay for the service (e.g., this can be done if the owners—humans—of AAAL and AAAH agree that the owners of AAAH will pay for the service and bill the owner of the mobile node for it).

In this model, the mobile node requests access to the network by sending a message to a front-end AAA server. The request includes the identity of the mobile node. An example of such identity is the Network Address Identifier (NAI), which has the form *user@example.com*. The front-end AAA server has no prior security association with the mobile node. It relays the message to AAAL, which checks if the mobile node belongs to its domain based on the

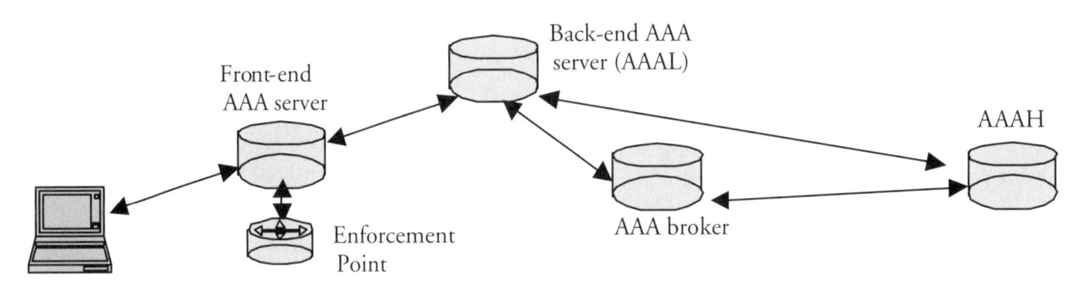


**Figure 8–1** *AAA model.*

mobile node's identifier. If AAAL were not serving *example.com*, it locates AAAH, which serves that domain. This can be done, for example, by sending a DNS request for *AAA.example.com*, which returns the IP address of one or more AAAH servers. The mobile's request for access is then sent to AAAH. AAAH challenges the mobile node (e.g., by sending a nonce) through AAAL. In this scenario, AAAL simply relays the message to the mobile node via the front-end AAA server. The mobile node encrypts the nonce with a key, which is known only by AAAH. The mobile node's reply is forwarded along the chain of trust from the front-end AAA server to the back-end AAA server and through the broker (if AAAL and AAAH do not have a secure relationship) to AAAH. AAAH can authenticate the mobile node. Furthermore, it can generate *session keys* (recall the session keys generated after an IKE exchange) to allow for future communication between the mobile node and the front-end AAA server.

After successfully completing the authentication process, the operator in the visited network authorizes the user (represented by the device) for network access. This can be done by sending a message from the front-end AAA server to the EP, requesting that it allow a particular device's traffic to be forwarded. The EP is a router that sets filters in its ingress interface. Before the device receives the necessary authorization, the EP will not include its address (IP address or MAC address) in its Access Control List (ACL), which lists the addresses allowed to gain access through the router's ingress interface; hence, it will not forward the device's packets. When the EP is informed that the device is authorized to access the network, the EP's ACL will be configured dynamically to allow access for that device. In Figure 8–1, we show the front-end AAA server and the EP as two separate entities; however, physically they can coexist in the same box. This box is called a Network Access Server (NAS).

AAA brokers are included in this architecture to allow it to scale for a large number of AAA servers (local and home servers). If 100,000 AAA servers (belonging to different operators) exist on the Internet, we need $10^{10}$ security associations between the servers to allow mobile nodes associated with any operator to roam into any other operator's network. Therefore, we need AAA brokers that act as trusted (by back-end AAA servers) third parties, and therefore allow for secure communication between different back-end AAA servers. Such communication must be secure because it may contain sensitive data about the users' subscription and expose data encrypted by the user's and AAAH's secret key. Ideally, this data should not be exposed frequently because it may aid Bad Guy in identifying the secret key and later pretending to be a legitimate user.

So why is AAA needed for mobility? In theory, all nodes should be authenticated before being authorized to access the Internet or an operator's network. When nodes are stationary and connected through wired links, network operators may be content with physical security (e.g., the existing telephone network). However, when nodes are connecting to networks through wireless links and are free to move anywhere on earth, this assumption does not hold and AAA infrastructure is needed. Therefore, the AAA concept is always needed, but wireless networks and mobility strengthen the need for AAA infrastructure.

Currently, the Remote Authentication Dial-In User Service (RADIUS) is used on the Internet to allow AAA servers (front and back end) to communicate. The protocol used between the mobile node and the front-end server varies depending on the type of access. However, since this protocol need not be access-dependent (this dependence on the type of access is due to historical reasons and a lack of an access-independent protocol), the PANA Working Group was chartered in IETF to define a new access-independent protocol to be used between the mobile client and the front-end AAA server. In addition, the AAA Working Group is currently standardizing a new-generation AAA protocol called Diameter to replace RADIUS.

## 8.2 Achieving Seamless Mobility

When people talk about seamless IP mobility, they usually mean that the impacts of IP mobility should be confined to the IP layer (i.e., it should be transparent to upper layers). Mobile IPv6 provides applications with a stable (home) address, which allows connections to survive a handover. However, the **impacts** of the handover still affect upper layers when we consider packet losses, delays, or actions that need to be taken after movement. Ongoing research addresses a number of areas that will allow mobility to become as seamless as possible. In this section, we discuss some of these areas and ongoing work.

### 8.2.1 Link-Layer Agnostic Interface to the IP Layer

In Chapter 7, we saw how the movement detection time can be eliminated by anticipating movement. We also saw how the amount of time needed for anticipation can be calculated. However, it is crucial to observe that anticipation relies on "hints," or triggers, from the link layer that inform the IP layer about the status of the layer 2 handover. These triggers need to arrive at the right time to allow the IP layer enough anticipation time to perform its handover with minimal or zero packet losses.

Several link layers are used by hosts and routers on the Internet today. How can we ensure that they will deliver these triggers? An analogy can be made with applications running on IP stacks. Thousands of applications, developed by different vendors, are currently used on the Internet. How do they all manage to run on different IP implementations? The sockets API, a pseudo-standard API implemented by applications running in user space, allows them to communicate with the IP stack (obviously not all applications run on all operating systems; however, many operating systems implement the pseudo-standard API, which allows an application to run on any of them without significant modifications). Hence, if we can introduce a generic API that receives certain events, specific to each link layer, and present them in an abstract form to the IP layer, we can allow anticipation to take place independently of the underlying link layer. For instance, the functions in this API can request that a link-layer driver inform them when the signal-to-noise ratio (SNR) on their link drops by 6 dBm. In this example, 6 dBm will mean that a handover is about to take place; therefore the IP layer needs to be informed.

This task is not as simple as it sounds; it requires knowledge (within the API) of the different link layers and of how much time is sufficient for anticipation to work on each one. Some work has been done to specify the abstract information that will be presented to the IP layer during handovers (the easy part) [4]. More work is needed in order to synchronize the IP layer handover with the link-layer handover. This is likely to require that link layers prepare this information and support a pseudo-standard interface that allows them to send this information to an IP layer that is unaware of the underlying link's characteristics.[1] The same API can be used to provide some "static" information about the link layer (e.g., available bandwidth), whether the link layer supports encryption, and other useful information that allows upper layers to make smart decisions, especially when selecting the source address to be used by each application.

### 8.2.2 Context Transfer

Access routers are likely to have some information related to mobile nodes attached to their links. For instance, if we consider the earlier section on AAA, we can see that an EP (also an AR) must know which mobile nodes are allowed access to the network and include this information in its ACL. A mobile node may also be a member of one or more multicast groups; if so, it must inform the AR in order for it to be able to receive packets addressed to that multicast group. A mobile node may also have particular bandwidth requirements for its applications and must reserve this bandwidth, at least on the wireless interface if not end to end. It is clear that an AR must keep some information about the mobile node. We refer to this information as the mobile node's *context*.

An integral part of achieving seamless mobility is to ensure that the mobile node does not need to inform the new AR about all of its context after movement, as these actions add further delays and are likely to disrupt ongoing connections. It would be very useful if the old AR could transfer this information to the new AR when it knows that the mobile node has moved. This knowledge is gained upon the reception of a binding update requesting that traffic be redirected to a new location. The context transfer must be done in a

---

[1] This work is likely to raise some concerns about layering violations; however, optimizations are likely to require some exchange of information between layers. The best we can do is make sure that this exchange is done in an abstract manner to avoid having IP layer implementations knowing about the details of each link layer to be able to interpret the information given.

secure and timely manner to stop Bad Guy from modifying the mobile node's context while making sure that the context arrives at the new AR before any disruption is experienced by ongoing connections.

It is important to decide on the type of information that will be transferred during a handover. For instance, if some parts of the mobile node's context are changing rapidly, they may not be useful when received at the new AR, because by the time they are transferred, they have already changed. In these rare cases, the mobile node must reestablish these parts of the context at the new AR.

The above discussion illustrates the need for a protocol that transfers the context of the mobile node from the old AR to the new AR. These protocols are already defined for some cellular systems (e.g., UMTS). However, there is currently no generic protocol defined for IP networks. On the bright side, standardization work is ongoing within the Seamoby Working Group in IETF and can be found in [5] and [6].

### 8.2.3 Candidate Access Router Discovery (CARD)

When we discussed fast handovers, it was clear that the current AR needed to have a table that includes the neighboring APs, their MAC addresses, and their corresponding AR (see Chapter 7, Table 7–1). We assumed that this information is manually configured in the AR.

One obvious disadvantage of manual configuration is that it cannot handle dynamic changes in the network. For instance, suppose a mobile node is moving away from AR1 and it discovers a new AP, say APx, which is known by the current AR to be connected to AR2. When the mobile node sends a proxy solicitation asking for the prefix and AR that APx is connected to, AR1 will send it a proxy advertisement for AR2. However, if AR2 is too busy at the time to accept new mobile nodes, AR1 has no way of knowing that. On the other hand, if AR2 were discovered dynamically and part of the discovery process included information about the availability of the link served by AR2, AR1 might recommend another AR or increase the power transmitted to the mobile node (this is possible on CDMA links) to keep it in the same link if no other AR was available.

Again, such discovery protocol does not exist in IP networks; however, work in this area has started. Reference material can be found in [2].

## 8.3 Network Mobility

Mobile IPv6 solves the host mobility problem by allowing hosts to roam within the Internet while keeping ongoing connections and maintaining reachability. However, there are several mobility scenarios that involve a moving network as opposed to a host. For example, a user can be mobile while carrying a number of devices, such as a mobile phone, a computer, and a Personal Digital Assistant (PDA); this configuration is commonly called a Personal Area Network (PAN). Each device is likely to need Internet connectivity at some point in time. Furthermore, in the near future we will likely have new devices, like cameras and CD players, implementing the TCP/IP protocol suite. A device can connect directly to the Internet if it can find wireless coverage corresponding to its capability or indirectly via another router. For instance, a network consisting of a PDA, a CD player, and a mobile phone can connect to the Internet via the mobile phone when cellular coverage exists. In this case, the mobile phone acts as a router, providing connectivity to the rest of the devices. This leads to a definition for mobile networks:

> A network containing one or more devices that connect to the Internet through one or more Mobile Routers (MRs). A mobile router has an ingress interface connected to nodes inside the mobile network, and one or more egress interface connected to the Internet through an access router.

When the mobile router has more than one egress interface or the mobile network contains more than one mobile router connecting to the Internet, the mobile network is *multihomed*. Figure 8–2 shows an example of a multihomed mobile network.

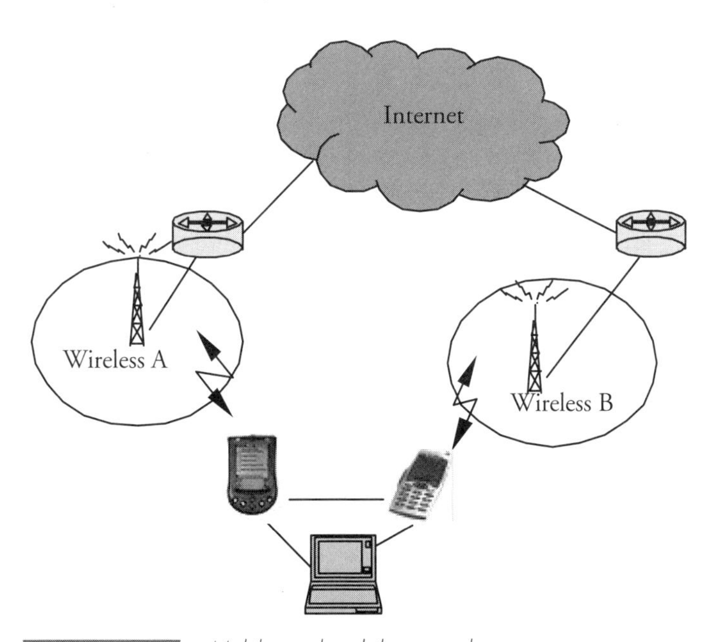


**Figure 8–2** *Multihomed mobile network.*

In Figure 8–2, a mobile network consists of two different routers and a host. Each router is connected to a different access network that runs a different wireless technology. For instance, the mobile network could be located in an area where there is IEEE802.11b and cellular coverage. One of the devices happens to have an IEEE802.11b interface and the other has a cellular interface; hence, they can both act as mobile routers for the devices in the mobile network. Suppose the mobile network moves and the device with IEEE802.11b loses coverage; it will no longer be multihomed and will have only a single interface connecting it to the Internet (the cellular interface). The PDA can certainly connect to the Internet via the other mobile router connected to the cellular interface.

A mobile network can also join another mobile network, forming a *nested mobile network*. Suppose a transportation company decided to offer Internet connectivity to its passengers traveling on trains and buses. To do this, the company puts one or more mobile routers on each train. Suppose passengers get on the train with their own PANs. Each PAN will have a mobile router that attaches to the mobile router on the train. However, the entire train's network will look like a single mobile network connected to the Internet through the train company's router. *Nested mobile network* refers to a mobile network that contains a hierarchy of mobile routers, with a Top-Level Mobile Router (TLMR) connecting devices in the mobile network to the Internet. A nested mobile network can certainly be multihomed; for instance, if one of the passengers on the train decided to offer Internet connectivity to the rest of the passengers through her mobile router (and suffer the costs!), the train will have two mobile routers offering connectivity to the Internet.

To provide mobility for mobile networks in a similar form as that provided to mobile hosts, nodes inside the network need to be reachable. In addition, it should be possible to allow these nodes' connections to survive while the mobile network moves within the Internet. Making the nodes reachable involves providing them with a stable address. When dealing with mobile networks, this means that the mobile router needs a stable prefix to advertise on its ingress interface. When at home, the mobile router forwards packets to nodes inside its network as usual. The mobile router can also act as a home agent to nodes inside the mobile network. When the entire network moves, the mobile router acquires a care-of address on its egress interface (just like a normal mobile node). Hence, it can establish a tunnel to its home agent and inform it that all packets destined to any address derived from the mobile router's prefix (advertised on the mobile router's ingress interface) should be forwarded to the mobile router's care-of address. The mobile router can inform the home agent of its prefix explicitly by extending the binding update message or by simply running a routing protocol inside the tunnel to the home agent. Hence, the mobile router seems like a normal mobile node when viewed by the AR on the visited link. Every node inside the mobile network continues to send and receive packets as if they never left their home network. All sent and received packets are tunneled to and from the home agent respectively. As a result, nodes inside the mobile network continue to be reachable and their connections survive mobility. The details of this scenario are described below and shown in Figure 8–3.

Route optimization becomes more pertinent when considering mobile networks. To illustrate this, let's consider the nested mobile network example discussed earlier and shown in Figure 8–3. The TLMR (e.g., the train's mobile router in our example) is advertising Prefix_X on its ingress interface; this is a stable prefix, which does not change as the train moves. The TLMR has a tunnel established with its home agent (HA_X), requesting that all packets destined to any address derived from Prefix_X be tunneled to the TLMR. The mobile router (e.g., a PAN's mobile router belonging to one of the passengers) is advertising Prefix_Y on its ingress interface. On its egress interface, the PAN's mobile router configures an address based on Prefix_X and uses it as a care-of address when updating its home agent (HA_Y). As a result of this configuration, packets sent from correspondent nodes to a node inside the PAN are sent to HA_Y, which tunnels them to HA_X, which in turn tunnels them to the TLMR. The same path is traveled in the opposite direction for packets leaving the PAN.

From this example, we can see how nesting adds further delays and tunneling overhead to all packets sent and received from the mobile network.

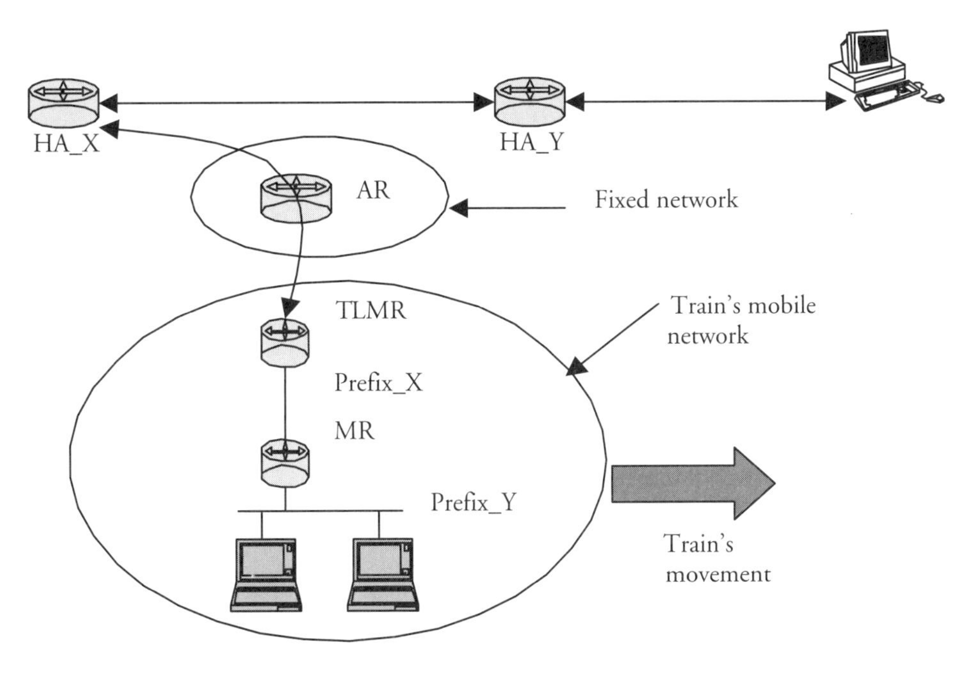


**Figure 8–3** *Mobile networks' need for route optimization.*

Furthermore, nodes inside the network cannot optimize routes with correspondent nodes because they are not aware of their movement. Optimizing routes to the nodes inside the mobile network requires a mechanism that allows these nodes to configure a care-of address that belongs to the topology of the visited network. Currently, there is ongoing work in the nemo (network mobility) Working Group in IETF to define possible solutions for problems associated with network mobility.

## 8.4 Summary

This chapter presented some of the enablers for deploying a mobile communication system based on IP. We showed an overview of AAA systems and explained why they are important for wireless networks in particular.

Achieving seamless mobility adds several components to the mobility management protocol. We discussed the need for a generic API between the IP layer and the link layer. This API allows IP layer mobility management to work independently of the underlying link layer. It is also an enabler for better integration of IP layer and link-layer mobility. Next, we saw that context transfer is an integral component in achieving seamless mobility, as it allows applications to continue to work after movement without reestablishing their needs with the new AR.

Candidate Access Router Discovery is also very helpful when it comes to assisting mobile nodes to anticipate their movement while minimizing manual configuration in ARs.

Finally, we discussed network mobility, which is a relatively new topic in IPv6. We defined *mobile network* and saw various scenarios where mobile networks are likely to exist. We also discussed the importance of route optimization in the context of network mobility.

Chapter 9 presents some of the challenges associated with widely deploying IPv6 on the Internet and how mobility adds a new angle to those issues.

## Further Reading

[1] Calhoun, P., J. Loughney, E. Guttman, G. Zorn, and J. Arkko, "Diameter Base Protocol," draft-ietf-aaa-diameter-17, work in progress, December 2002.

[2] CARD Design team, M. Liebsch, A. Singh, H. Chaskar, and D. Funato, "Candidate Access Router Discovery," draft-ietf-seamoby-card-protocol-01, work in progress, March 2003.

[3] Devarapalli, V., R. Wakikawa, A. Petrescu, and P. Thubert, "Nemo Basic Support Protocol," draft-ietf-nemo-basic-support-01, work in progress, September 2003.

[4] Ernst, T., "Network Mobility Support Requirements," draft-ietf-nemo-requirements-00, work in progress, February 2003.

[5] Kempf, J. (Ed.), "Requirements for Layer 2 Protocols to Support Optimized Handover for IP Mobility," draft-manyfolks-l2-mobilereq-00.txt, January 2002.

[6] Kenward, G. (Ed.), "General Requirements for Context Transfer," draft-ietf-seamoby-ct-reqs-05, work in progress, October 2002.

[7] Loughney, J., M. Nakhjiri, C. Perkins, and R. Koodli, "Context Transfer Protocol," draft-ietf-seamoby-ctp-01, March 2003.

[8] Ohba, Y., S. Das, B. Patil, H. Soliman, and A. Yegin, "Problem Statement and Usage Scenarios for PANA," draft-ietf-pana-usage-scenarios-06, work in progress, April 2003.

# PART FOUR

# IPv6 AND MOBILE IPv6 DEPLOYMENT


- Chapter 9 IPv6 in an IPv4 Internet: Migration and Coexistence
- Chapter 10 A Case Study: IPv6 in 3GPP Networks

NINE

# IPv6 in an IPv4 Internet: Migration and Coexistence

The introduction of IPv6 into the existing IPv4 Internet has often been compared to trying to change engines on a flying jet plane! The process by which IPv6 is introduced into the Internet is a key factor in the success of the technology. IPv4 has been used for a long time and is expected to be used for a very long time to come, perhaps for the lifetime of the Internet. For these reasons, it is important that IPv6 is introduced with minimal (or zero) disturbance to the existing Internet.

When IPv6 becomes widely deployed on the Internet, a number of different communication scenarios between IPv6 and IPv4 nodes will exist. It is important to ensure that such scenarios will work in a manner that is seamless to users and requires minimal effort by operators. Otherwise, deployment will seem cumbersome and will be delayed. Mobility will add a new dimension to the deployment problems and therefore should be carefully considered. While we cannot be certain about the way IPv6 will be introduced and certainly cannot claim a "one size fits all" policy, we can anticipate the most likely ways of introducing IPv6 and try to make these scenarios as easy to manage as possible. For this reason, you will find several mechanisms in this chapter, and more in IETF that address slightly different requirements and deployment cases. These mechanisms form a toolbox from which network operators and vendors can choose to best meet the requirements of their deployment scenarios. Eventually, a few mechanisms are likely to be widely deployed and will cover most of the cases.

Several mechanisms are defined in IETF to allow a smooth introduction of IPv6. In this chapter we present some of these deployment scenarios and the problems associated with each one, and we summarize existing proposals that can address these problems.

## 9.1 How and When Will IPv6 Be Deployed?

The main driver behind IPv6 development is the expected depletion of IPv4 addresses. Its deployment is expected to start where this benefit is most needed. That is, IPv6 is likely to be deployed first in networks where devices are unable to acquire globally unique addresses, which allow them to be reachable.

The deployment of IPv6 is expected to take place in "islands" (or access networks). Access networks are those networks located between Internet users in corporate networks, homes, universities, cellular networks, and Internet backbones containing transit networks that enable users within one domain to communicate with other users in a different domain. This is due to the difficulty associated with upgrading the entire Internet backbone to support IPv6. It is easier to update routers in the access network and allow for the greater benefits of IPv6 (offering globally unique addresses to end hosts) than to upgrade all routers in the Internet backbone network.

There is a lot of speculation about the exact time of IPv6 deployment. This question (perhaps to your disappointment) is not addressed in this book. Ultimately, deployment depends on several factors, most of which are not technical. We make the case for the technical motives for IPv6 deployment and leave the speculation to you and other technology observers. However, it suffices to say that all current major router and host vendors do support IPv6 in their products.[1] Furthermore, commercial networks are running IPv6 in Japan, Australia, and Italy (to mention a few), and prelaunch trial networks have been set up in the Asia-Pacific region and Europe.

## 9.2 What Are the Problems?

The crux of the problem associated with IPv6 deployment is incompatibility—the inability of IPv4 nodes (hosts or routers) to understand the IPv6 header—as illustrated in Figure 9–1. A router with only an IPv4 implementation will not be able to forward packets containing an IPv6 header. Similarly, hosts with only IPv4 implementation cannot communicate with others using IPv6 because they will not be able to parse the IPv6 header.

Figure 9–1 illustrates different scenarios in which the incompatibility between IPv6 nodes and IPv4 nodes will prevent IPv6 nodes from communicating. Three different cases are presented. The first shows an incompatibility between an end host and the first-hop router. The host could be an IPv6 host, while the router may understand only IPv4. In this case, the host cannot acquire

[1] Consult with the IPv6 forum (www.ipv6forum.com) Web page to get the latest on IPv6 support in different product platforms (directly or through links to vendors' pages).

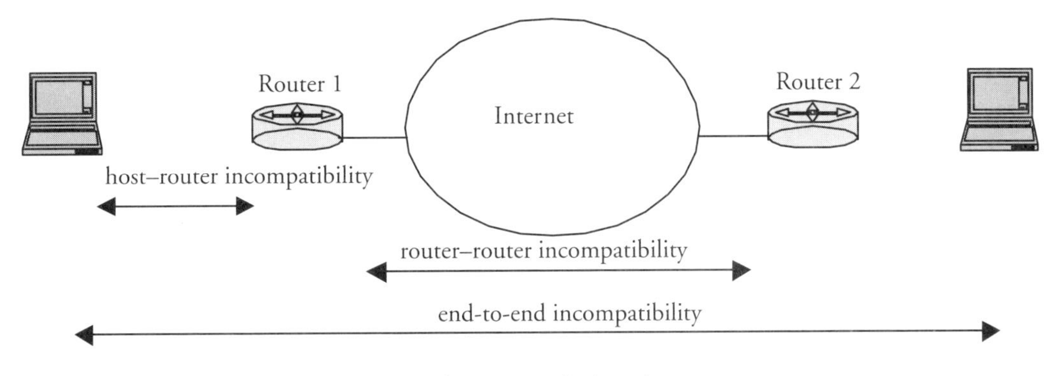


Figure 9–1 *Possible incompatibilities when IPv6 is deployed.*

an address that is valid beyond its link and consequently cannot communicate with other hosts outside the link. The second scenario shows an incompatibility between two routers, where one of these routers does not understand IPv6 packets and therefore cannot forward them to their destination. Finally, if both scenarios are solved, there is still a strong possibility that tens of millions of existing hosts will not be upgraded to support IPv6 for a long time (or they might be upgraded but without IPv6 enabled because it is not needed on the network). The nodes that enable only IPv4 will clearly not be able to communicate with other nodes using IPv6.

Host–router and router–router incompatibility can be overcome by tunneling, as described in the following section. On the other hand, end-to-end incompatibility requires the use of translators, as shown in Section 9.4.

**Note** Incompatibility could also occur within a host when old applications are run over an IPv6 protocol stack. This is because IPv6 introduces a different socket API toward applications in user space. Hence, existing applications must be upgraded to use the new API. In this chapter, we do not focus on this particular case but on the difficulties seen "on the wire" when IPv6 is deployed. If you are interested in a solution that allows existing applications to run over IPv6 stacks, refer to [4].

## 9.3 Tunneling

We saw in Chapter 2 how tunneling IPv6 packets in IPv6 headers can be useful for different reasons and that it is used by home agents to forward packets to mobile nodes in a foreign network. IPv6 packets can also be tunneled in IPv4 headers. Why? To avoid problems associated with intermediate routers being unable to forward packets containing IPv6 headers. When an IPv6 packet is encapsulated into an IPv4 header, only the outer (IPv4) header is visible to intermediate routers, hence, eliminating possible problems resulting from IPv4 routers being unable to forward IPv6 packets. Tunneling can be useful when two IPv6 hosts wish to communicate over an IPv4 Internet, which is likely to be a very common scenario in the short and medium term, as ISPs are not expected to upgrade every router in their potentially large networks to be able to introduce IPv6.

Tunneling can be done on an end-to-end basis, with the tunnel entry point being the host originating packets and the tunnel exit point being the ultimate destination, or between intermediate routers (e.g., between Router1 and Router2 in Figure 9–1), or between a host and a router. Independently of where the tunnel entry and exit points are, both points must have an IPv4 address (for the source and destination fields in the IPv4 header) as well as an IPv6 address (for the source and destination fields in the original packet if tunneling is done from the originating host to the final destination). Hence, in all cases, the tunnel entry and exit points must enable both IPv4 and IPv6 protocol stacks.

A very likely tunneling scenario is expected to take place between two different IPv6-enabled islands that communicate over an IPv4 Internet (Router1 and Router2 in Figure 9–1). The crucial issue is how the tunnel entry point discovers the tunnel exit point's IPv4 address. For instance, how does Router1 know that an IPv6 packet, addressed to host_A (which is located in the network behind Router2), should be tunneled to Router2's IPv4 address? One obvious way to achieve this involves manual configuration of routers. While manual configuration is simple and does work in many cases, it has the disadvantage of being less robust when faced with changes in the topology (e.g., due to network failures).

In the following sections, we present different methods that allow for such tunneling to take place using both manual and dynamic tunnel configuration.

## 9.3.1 Configured Tunnels

Configured tunnels are simple to use; an operator trying to allow communication between different IPv6 islands over an IPv4 network must configure each router on the edge of each island to tunnel traffic to the IPv4 address for the tunnel exit point corresponding to the IPv6 destination address in the original IPv6 header. To do this, an operator must know the corresponding IPv4 address for a particular IPv6 address, or prefix. For instance, suppose that Router2 (Figure 9–1) had an IPv4 address of `1.2.3.4` and it served a network that was assigned the IPv6 prefix `3ffe:200::/32`. If an operator running Router1 wants to make sure that IPv6 nodes inside its network can communicate with other IPv6 nodes in Router2's network, she can configure Router1 with a static route that causes Router1 to tunnel any packet addressed to any address starting with

`3ffe:200::/32` to the IPv4 address `1.2.3.4`. Hence, when Router1 receives a packet addressed to an IPv6 node with an address `3ffe:200:8:1::1`, it will encapsulate it in an IPv4 header containing its own IPv4 address as a source address and the destination address `1.2.3.4`. This configuration will enable communication between IPv6 nodes in the Router1 → Router2 direction. To enable communication in the reverse direction, Router2 will need to have a configuration similar to Router1's.

This mechanism is quite simple and effective. At the same time, it is clear that this mechanism can have scalability problems when considering a very large number of sites that need to communicate to each other. That is, if 1,000 sites need to communicate with each other, each site must be manually configured with the other 999 tunnel exit points for each IPv6 prefix. If the tunnel exit point (border router, e.g., Router1) fails, manual intervention is needed to modify the tunnel exit point (in case more than one is available) in the other 999 sites. Clearly, a dynamic tunneling mechanism will allow tunneling to be more robust to network failures. The following sections show different ways of achieving dynamic tunneling between IPv6 nodes.

### 9.3.2 6-to-4 Tunneling

The motivation behind 6-to-4 tunneling is to allow different IPv6 domains (islands within the Internet) to communicate over an existing IPv4 backbone without the need for explicit tunnel configuration in the border router of those domains. Hence, 6-to-4 was designed to allow border routers to automatically discover the IPv4 address of the tunnel end point (border router) for a particular domain. As a result, a packet arriving at the border router of one domain will be encapsulated within an IPv4 header containing the border router's IPv4 address in the source address field and the IPv4 address of the border router of the receiving domain in the destination address. The IPv4 address of the tunnel exit point is known because it is included in the IPv6 address of the destination; 6-to-4 defines and reserves a special prefix for sites that wish to be reachable via a 6-to-4 address. A 6-to-4 prefix is represented as follows:

```
2002:v4_addr::/48
```

where `2002::/16` is a reserved prefix to be used by sites wishing to be reachable via 6-to-4, and `v4_addr` is the 32-bit IPv4 address of the border router in the 6-to-4 site. The rest of the address (16 bits for the subnet identifier and the 64-bit interface identifier used by hosts) is used in the normal manner, like any other IPv6 unicast address. Including the tunnel exit point's IPv4 address in the IPv6 address allows border routers to dynamically discover the tunnel exit point without keeping any state, and consequently no manual configuration will be required.

The tunneling router is called a 6-to-4 router because it implements the 6-to-4 tunnel interface. That is, it understands the 6-to-4 address format and can extract the tunnel exit point from the address.

Figure 9–2 illustrates how 6-to-4 can be used to allow communication between different IPv6 hosts within three different IPv6 sites over an IPv4 Internet. Sites A and C are called 6-to-4 sites because they are allocated addresses that start with the `2002::/16` prefix. Hence, RouterA has an IPv4 address of `9.254.253.252` (represented by the Hexadecimal `09fe:fdfc:` in its 6-to-4 address) and RouterC has an IPv4 address of `192.1.2.3` (represented by the Hexadecimal `c001:0203` in its 6-to-4 address). In addition to the 6-to-4 prefix, Site A has a native IPv6 unicast prefix (`3ffe:100::/32`) and a direct connection to a native IPv6 backbone. Site B also has a native IPv6 unicast prefix and is connected to the IPv6 Internet through another router, a *6-to-4 relay* router. The 6-to-4 relay router is a normal IPv6 router that also understands 6-to-4. That is, it is able to understands the 6-to-4 address format and can encapsulate IPv6 traffic destined to 6-to-4 addresses. Like all 6-to-4 routers, a 6-to-4 relay is a dual stacked router with both IPv4 and IPv6 enabled.

Site B might be connected to the 6-to-4 relay through a native IPv6 link or a configured tunnel (e.g., a default route pointing to the 6-to-4 relay). If Site B is connected to the 6-to-4 relay router through a native IPv6 link, and a routing

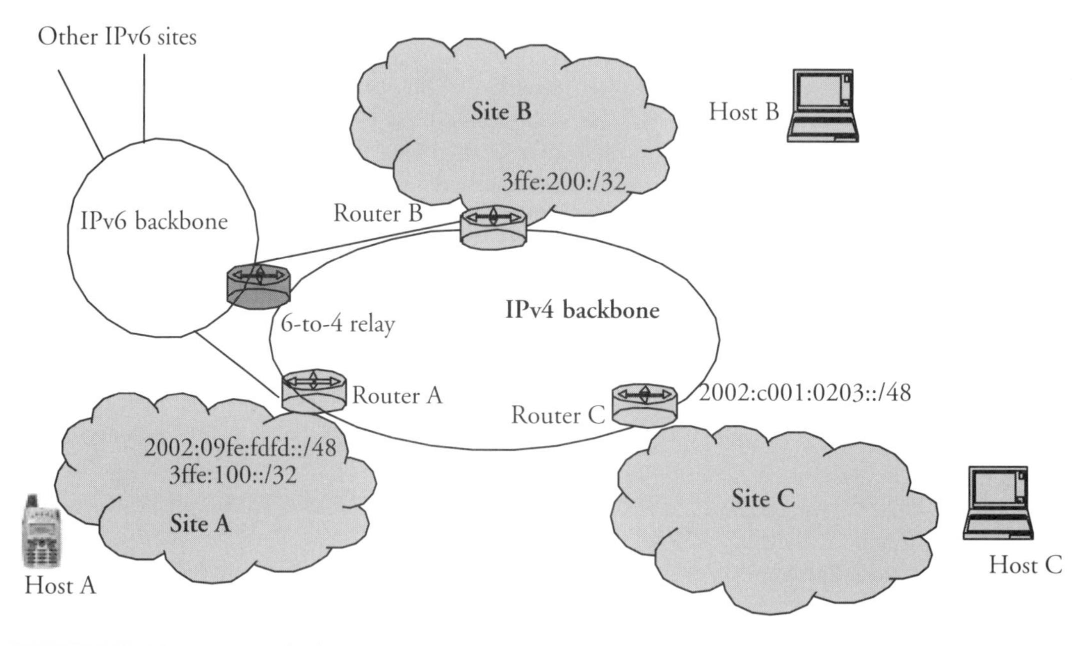


**Figure 9–2** *A 6-to-4 deployment scenario.*

protocol runs between RouterB and the relay router, the relay router will inject a route for 2002::/16 toward RouterB so that it can receive all the 6-to-4 traffic and encapsulate it to the right IPv4 address (i.e., in case RouterB is not a 6-to-4 router). To understand how 6-to-4 works in detail, we consider communication cases between Hosts A, B, and C.

First we consider a case where Host C is trying to communicate with Host A. Host C queries the DNS to get Host A's IPv6 address. Two addresses are returned by the DNS: one 6-to-4 address (e.g., 2002:09fe:fdfc:10::20:30) and another IPv6 unicast address (e.g., 3ffe:100:8:10::20:30). Since Host C has only one 6-to-4 address (e.g., 2002:c001:0203:1::10:50), it chooses to communicate with Host A's 6-to-4 address. Packets sent from Host C to Host A arrive at RouterC, a 6-to-4 router, which extracts RouterA's IPv4 address and uses it as a destination address in the IPv4 header encapsulating the original IPv6 packet. When Host A replies with an IPv6 packet addressed to Host C, RouterA, being a 6-to-4 router, extracts the IPv4 address of RouterC and uses it as a destination address in the IPv4 header encapsulating the original IPv6 packet. In this manner, all nodes in sites A and C can communicate over the IPv4 backbone, separating the two sites using 6-to-4.

Now let's consider a scenario where Host C is trying to communicate with Host B. A DNS reply returns Host B's IPv6 address (e.g., 3ffe:200:1:2::10:20) to Host C. When Host C sends an IPv6 packet to Host B, it arrives at RouterC to be forwarded to the next hop. Since 3ffe:200:1:2::10:20 is not a 6-to-4 address, and Site C has no native IPv6 connectivity to Site B, RouterC must find some way of locating a router that can forward packets to Site B. The 6-to-4 relay router can do that; however, RouterC needs to know its IPv4 address. RFC 3068 reserves an IPv4 subnet prefix and an IPv4 anycast address within that subnet for 6-to-4 relays. The reserved IPv4 anycast address for 6-to-4 relays is 192.88.99.1, and the reserved prefix is 192.88.99.0/24. The rest of the prefix is not actually used. The reason for reserving an entire prefix is that transit network operators usually stop their routers from accepting a specific route to a single address, as it defeats the point of route aggregation and exhausts the scarce routing table resources. This address can be used by RouterC to send packets to any relay on the Internet, which by definition will have connectivity to a native IPv6 backbone. In this manner, RouterC can encapsulate Host C's packets to the 6-to-4 relay router, which in turn decapsulates packets and forwards them to RouterB.

IPv4 anycast was proposed to allow for multiple 6-to-4 relay routers on the Internet while being able to route packets to the closest possible relay. Furthermore, an anycast address will help in overcoming router failures, as normal routing will reroute packets to the nearest possible router. However, operators are likely to need unicast addresses as well to identify each 6-to-4 relay router for troubleshooting purposes. RFC 3068 discusses possible ways of including such unicast addresses inside BGP advertisements, but there is no current concrete method that allows border routers to discover the unicast address of each 6-to-4 relay router.

Clearly, the concept of 6-to-4 relays has several open issues. For instance, there is no guarantee that the relay is close to either the originating or the destination sites. Site C might be topologically close to Site B, while the relay router is very far from both sites, resulting in significant delays. Another concern is related to the security hazards associated with open relays (i.e., relays accepting packets from anyone). Any node can send encapsulated packets to the relay with a spoofed (a victim's) IPv6 source address; the relay will forward those packets to their intended IPv6 destination, causing it to reply to the victim. This is similar to the reflection attacks discussed in Chapter 5. Currently, there are no standard solutions that prevent these reflection attacks through 6-to-4 relays. The concept of 6-to-4 relay routers is often met with skepticism from an operational point of view due to the security concerns associated with it. Without 6-to-4 relay routers, this mechanism cannot allow bidirectional communication unless both hosts have a 6-to-4 address (i.e., communication between Sites B and C would not be possible). This is a significant disadvantage of 6-to-4 tunneling.

### 9.3.3 Routing Protocols-Based Tunnel End Point Discovery

Another way to discover the IPv4 tunnel end point is to propagate it inside the routing protocol messages. Using this mechanism, a border router can advertise the reachability of a particular IPv6 prefix through an IPv4 address. The IPv4 address is embedded inside an IPv4-mapped IPv6 address to maintain the format of an IPv6 address. This mechanism was designed to work for OSPF, IS-IS, and BGP. Hence, it can be used inside a site or between autonomous systems. The information regarding the tunnel exit point discovery is encapsulated inside options specific to the routing protocol being used. Routers that do not understand these options will ignore them but will still propagate them through the routing infrastructure. Therefore, this mechanism is suited to IPv6 sites separated by large IPv4 networks or to connecting small IPv6 islands within a single site that has not been fully upgraded to support IPv6.

### 9.3.4 Intrasite Automatic Addressing Protocol (ISATAP)

Allowing native IPv6 connectivity within a site involves upgrading all routers, hosts, and some essential servers within that site. Upgrading all these nodes would involve a large amount of work for the site administrator, especially when considering large sites (e.g., large companies or large ISPs providing connectivity to home networks). The aim of ISATAP is to allow IPv6 connectivity to hosts within a site without having to enable IPv6 in all routers within such site. Hence, ISATAP allows site administrators to gradually upgrade their sites. We use the term *site* loosely to mean a network consisting of multiple links (e.g., a company or a university campus), and it should not be confused with IPv6 sites, which are bound by site border routers and use a site-local prefix.

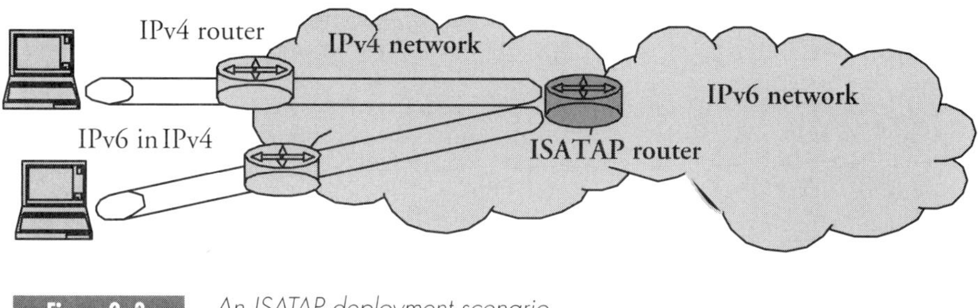


**Figure 9–3** *An ISATAP deployment scenario.*

The scenario addressed by ISATAP is one where a dual stacked host wishes to use IPv6 but is unable to do so because the first-hop router is not IPv6-enabled. Normally, this would mean that the host cannot configure an IPv6 address with a scope larger than link-local; furthermore, packets will not be forwarded by its default router, which is not IPv6-enabled. ISATAP solves this problem by allowing administrators to enable one or more ISATAP routers within a site, as shown in Figure 9–3.

ISATAP routers act as tunnel exit points for hosts. A host wishing to get IPv6 connectivity sends a router solicitation to an ISATAP router. ISATAP routers reply with a router advertisement unicast to the host. Both router solicitations and router advertisements are tunneled in IPv4 headers. A solicitation is sent with the host's IPv4 address as a source address in the IPv4 field and the ISATAP router's IPv4 address in the destination field. To do this, the host must somehow discover the ISATAP router's IPv4 address. This is done by sending a DNS request for *isatap.domainname* (e.g., *isatap.example.com* to find an ISATAP router in domain *example.com*). The DNS reply contains one or more IPv4 addresses that can be used as destination addresses for the IPv4 header encapsulating future router solicitations. The host stores these addresses in a list called the Potential Router List (PRL). However, the actual router solicitation message needs IPv6 source and destination addresses. These addresses are called *ISATAP addresses* and are formed as shown in Figure 9–4.

The prefix part of the address is the normal IPv6 prefix of the ISATAP link. The first 32 bits of the interface identifier are set to `0000:5EFE`, and the

| 64 bits | 32 bits | 32 bits |
|---|---|---|
| Prefix | 0000:5EFE | V4ADDR |

**Figure 9–4** *ISATAP addresses.*

last 32 bits contain the node's IPv4 address. When a host sends a Router Solicitation, the prefix contains the well-known link-local prefix; the interface identifier can be formed for the host and the router addresses using the host's IPv4 address, and for the router's address in the PRL, respectively. When a host receives a router advertisement, it can configure a global ISATAP address based on the announced prefix and the same IPv4 address it used earlier to form its link-local address. All packets sent by the host are tunneled in IPv4 to the ISATAP router, which decapsulates and forwards them as usual. Hence, IPv6 uses the IPv4 network as a link layer (this is no different than any other form of tunneling).

When using ISATAP, hosts cannot send router solicitations to the well-known all-routers multicast address, because such an address will need to have an equivalent in the IPv4 destination address. In other words, if router solicitations and advertisement were to be sent to IPv6 multicast addresses, there would be a need for IPv4 multicast routing within the site to deliver these packets to multiple routers (just like a multicast address is mapped to an Ethernet multicast address on a shared link). Since IPv4 multicast routing is not widely deployed, this assumption could not be made and was seen as an obstacle to the use of ISATAP. For this reason, ISATAP treats the IPv4 network as a Non-Broadcast Multiple Access (NBMA) link layer—that is, a point-to-point link layer with no multicast capability. This assumption has led ISATAP to rely on DNS records to discover the ISATAP routers' IPv4 (and consequently IPv6) addresses as opposed to the standard multicast mechanisms introduced in Neighbor Discovery.

## 9.4 Translation

So far, we have presented mechanisms that aim at "hiding" the IPv6 header from IPv4-only routers using tunneling in order to be able to forward IPv6 packets through an IPv4-only network. These mechanisms are aimed at allowing IPv6 nodes (or dual stacked nodes) to communicate with each other independently of the underlying network layer. However, we have not yet addressed the end-to-end incompatibility problems that arise when an IPv6 node wishes to communicate with an IPv4 node. There are different cases that can lead to this scenario. To understand them, we need to distinguish between *IPv4-capable* and *IPv4-enabled* nodes. An IPv4-capable node is a node that contains an IPv4 protocol stack. For this stack to be enabled, the node must be assigned one or more IPv4 addresses. An IPv4-enabled node is clearly IPv4-capable, but the reverse statement is not true. The same definitions apply to IPv6 nodes. At any point in time, a given node can be only IPv4-enabled (IPv4-only node) or only IPv6-enabled (IPv6-only node) or IPv4 **and** IPv6 enabled.

Based on the above definitions, we can easily anticipate that end-to-end incompatibility will take place if:

- An IPv6-only node attempts to communicate with an IPv4-only node.
- An IPv6-enabled, IPv4-capable node attempts to communicate with an IPv4-only node.
- An IPv6-enabled, IPv4-capable node attempts to communicate with an IPv4-enabled, IPv6-capable node.

In other words, if all nodes on the Internet were both IPv4-enabled and IPv6-enabled there would be no end-to-end incompatibility problem. However, this is clearly not the case for the hundreds of millions of IPv4 nodes that are currently deployed on the Internet. One way of avoiding this problem could be achieved by mandating that all IPv6-enabled nodes must also be IPv4-enabled. After all, these nodes are new and there is no reason for removing IPv4 protocol support from existing operating systems that are updated with IPv6 support. However, enabling IPv4 in all hosts would require all networks to have enough IPv4 addresses for all of their hosts. Since the main reason for introducing IPv6 is the lack of IPv4 addresses, this assumption may not be realistic in many cases, especially with the expected proliferation of cellular devices on the Internet. In the next chapter, we also see that IPv6 is exclusively used by UMTS networks for some applications. For these reasons and others, end-to-end incompatibility poses a problem that requires some attention and solutions.

The only way to solve this problem is to introduce a function that resides somewhere between the two communicating nodes that transforms one header format to another. This will allow IPv6 nodes to send and receive only IPv6 packets, while allowing IPv4 nodes to send and receive IPv4 packets. This translating function comes at a price, discussed in the following sections. In general, translation introduces problems and should be avoided if possible (e.g., by allowing IPv6-enabled nodes to be IPv4-enabled and using IPv4 to communicate with IPv4 nodes); however, there are circumstances in which translation is necessary, as discussed earlier.

### 9.4.1 Stateless IP ICMP Translator (SIIT)

SIIT provides a stateless algorithm that can be used to translate IPv6 headers into IPv4 headers, and vice versa. As the name suggests, the translator also works for ICMP (i.e., it translates ICMPv6 into ICMPv4, and vice versa). The translation function is implemented into dual stack routers, which are called *SIIT routers*. Figure 9–5 shows the operation of SIIT routers.

SIIT assumes that an SIIT router will be allocated a pool of IPv4 addresses. When an IPv6 host, implementing SIIT (hereafter called the *SIIT host*), attempts to communicate with an IPv4 host (it knows so because the DNS will only return an IPv4 address for the correspondent), it requests a temporary IPv4 address. Using this address, it forms an IPv4-translated IPv6 address (see section 2.5.6). Using the correspondent's IPv4 address, it forms an IPv4-mapped IPv6

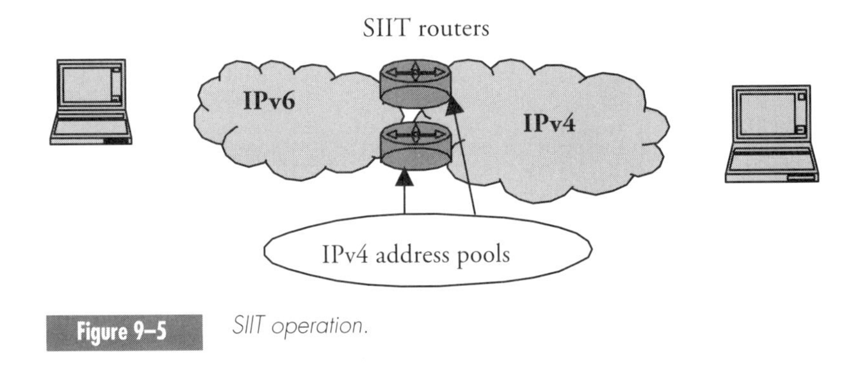


Figure 9–5 SIIT operation.

address. From this point onward, the application on the IPv6 host will see that it is communicating with an IPv4 host whose address is included in the IPv6 address format. Packets sent from the IPv6 host will include the IPv4-mapped IPv6 address in the destination address field and the IPv4-translated IPv6 address in the source address field.

All SIIT routers inject routes (using an IGP, e.g., OSPF) within the IPv6 cloud to announce their ability to forward packets destined to IPv4-mapped IPv6 addresses. Hence, packets sent from IPv6 hosts to IPv4-mapped addresses will be received by one of the SIIT routers. Note that any SIIT router can receive these packets, since no state is required to be able to translate the packet. When packets are received at the SIIT router, it will know that translation is required, since these packets are destined to mapped addresses and originate from translated addresses. The IPv4 header is formed using information from the IPv6 header. For instance, the source and destination addresses are taken from the last 32 bits in the translated and mapped addresses respectively. The rest of the fields in the IPv4 header are either well-known values (e.g., the IP version field) or have corresponding fields in the IPv6 header (e.g., the hop limit in the IPv6 header has the same range and meaning as the Time to Live, TTL, field in the IPv4 header). The reverse is not true; some IPv6 values will not be translatable to a corresponding IPv4 field. For instance, there is no field in the IPv4 header or options that can carry the value and meaning of the IPv6 flow label field. Hence, some of the IPv6 functions will be lost during the translation. This is not a surprise, as IPv6 introduces more functions than IPv4.

After the packet is translated, packets are sent on the IPv4 network to the IPv4 correspondent. Note that we chose to separate the IPv4 and IPv6 clouds in Figure 9–5. This is clearly a logical separation. IPv4 and IPv6 can be running on the same link. However, in this case, packets will still be routed through an SIIT router. The SIIT router itself could be located on the same link as the hosts or further up in the routing hierarchy.

SIIT routers will also inject routes within the IPv4 cloud to announce their ability to forward packets destined to any of the IPv4 addresses included in their IPv4 address pools. For instance, an SIIT router allocated an address pool represented by the IPv4 prefix `192.1/16` will inject a route for that subnet, causing packets addressed to any IPv4 address derived from that subnet to be forwarded to the SIIT router. Upon receiving any of these packets, SIIT routers will know that they should be translated (because they include destination addresses from the address pool) and perform the SIIT algorithm to construct the IPv6 header. At this point, the SIIT router must forward the translated packet to the IPv6 host. If the SIIT router is on the same link as the IPv6 host, this will not pose a problem, and normal neighbor solicitation and neighbor advertisement messages can be exchanged to determine the link-layer address corresponding to the IPv4-translated IPv6 destination address. However, if the SIIT router is more than one hop away from the host, it must determine the host's unicast IPv6 address in order to encapsulate the packets to that host. Encapsulation is needed to keep the translation process transparent to upper layers (since applications see only the translated and mapped addresses and TCP uses those addresses to calculate its checksum). RFC 2765 does not explain how the SIIT router will know the IPv6 unicast address of the IPv6 host, nor does it explain how the IPv6 host can be assigned with an IPv4 address that can be used to construct the translated address. This problem can be solved in two ways:

- Always keep SIIT routers on the same link as the IPv6 host; that is, enable the SIIT function in all default routers and assign each link a pool of IPv4 addresses that is given through DHCPv6.
- Find a new protocol that can inform all SIIT routers of the IPv6 unicast address of the IPv6 host that is assigned a particular IPv4 address for this purpose (e.g., the address can be assigned by a DHCPv6 server and the server can update SIIT routers). This approach was proposed in [6].

The advantage of the second option is that it allows a more efficient utilization of IPv4 addresses (because they are shared by more nodes). However, the first option is a lot simpler to implement.

The main benefit of SIIT is that it is stateless; hence, an SIIT router's failure will be no different than any other router's failure, as another SIIT router can perform the same task. Furthermore, since the translation is performed on the IP layer, it is independent of the application being used. Some applications send IP addresses in their payload. While this behavior violates the layering model, it is common in some popular applications like FTP. When these applications communicate through SIIT (they know that because they use translated and mapped addresses and run on an IPv6-only host), they should include only the IPv4 address part of the translated address when sending the host's IP address to the correspondent. In this manner, the receiver (an IPv4 host) will not know that it is communicating with an IPv6 host.

Translation, by definition, transforms the content of the IP header; therefore, end-to-end integrity is not maintained. As a result, Authentication Header cannot be used to protect communication between the IPv6 and IPv4 hosts when communicating through a translator, since Authentication Header covers the IP header as well as the payload. On the other hand, transport mode ESP can be use as it is independent of the content of the IP header. However, when negotiating an SA, the IKE implementation in the IPv6 host should send its IPv4 address (not the IPv4-translated IPv6 address) to its IPv4 peer. IKE is just another example of an application that communicates IP addresses in its payload. Furthermore, if ESP were used, the IPv6 host cannot include a destination options header after the ESP header as the destination options header cannot be translated (since it's not visible to the SIIT router) and will not be understood by the final IPv4 destination.

### 9.4.2 Network Address Translator and Protocol Translator (NAT-PT)

NAT-PT uses the SIIT algorithm and a centralized address allocation function to allow communication between IPv6 and IPv4 hosts. The aim of NAT-PT is to provide an address and protocol (IP and ICMP) translator that is transparent to end hosts. Unlike SIIT hosts, IPv6 hosts whose traffic is translated by NAT-PT are completely unaware of the translators' existence.

The NAT-PT functions are implemented on routers that are typically located near the site border. When an IPv6-only host attempts to communicate with an IPv4-only host, the local DNS returns an IPv6 address (representing the IPv4 host), hence deceiving the IPv6 host into thinking that it is communicating with an IPv6 host. The returned AAAA record will contain an IPv6 address that has the real IPv4 address embedded inside the least significant 32 bits. The returned address can be represented as follows:

```
PREFIX:V4ADDR
```

where `PREFIX` is a 96-bit prefix assigned to the NAT-PT, and `V4ADDR` is the IPv4 address of the correspondent. `PREFIX` has no special format and looks exactly like any other IPv6 unicast prefix. Hence, an IPv6 host will start communicating with that address. Packets sent from the IPv6 host are routed to the NAT-PT router per normal routing. NAT-PT translates the header using the SIIT algorithm with one exception: the IPv4 source address used to represent the IPv6 host will be one of the addresses in the IPv4 pool assigned to the NAT-PT. NAT-PT also keeps state to be able to translate the inbound IPv4 packets addressed to the same IPv6 host. For instance, if NAT-PT uses `192.1.2.3` to represent the IPv6 address `3ffe:200:8:1::1`, it must store this information so that it can translate incoming packets (from the IPv4 correspondent) addressed to `192.1.2.3`. NAT-PT also needs to inspect the packets to see if the application in use embeds any IP addresses, and if it does, these addresses also need to be modified accordingly. Hence, NAT-PT must include Application Level Gateways (ALGs) for those applications that embed IP addresses in their payloads. The ALGs understand the application-level protocols (e.g., FTP) and modify the payload to make sure that the receiving host understands the addresses included in the application's payload.

Another flavor of NAT-PT includes port translation. When this function is included, the translator is called NAPT-PT. Port translation involves multiplexing a number of connections between different hosts on the same IPv4 address. For instance, hosts with IPv6 addresses `3ffe:200:8:1::1`, `3ffe:200:8:1::2`, `3ffe:200:8:1::3`, and `3ffe:200:8:1::4` can all be represented by the same IPv4 address `192.1.2.3`, which is assigned to the NAT-PT box if different source port numbers are used. Effectively, NAPT-PT terminates the connection and reestablishes it with the IPv4 host using different port numbers from those chosen by the IPv6 host.

NAT-PT is a centralized stateful translator, whereas SIIT is a distributed stateless translator. Because NAT-PT is transparent to communicating hosts, it includes ALGs, which parse the application payload and modify it. This implies that end-to-end security is not achievable when NAT-PT is used. Hosts do not get explicit notifications when NAT-PT is used and hence cannot know which applications can be secured and which cannot. Furthermore, since NAT-PT is stateful, a failure in the router will result in losing all ongoing connections unless another router can take over (i.e., a protocol is used to transfer state to a standby NAT-PT router). On the other hand, NAT-PT is useful for sites that do not have enough IPv4 addresses to allocate them to IPv6 hosts on demand (e.g., if SIIT were used). Port translation allows NAT-PT to use IPv4 addresses in an efficient manner. For this reason, NAPT-PT is expected to be the most commonly deployed form of translator, especially in regions where IPv4 addresses are considered a scarce resource.

## 9.5 Other Deployment Scenarios and Considerations

We have discussed the essential problems that need to be solved in order to introduce IPv6 gradually within sites and to allow communication between different sites and hosts. Other optimizations may also be needed in some cases to simplify network management. While the lack of these optimizations does not present a show stopper for IPv6 deployment, their presence may encourage some network administrators to adopt IPv6 earlier. Some claim that operators can save network management costs and resources if they run an IPv6-only network, which can make life easier for the network manager. Some proposals were also made to allow dual stacked mobile nodes to use one mobility management protocol: Mobile IPv6. The following sections discuss these ideas and summarize some of the existing solutions in this area.

### 9.5.1 IPv6-Only Networks

The Dual Stack Transition Mechanism (DSTM) was designed to allow operators to deploy an IPv6-only routing infrastructure. Hosts, on the other hand, can be IPv4- and IPv6-enabled to allow them to communicate with any host on the Internet without the need for translation. DSTM operation is quite simple and is, in many ways, the reverse of the ISATAP mechanism described earlier. DSTM introduces a dual stack router that acts as a Tunnel End Point (TEP), connected on the edge between the IPv6 and an IPv4 network. The TEP's scope is typically a small to medium site (e.g., enterprise network). DSTM assumes that the network between the hosts and TEP is running IPv6 only.

When hosts attempt to communicate with other IPv6 hosts, communication works using normal IPv6 standards and routing. When a DSTM host attempts to communicate with an IPv4 host, it requests an IPv4 address from a server (e.g., DHCPv6). The host provides this address to applications, which start normal communication using the IPv4 protocol stack. Outgoing packets are tunneled in IPv6 headers to the TEP, which decapsulates the received packets and forwards them on the IPv4 network. Incoming packets (from the IPv4 host) are tunneled back to the IPv6 host that is assigned the IPv4 address in the destination field of the IPv4 header. DSTM assumes that the TEP will cache the IPv4 address corresponding to each IPv6 address when an outgoing packet is received. Hence, the TEP will be able to find the correct IPv6 address upon receiving an IPv4 packet. This assumption can be valid in a trusting network (e.g., a closed enterprise network where the administrator trusts that employees will not launch DoS attacks on other employees). However, in a public network, Bad Guy could easily change the TEP's cache by sending a fake packet containing an IPv4 address (in the source address field of the encapsulated header) that is assigned to another device, causing the TEP to change the association between that IPv4 address and the corresponding IPv6 address and possibly connection hijacking.

### 9.5.2 Mobility Considerations

A given mobile node, by definition, can be located in different places on the Internet. Different networks may support IPv6, IPv4, or both. Mobile nodes may also communicate with any other node on the Internet. The transition mechanisms shown earlier will help mobile nodes communicate with other nodes on the Internet, just as they help stationary nodes in doing so. Tunneling techniques are completely transparent to mobility management signaling and the content of the IPv6 header, including the mobility header, the home address option, and the routing header type 2. Translation, on the other hand, introduces different challenges. When translation is used, an IPv6 mobile node clearly cannot communicate with an IPv4 node using an optimal route, which involves a return routability test and signaling specific to Mobile IPv6, which depends on end-to-end security. None of the Mobile IPv6 route optimization mechanisms can work if packets are intercepted and modified by a translator or ALG. This is not a serious problem, as Mobile IPv4 does not provide a route optimization function; hence, when IPv6 nodes communicate with IPv4 nodes in a nonoptimal manner, this is no worse than any IPv4 mobile node communicating with another IPv4 correspondent node.

Dual stacked mobile nodes with both IPv4 and IPv6 enabled could use Mobile IPv6 only to manage their mobility and make them reachable even while communicating with IPv4 nodes using the IPv4 protocol stack. One could imagine some extensions to Mobile IPv6 (see [7] for example) that allow mobile nodes to inform a dual stack home agent to forward its IPv4 and IPv6 traffic, as shown in Figure 9–6.

A mobile node could be assigned two home addresses, an IPv6 and an IPv4 one. A single binding update (including a new option) could request the home agent to act as a proxy for the mobile node's IPv4 and IPv6 home addresses and tunnel all packets to the mobile node's current care-of address. The care-of address can be an IPv4 or an IPv6 address. If it is an IPv4 address, it can be represented as an IPv4-mapped IPv6 address inside the Mobile IPv6 mobility header. In this case, the home agent could tunnel the mobile node's

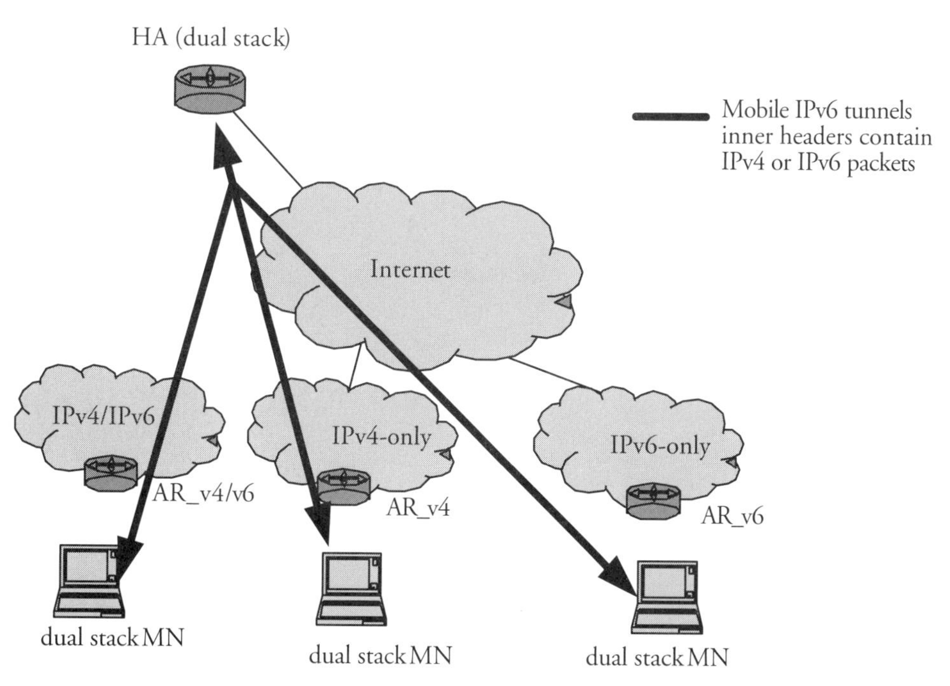


**Figure 9–6** *Dual stack mobile nodes using Mobile IPv6 only.*

IPv6 traffic in an IPv4 header, where the destination address field contains the mobile node's IPv4 care-of address. Effectively, the home agent can act as a remote tunnel entry and exit point (for outbound and inbound traffic, respectively), which guarantees both IPv4 and IPv6 connectivity to the mobile node independently of its current location and the IP version support within the visited network. That is, a dual stack (IPv4 and IPv6 enabled) mobile node using the mechanism presented above can be reachable and maintain ongoing sessions independently of the IP version supported by its default router, or correspondent node.

Furthermore, this mechanism allows mobile nodes to reuse the handover optimizations presented in Chapter 7 for its IPv4 and IPv6 addresses. This is useful because it allows mobile nodes to use Mobile IPv6 optimizations independently of the address used by its applications. The alternative would be to use both Mobile IPv4 and Mobile IPv6 handover optimizations, which is clearly inefficient.

## 9.6 Summary

In this chapter, we discussed the main problems associated with introducing IPv6 in the current IPv4 Internet. We discussed two different types of problems: incompatibility on the connectivity level (i.e., between routers) and end-to-end incompatibility (between hosts).

Tunneling of IPv6 in IPv4 was introduced as a solution to incompatibility between intermediate routers. Different mechanisms were discussed, each addressing different scenarios. We also showed that it is possible to tunnel IPv4 in IPv6 (e.g., DSTM).

Translation of IPv6 packets into IPv4 packets, and vice versa, was discussed in the context of enabling IPv6 hosts to communicate with IPv4 hosts. Both SIIT and NAT-PT were discussed.

Finally, we saw that mobile nodes can benefit from enabling both IPv4 and IPv6 home addresses and using Mobile IPv6 to be reachable on both addresses. This is useful for mobile nodes, since they can roam within different networks, which provides different levels of support for IPv6. Hence, minimizing the reliance on visited networks allows for a more robust operation of IPv6.

In Chapter 10, we use the information gained throughout this book and apply it to the third Generation of cellular networks to see how IPv6 and Mobile IPv6 can be used. In addition, we see how the transition mechanisms shown in this chapter can be applied in that context.

# Further Reading

[1] Carpenter, B., and K. Moore, "Connection of IPv6 Domains via IPv4 Clouds," RFC 3056, February 2001.

[2] Gilligan, R., and E. Nordmark, "Transition Mechanisms for IPv6 Hosts and Routers," RFC 2893, August 2000.

[3] Huitema, C., "An Anycast Prefix for 6to4 Relay Routers," RFC 3068, June 2001.

[4] Lee, S., M-K. Shin, Y-J. Kim, E. Nordmark, and A. Durand. "Dual Stack Hosts Using 'Bump-in-the-API' (BIA)," RFC 3338, October 2002.

[5] Nordmark, E., "Stateless IP/ICMP Translation Algorithm (SIIT)," RFC 2765, February 2000.

[6] Soliman, H., and E. Nordmark, "Extensions to SIIT and DSTM for Enhanced Routing of Inbound Traffic," draft-ietf-ngtrans-siit-dstm-02, work in progress, February 2001.

[7] Soliman, H., and G. Tsirtsis, "Dual Stack Mobile IPv6," draft-soliman-v4v6-mipv4-00, work in progress, August 2003.

[8] Templin, F., T. Gleeson, M. Talwar, and D. Thaler. "Intra-Site Automatic Tunnel Addressing Protocol (ISATAP)," draft-ietf-ngtrans-isatap-13, work in progress, March 2003.

[9] Tsirtsis, G., and P. Srisuresh, "Network Address Translation: Protocol Translation (NAT-PT)," RFC 2766, February 2000.

TEN

# A Case Study: IPv6 in 3GPP Networks

In the early 1990s, while GSM deployment was starting in many parts of the world, research had begun to look into advanced radio technologies for the third generation of cellular networks. Generations of cellular networks are tied to the family of radio technology being used; the first generation used analog radio technology (e.g., NMT and AMPS), and the second used digital radio technology (e.g., GSM, TDMA). For third Generation cellular networks, one of the aims of the cellular vendors was to globally unify the radio technologies used. A single radio technology allowed vendors to manufacture the same devices in larger numbers, resulting in lower costs for operators and ultimately for users. More importantly, having a single cellular technology allowed global roaming using the same device. This could not be done with earlier generations, as some countries use unique cellular technologies (e.g., the Japanese PDC system), and allocated radio spectrum in various countries is not always compatible.

In order to achieve this grand vision of total global roaming without the need for changing devices, the International Telecommunications Union *(www.itu.org)* developed a set of recommendations on how this should be achieved, with the main aim of supporting services requiring significantly higher bandwidth than in current cellular networks. These are captured in the International Mobile Telephone-2000 (IMT-2000) standards and form the definition of what we call the third generation of cellular networks, or more commonly, 3G.

The IMT-2000 standard consists of a family of systems, each family representing a migration from current cellular network standards toward 3G. The two main families of systems are the Universal Mobile Telephony System (UMTS), based on the evolution of the widely deployed GSM standard (European), and cdma2000, based on the evolution of cdmaOne (IS-95, North American) standard.

During 1998, intense debate was ongoing in European Telecommunications Standards Institute (ETSI) and The Association of Radio Industries and Businesses (ARIB) standards body of Japan to settle on a radio technology suitable to meet the IMT-2000 requirements. In July of that year, two proposals out of the five that were submitted were selected to be part of the family of IMT-2000 systems: Wideband Code Division Multiple Access (WCDMA) in frequency division duplex (FDD) mode and WCDMA in time division duplex (TDD) mode.

The 3$^{rd}$ Generation Partnership Project (3GPP) was formed in December 1998 from members of ETSI, ARIB, the Telecommunication Technology Committee (TTC) in Japan, the Telecommunication Technology Association (TTA) in Korea, the China Wireless Telecommunication Standard group (CWTS), which joined later in 1999, and the American National Standards Institute T1P1 group (ANSI T1P1).

The original aim of 3GPP was to produce consistent, open, global standards for 3G based on the evolution of GSM cellular networks and to use the new WCDMA radio access technology.

This new 3G cellular network was to be called Universal Mobile Telephony System to emphasize that WCDMA is really the radio (air) interface technology that is being used and not the actual cellular network in general. The scope was subsequently amended to include the maintenance and development of the GSM communication Technical Specifications and Technical Reports, including evolved radio access technologies such as General Packet Radio Service (GPRS) and Enhanced Data Rates for GSM Evolution (EDGE).

For the evolution to the cdma2000 3G system, the 3$^{rd}$ Generation Partnership Project 2 (3GPP2) was formed in January 1999 from members of ARIB (Japan), CWTS (China), Telecommunications Industry Association (TIA, North America), and TTA (Korea).

In this chapter, we present an overview of the system proposed by 3GPP (i.e., based on WCDMA). We chose a 3GPP system (as opposed to 3GPP2) because it defines the use of IPv6 and has in fact mandated its use for some applications, which made it more relevant to this book! Following the system overview, we show how IPv6 is defined in this system, discuss possible options for migrating the network toward IPv6, and finally explain how Mobile IPv6 can be used to integrate UMTS networks with WLANs.

## 10.1 3GPP Background

3GPP is based on collaboration between different standards organizations. Its participants are members of these standards organizations. Members are companies that pay yearly fees to obtain one or more voting rights and for their delegates to attend 3GPP meetings. Fees are paid to the standards organization (e.g., ETSI) and not to 3GPP (note the company-based membership, which is different from the individual-based participation in IETF). A company can be a member of several standards organizations (e.g., ETSI and ARIB). In addition to members, 3GPP includes Market Representation Partners (MRPs) like the GSM association, UMTS forum, IPv6 Forum, Global Mobile Suppliers Association. and Universal Wireless Communications Consortium (UWCC). All 3GPP documentation is open and can be downloaded from *www.3gpp.org*.

3GPP is divided into five Technical Specification Groups (TSGs):

- Radio Access Network (TSG-RAN)
- Core Network (TSG-CN)
- Service and System Aspects (TSG-SA)
- GPRS/EDGE Radio Access Network (TSG-GERAN)
- Terminals (TSG-T; i.e., handsets or end host in IP language)

Each TSG has several Working Groups (WGs) developing standards for their respective areas (note the systems approach taken in dividing the TSGs compared to the layered or functional approach done in IETF). For example, the WCDMA physical layer (layer 1) radio interface protocol is standardized by TSG-RAN WG1, while the WCDMA layer 2 and layer 3 radio interface protocols are standardized by TSG-RAN WG2.

Standards are developed in different releases. Each release adds new functions or features (and sometimes fixes bugs in previous releases) to the previous release.

The first 3GPP release was known as Release 1999. As time progressed, it was clear that issuing releases each year would not allow sufficient time to standardize new features. An alternative approach was adopted after 3GPP Release 1999 to standardize releases based on version numbers, thereby not being linked to a year. Features, once completed, would be added to the upcoming release. Due to the standards process within 3GPP during 1998 and 1999, once approved, 3GPP Release 1999 (R99) specifications had version numbers in the range 3.*x*.*x*, so the next release was dubbed 3GPP Release 4.0 (Rel-4). This was finalized in March 2001, and 3GPP Release 5 (Rel-5) was finalized in March 2002. Work is currently ongoing to standardize 3GPP Release 6 (Rel-6), which is expected to be finalized in early 2004.

Once standards are frozen, changes are made to a release in order to correct, clarify, or tidy up the specifications. However, functionality may not be added or removed after a release has been frozen.

## 10.2 3GPP UMTS Network Architecture

The UMTS network is divided into three components: the User Equipment (UE), UMTS Terrestrial Radio Access Network (UTRAN), and the Core Network (CN). Standardization of each part of the network corresponds to the 3GPP TSGs described in section 10.1 and is shown in Figure 10–1.

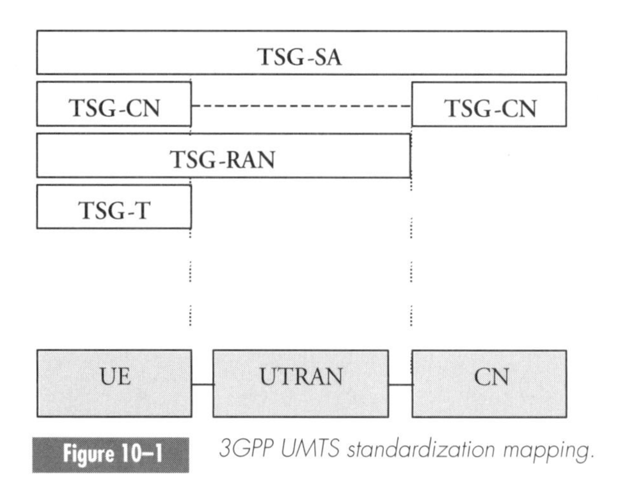


**Figure 10–1** 3GPP UMTS standardization mapping.

A defined standard interface is used to allow communication between the UE, UTRAN, and CN, as shown in Figure 10–1. An *interface* (in 3GPP language) specifies the set of protocols used by the entities separated by this interface. The Uu interface is the WCDMA radio interface used to communicate between the UE and UTRAN. This includes the physical and link layers, which are each terminated in different nodes in UTRAN.

The CN is further divided in two parts: the first is the circuit switched part based on the evolution of the current GSM core networks, and the second is the packet switched network based on the Generic Packet Radio Service (GPRS) core network. The GPRS core network was designed to allow a migration path between today's GSM networks and UMTS. The idea was to allow gradual migration from GSM by adding a new core network that is nearly independent from the radio access technology, then changing the radio interface to WCDMA while keeping the same GPRS core network. The Iu interface communicates between the CN and UTRAN. The Radio Access Network Application Protocol (RANAP) carries signaling between the UTRAN and the CN. The 3GPP network components are shown in Figure 10–2.

The UE has two functions: the mobile termination (MT), which terminates the WCDMA interface, and the terminal equipment (TE) which contains an IP stack and runs the user's applications. The MT and TE can be combined in one device, (e.g., a telephone) or can be split into two (e.g., a laptop being a TE connected to an MT; in this case the MT can be an interface card or a telephone acting like a modem).

## 10.2.1 Packet-Switched Core Network

The packet-switched core network (see Figure 10–3) consists of three main nodes: the Serving GPRS Support Node (SGSN), the Gateway GPRS Support Node (GGSN), and the Home Location Register (HLR). Several SGSNs can be

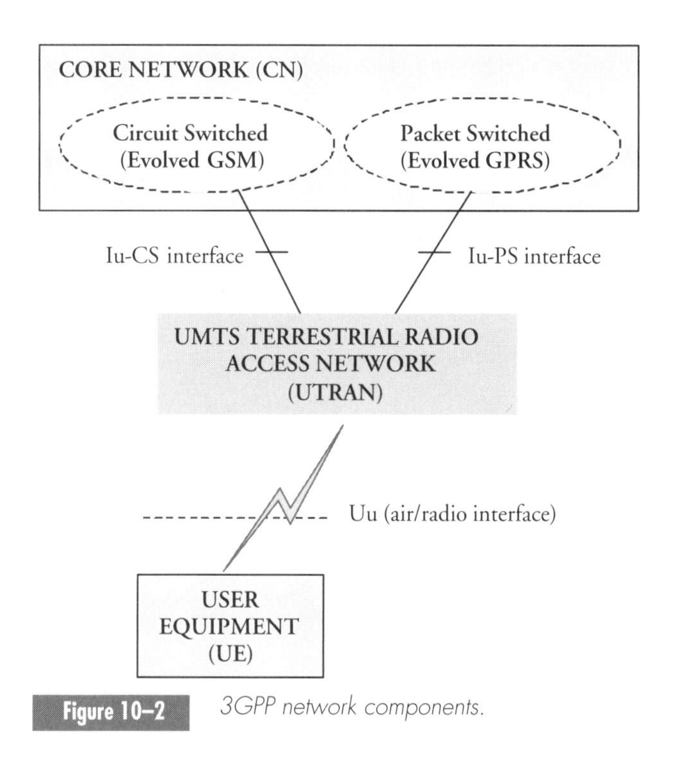


**Figure 10–2** *3GPP network components.*

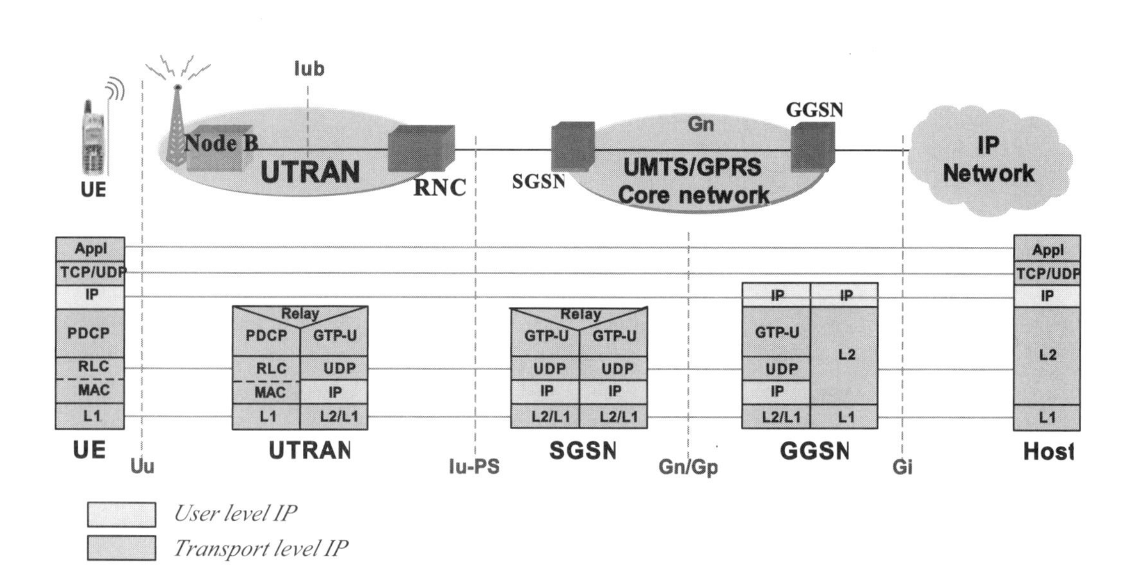


**Figure 10–3** *The UMTS packet-switched (PS) domain network and user plane protocols.*

connected to the same GGSN. The SGSN interfaces to UTRAN through the Iu interface and is responsible for some radio resource management functions, client authentication/authorization, access to the UMTS network, and locating the appropriate GGSN for each connection request from the UEs. In order to authenticate clients, the SGSN uses the Mobile Authentication Protocol (MAP)[1] to communicate with the HLR, which acts like an AAA server for UMTS subscribers. The subscribers' credentials are stored on the Subscriber Identity Module (SIM), which is implemented on a smart card that is accessible by the UE through the Cu interface (a standard electrical interface that allows the UE to read information from the SIM card).

The GGSN is connected to the SGSN through the Gn interface. The GGSN receives IP traffic from the UE and forwards it on the Gi interface. The Gi interface has no special protocol associated with it; in many cases the Gi interface is simply an Ethernet interface connecting the GGSN to another router.

The SGSN and GGSN can be separated by an IP backbone that belongs to the UMTS network operator or another ISP. The idea is to enable UMTS operators to locate SGSNs and GGSNs anywhere on the Internet and have a normal, routed IP backbone between them. However, the downside to this design is that delays can be experienced when forwarding packets between the two nodes, which could degrade the user's experience. This issue becomes clear when we consider the traffic path within the UMTS network.

When work started on the design of the GPRS core network in the early 1990s, before the Internet grew so rapidly, it was not clear to the designers which network layer protocol would be dominant in the future; IP was a strong candidate, but other protocols, like X.25 and Asynchronous Transfer Mode (ATM), were also potential network layer protocols at the time. While the designers aimed at having one network layer technology that could be used by the operator (for ease of network deployment and management), they did not want to limit any UE from attaching to the GPRS network. This is one of the factors that led to the underlying IP (transport) level in Figure 10–3. The intention was to allow a single network layer technology within the core network (IP) to allow the SGSN and GGSN to be separated by a routed IP network while allowing any UE with any network layer to connect to the GPRS core network. Hence, the *transport level IP* layer is completely independent from the *user level* IP (or network layer). We now know that IP will also be used on the user level (in the UE), and this design may now seem inefficient.

The UE implementation includes the standard TCP/IP protocol suite. The Packet Data Convergence Protocol (PDCP) is used as an upper-layer agnostic framing protocol that receives IPv4 or IPv6 packets and places them on their respective channels. It is the first point of contact of the IP packet into the WCDMA radio protocol stack. One purpose for PDCP is to put IP packets into their

[1] The author takes full responsibility for overloading this acronym in HMIPv6!

respective radio channels. The IP header compression is also implemented on this layer. Packets sent over the Uu interface are received in Node B, which then relays them (after terminating the physical layer) onto the Iub interface to the Radio Network Controller (RNC). Some of the data received on the physical layer is sent on the Iub to the RNC for the purpose of power control.

The RNC encapsulates the user's IP traffic into the GPRS Tunneling Protocol's (GTP) header and forwards them to the SGSN. GTP is an application layer-tunneling protocol that runs over UDP. As a result of using GTP (as opposed to the IP in IP tunneling used in Mobile IP), the user's IP packets will be transported on the application layer (GTP), which runs over UDP, which runs over the transport-level IP network. Hence, the original IP header will become invisible for all routers between the SGSN and the GGSN. When packets are received at the GGSN, GTP is terminated and the original IP packets are forwarded on the Gi interface.

The above architecture implies that only the GGSN will be able to route the UE's IP packets. From the UE's IP stack viewpoint, the GGSN is the default router. All other nodes (SGSN, RNC, and Node B) are not visible to the UE's IP stack and can be treated as link-layer devices as far as the UE is concerned.

### 10.2.2 Circuit-Switched Core Network

To allow for the gradual introduction of Packet Switched (PS) services, UTRAN connects to the legacy Circuit Switched (CS) network through the Iu-cs interface. This allows users to have IP and circuit switched applications running on the same device. When UMTS is deployed, traditional circuit switched applications, like voice telephony, are not possible on the packet switched domain. They must run over the circuit switched domain (the intention is to run everything on the packet switched domain in the future, but this is not possible today). If UTRAN did not allow for circuit switched services, users would have to buy two devices—one for each network—which would clearly delay the deployment of the PS domain.

The circuit switched domain is not relevant to this book. It is mentioned here for completeness and is not considered in later sections.

## 10.3 UTRAN Architecture

UTRAN consists of several Radio Network Subsystems (RNSs), each consisting of two entities: Node B, essentially a radio base station terminating the physical layer, and a Radio Network Controller (RNC) terminating the remaining radio protocols. Several Node Bs are connected to an RNC to form an RNS. An operator's network would typically contain more than one RNS. A more detailed representation of the UTRAN network architecture is shown in Figure 10–4.

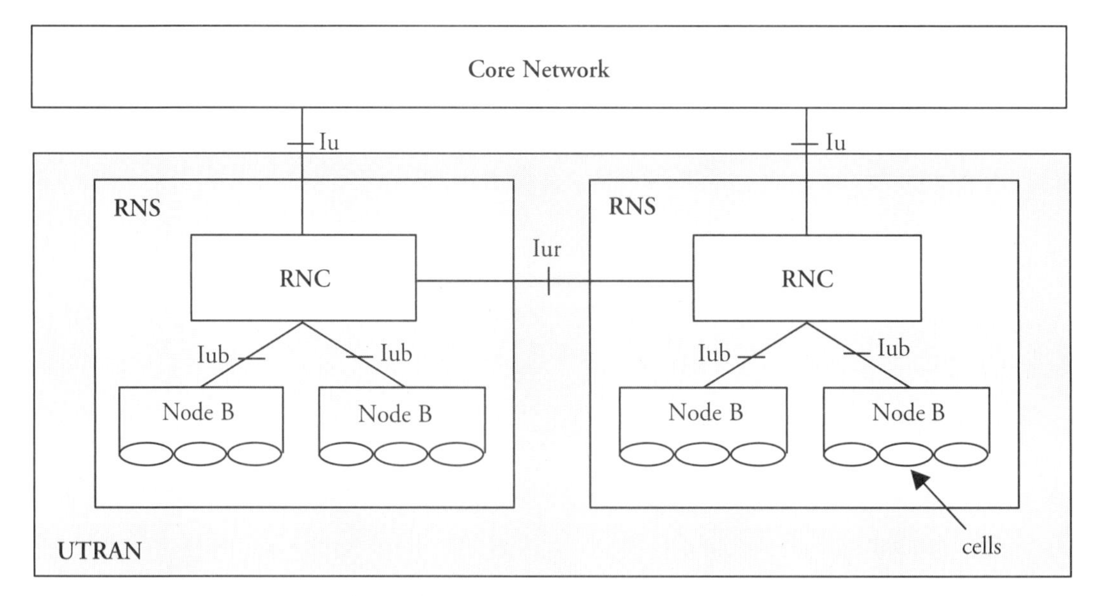


Figure 10–4 3GPP network components.

UTRAN consists of several RNSs, each containing an RNC and a number of Node Bs connected to it. A Node B can connect to only one RNC, using the Iub interface for encapsulating the user data (user plane) and signaling to establish a radio bearer. A bearer is merely a carrier for the user's packets. IP packets from the UE are encrypted before being sent the Uu interface and are decrypted at the RNC. Encryption is based on UMTS-specific algorithms and is performed by the Radio Link Control (RLC) layer. In this section we give a brief overview of the UTRAN architecture and functions. We do not delve into the radio-specific details, as they are not the focus of this chapter. If you are interested in the details of UTRAN, refer to [5]

## 10.3.1 Wideband Code Division Multiple Access

Imagine being at a restaurant or café in one of the large multicultural cities where people of different nationalities are conversing in their native languages. Suppose different people at different tables are talking in English, Chinese, German, Arabic, and so on. A Chinese person could tune in to the Chinese conversation and ignore the other languages that he does not understand. Someone who does not understand any of the languages would be unable to understand any of the conversations—they would sound to her like unintelligible noise. On a very high level, this is approximately the way CDMA encodes information over the Uu interface, and so does its wideband protocol, WCDMA.

Every UE is assigned one or more *channelization codes* (analogous to the languages in our example). A channelization code is a bit sequence whose length varies depending on the bandwidth allocated to the UE and the method used to generate the code. A sender (UE in the uplink or Node B in the downlink) multiplies the signal (data) by the code to spread it over the entire carrier rate. The signal maintains the same amount of overall power, but is spread into a larger spectrum. Signal spreading to a larger spectrum, using a particular code, allows multiple UEs to share the same spectrum for sending and receiving packets by maintaining separate codes for each UE. A spread over a large spectrum minimizes the interference between different users and simplifies the receiver's job in recovering the original signal.

When a spread signal is received, the same code that is used to spread it is used again to despread the signal and recover the original message. The despreading process is done in a receiver called the RAKE receiver, which is located in the UE and Node B.

If we observe a WCDMA channel, we would see a channel containing white noise with amplitude equal to (or slightly above) the level of ambient noise. Just like the person who did not understand any of the languages spoken in our restaurant example, the observer can only retrieve the original signal if she knows the code used to spread the signal. This is one of the reasons that made CDMA technologies attractive for military use.

Channelization codes are used to distinguish different UEs. To minimize the interference between different UEs, it is important to ensure that there is little or no ambiguity between the different codes. That is, if UE1 uses code1, and UE2 uses code2, it is pertinent to ensure that code1 will extract UE1's signal and treat UE2's signal as noise. *Orthogonal codes* will satisfy this requirement. Two codes are considered orthogonal when their cross-correlation is equal to zero. When two codes are not orthogonal, interference takes place and needs to be accounted for in the receivers.

To distinguish signals sent from UEs from those sent from Node Bs, Node Bs use a different code, the *scrambling code*, on the downlink. After the signal is spread with the channelization code, it is multiplied with the scrambling code. In this manner, UEs can separate signals received from Node Bs from those received from other UEs.

There are two different types of WCDMA transmission that are supported by 3GPP standards: frequency division duplex (FDD) and time division duplex (TDD). When FDD is used, two different frequencies are used in the uplink and the downlink. The frequency spacing is large enough to avoid possible interference. Therefore, in FDD the UE will transmit on a different frequency than that used to receive; this allows the UE and Node B to transmit and receive simultaneously. On the other hand, TDD implies that the same frequency is used for both the uplink and the downlink. As a result, a particular period of time is allowed for uplink transmission by the UE, and another period of time is allowed for downlink transmission. The time period for uplink and downlink transmission need not be equal. For instance, applications like Web browsing or FTP need more bandwidth on the downlink than the uplink. Current deployment plans rely predominantly on a WCDMA FDD operation.

### 10.3.2 Power Control and Handovers

One of the most important functions of WCDMA is Radio Resource Management (RRM). RRM can be divided into two main functions: power control and QoS. QoS-related functions include load control and admission control. Power control–related functions include different types of power control in WCDMA and handovers. In this section, we focus on power control–related functions. Understanding these functions will enable you to get a better insight into how handovers are managed in UTRAN.

To understand the importance of power control in WCDMA, we revisit our multicultural restaurant example. Suppose the Chinese customer was dining on a table surrounded by an English-speaking table, a German-speaking table, and another Chinese-speaking table. We previously assumed that the Chinese person could listen to the Chinese conversation at the next table and treat the other languages as noise. This is true—if the Chinese speakers were loud enough to be heard. If the English-speaking people were much louder than the Chinese, the noise would overcome the otherwise understandable Chinese language. The same problem occurs in WCDMA uplinks where multiple UEs are transmitting to a single Node B. If one UE starts transmitting with much more power than others, other UEs' data would not be recovered by the Node B's receiver. For instance, a UE located on the border between two cells and receiving a fading signal might want to increase its power so it can send a signal that can be recovered by the Node B. In the meantime, another UE, close to the same Node B would not see the need to increase its power. Hence, the far UE can overshoot the near UE and make it unable to communicate with the Node B. This near-far problem is one of the main reasons for having power control and soft handovers, as will be shown in the following sections.

The near-far problem is not a serious issue for the downlink due to the one-to-many nature of the transmission between the Node Bs and the UEs. However, the Node Bs must be careful to not transmit too much power to avoid interfering with neighboring Node Bs.

There are three different types of handovers that can take place in UTRAN: *softer handover*, *soft handover*, and *hard handover*. These are discussed in the following sections.

#### 10.3.2.1 CLOSED-LOOP AND OUTER-LOOP POWER CONTROL

Fast closed-loop power control is an extremely important feature in WCDMA for controlling power levels on the uplink and avoiding the near-far problem. In closed-loop power control, Node Bs measure the signal-to-interference Ratio

(SIR) in received signals from UEs; if the SIR is too high, the Node B commands the UE to reduce its power. If the SIR is too low, the UE is commanded to increase its power to reach a target SIR value set in the Node B. These measurements are done 1500 times per second, which makes them fast enough to handle any significant change in the radio conditions, for example, due to movement.

While closed-loop power control is done in the Node B, outer-loop power control is done, per radio link, in the RNC. The RNC infers information about each radio link quality from the BER received in each frame. The BER value is estimated from a *Cyclic Redundancy Check* (CRC) value added to each frame by the Node B. The reason for doing outer-loop power control in the RNC is to allow for power control when the UE is simultaneously attached to two different Node Bs—that is, during soft handover.

#### 10.3.2.2 SOFTER HANDOVER

Handovers within UTRAN are handled by the radio protocols without any involvement from upper layers. Softer handovers take place when a UE moves between two sectors on the same Node B. During softer handover, the UE communicates with the Node B via two different channels, one from each sector. However, in this case the UE must use a different code in the downlink for each sector. In the uplink, the UE continues to use the same code for both channels. When the Node B receives the UE's signals, it compares them and picks the better signal. Selection can happen in different ways; for instance, Node B can select on a frame-by-frame basis. Alternatively, per-bit selection is also possible.

#### 10.3.2.3 SOFT HANDOVER

Soft handover takes place when a UE moves between two different Node Bs. Just as in softer handover, in soft handover the UE communicates with both Node Bs in the uplink using the same code and two **different** codes for its RAKE receiver on the downlink. This is shown in Figure 10–5.

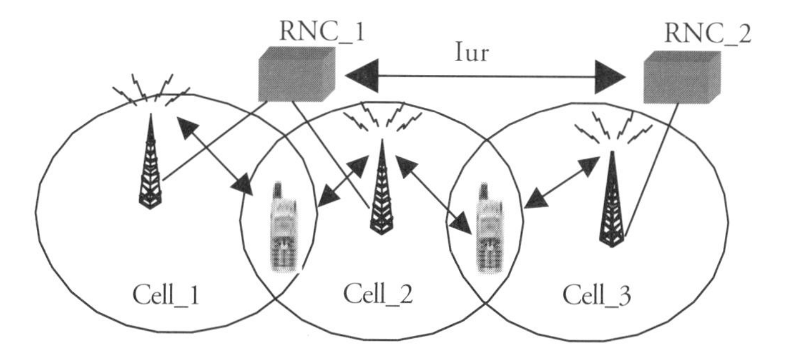


**Figure 10–5** *Soft Handover.*

To the UE, soft handover looks almost identical to softer handover. However, within UTRAN there is a large difference between the two handovers. Since soft handovers take place when the UE moves between different Node Bs (e.g., from Cell_1 to Cell_2), the combining of uplink traffic is done in the RNC. Therefore, power control also takes place in the RNC. The information needed to perform power control in the RNC is transmitted from Node B on the Iub interface. Hence, the RNC does the combination and selection of uplink frames.

It is possible for soft handover to occur when the UE moves between two Node Bs that are attached to different RNCs (e.g., Cell_2 and Cell_3). In this case, the *Serving RNC* (SRNC) performs power control and recombining on the uplink. Signals transmitted from the UE and received by the Cell_3's Node B are sent to the SRNC (RNC_1 in our example in Figure 10–5) via the *Drift RNC* (DRNC; RNC_2 in this example) where recombining and selection take place. The interface between RNC_1 and RNC_2 is called the Iur interface. When it becomes clear that the UE is receiving a better signal from the new Node B, the SRNC moves the traffic to the new Node B, and the old Node B will terminate the old link. When soft handover is taking place between two different RNSs (e.g., RNC_1 and RNC_2 above), the SRNC first starts an *SRNS relocation procedure*, which involves signaling to the core network and a 3GPP-specific context transfer protocol between the two RNCs. In this case, the context is associated with that particular UE's radio requirements.

Soft handover requires tight synchronization and has strict delay requirements to ensure that different Node Bs transmit the same radio frames at the same time in order for the UE to compare the received signals and pick the best quality signal. The same requirements need to be met on the uplink to allow the SRNC to select the best signals.

Soft handover is often confused with IP layer make-before-break handovers. It is pertinent to note that in soft handovers, the UE receives the **same** data through two different APs. The UE is unable to receive two different sets of data from the different APs, which is a basic requirement for an IP layer make-before-break handover. Receiving the same signal through two different APs makes soft handover completely transparent to the IP layer.

#### 10.3.2.4 HARD HANDOVER

Unlike the previous handovers, in hard handovers the UE detaches from one RNS, then attaches to a new RNS. In hard handover there is no SRNS relocation; the UE simply disconnects from the old RNC (connected to the old SGSN) and attaches to the new RNC (connected to the same or a new SGSN). When attaching to the new RNS, the UE includes the previous Routing Area Identifier (RAI). This term has nothing to do with IP routing; a routing area corresponds to the coverage area of an SGSN. From the RAI, the new SGSN can retrieve the old SGSN's IP address. The GTP protocol is used to ensure that traffic from the old SGSN is forwarded to the new one (inter-SGSN signaling) and that the GGSN is informed (by the new SGSN) that it should forward the UE's traffic to the new SGSN. All these actions are completely transparent to the UE.

Hard handover involves some packet losses during the break from the old RNS and reattachment to the new one, which would not take place in softer or soft handover. So why doesn't UMTS use soft handover all the time? There are two different cases where hard handover can take place instead of soft handover:

- Sudden loss of coverage: The UE might suddenly lose its connection to the current RNS. This is a rare case and is most likely due to an error.
- No Iur interface between the RNSs: Not all RNCs are connected with an Iur interface. In cases where they are not, soft handover is not possible between two different RNSs. Hard handover will take place instead.

Having introduced UTRAN's operation and mobility management, let's consider the core network operation.

## 10.4 UMTS Core Network

The UMTS core network is based on the GPRS core network originally standardized by ETSI in 1997. The main components of the core network are the SGSN, GGSN, and HLR. The SGSN and GGSN are involved in forwarding the UE's traffic, while the HLR is involved in authentication, authorization, and access control. Before we get into the details of the core network operation, we need to identify some of the major differences between this network and other IP networks that you might be familiar with. A few new terms and design philosophies are discussed below, followed by an overview of the core network's operation.

A fundamental design principle used in the GPRS core network is that a separation is needed between the UE's IP packets and those packets appearing on the wire within the core network. There are two possible reasons for this design:[2]

- Network layer independence: This means that the operator can deploy an IP network where the core network components can be located anywhere without assuming that the UE will use IP. This point was perhaps more relevant in the early 1990s when people were not sure that IP would become ubiquitous.

[2] This is the author's opinion on the subject based on discussions with people working with 3GPP networks and should not be interpreted as a representation of the consensus in the 3GPP community.

- Flexible location of the SGSN and the GGSN: If tunneling is used between the GGSN and the SGSN, the core network would seem like a VPN running within the ISP's backbone network, which would allow an operator to place the GGSN anywhere. In a typical GPRS deployment scenario, the GGSN need not be located in the same site or city as the SGSN. The obvious tradeoff here is between latency and location flexibility.

Within the entire UMTS network, the terms *control plane* and *user plane* are used to distinguish the protocols used to set up the links between UEs and the UMTS network and to manage the UEs' mobility from those protocols involved in forwarding the UEs' packets respectively. For instance, GTP-U is the protocol used to encapsulate the UE's IP packets and to forward them from the RNC to the SGSN and from the SGSN to the GGSN (i.e., User Plane GTP). GTP-C is the signaling protocol used between the SGSN and GGSN to manage the UE's mobility within the core network (i.e., Control Plane GTP).

Another fundamental design principle of the GPRS core network is the concept of a link. The GPRS link definition is drastically different from the definition used in RFC 2460 and presented in our terminology section. A GPRS link is a logical link mapped from the physical radio link in UTRAN to a GTP *tunnel identifier* that is used to forward the UE's packets from the RNC to the SGSN and from there to the GGSN. The tunnel identifier is established when the UE attaches to the core network and activates a Packet Data Protocol (PDP) context. The PDP context initiates a virtual link between the UE and the GGSN. In other words, the link, in this context, is formed over a number of relay devices, which are not visible to the UE.

At this point, if you are familiar with today's IP networks operation and design, you may be alarmed, confused, or both. Hopefully, the following sections will eliminate the confusion about the operation of the core network.

## 10.4.1 PDP Context Activation

In order for the UE to connect to the core network, get an IP address assigned, and become reachable, it needs to have a PDP context. The PDP context defines the link between the UE and the GGSN. When a UE boots and wishes to connect to the UMTS network, it must perform a *GPRS attach* procedure. The attach procedure involves sending an attach request to the RNC that is relayed over the Iu interface's control plane (RANAP) to the SGSN. The SGSN sends the UE's credentials to the HLR for authentication. If the UE is authenticated, the SGSN accepts the attach request and informs the UE.

After a successful GPRS attach, the UE can request a PDP context. The PDP context activation request is initiated by the UE and relayed to the SGSN by the RNC. A simplified message sequence diagram for the PDP activation process is shown in Figure 10–6.

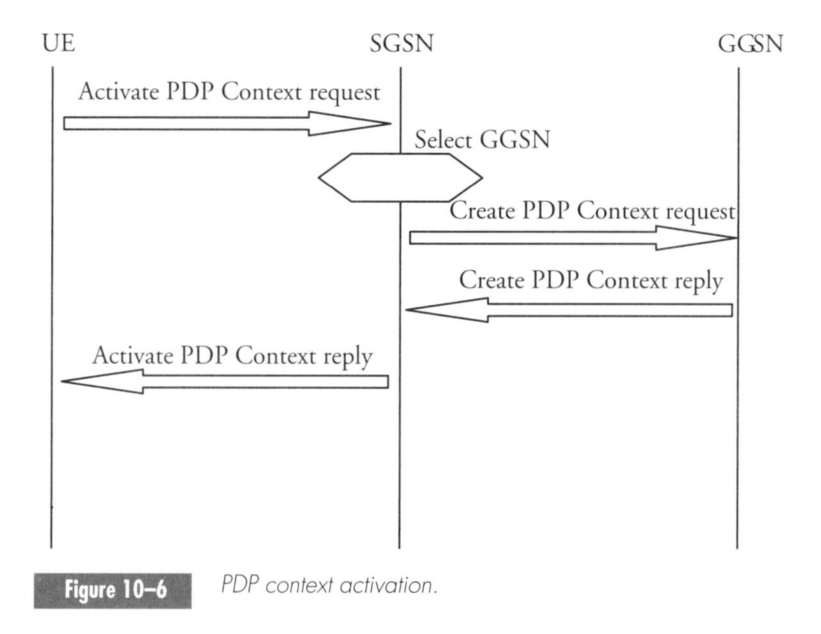


**Figure 10–6** *PDP context activation.*

All messages between the UE and SGSN are relayed through the RNC, as neither the SGSN nor the UE can send messages directly to each other. As we described earlier, the RNC and the SGSN can be considered link-layer bridges by the UE; therefore, messages from the UE must be reformatted and placed in the new protocol's format. In Figure 10–6, we show only the final destinations of the messages as opposed to the physical path that the messages travel to get to that destination.

The UE initiates the process by sending an *activate PDP context request* message to the SGSN. This message contains three main pieces of information: the *PDP type* (e.g., IPv6 or IPv4), the *Access Point Name* (APN), and QoS/filtering-related parameters.

The PDP type is either IPv4 or IPv6 for a single PDP context. IPv4 and IPv6 cannot exist simultaneously on the same PDP context. If a UE wishes to use IPv4 and IPv6 simultaneously, it needs to perform this procedure twice. Alternatively, tunneling may be used to hide IPv6 in IPv4, or vice versa.

The APN is yet another strikingly different concept from today's IP networking standards and practices. The APN is a specific service requested by the user and represented as a text string. The definition of a service here is quite vague, so we describe it by examples. One possible APN could be called "IMS" to indicate that the UE wishes to use 3GPP's IP Multimedia Subsystem (IMS), which defines a number of multimedia applications that the user wishes to use. If this APN is used, all the UE's traffic must be using the IMS. For instance, the

UE cannot request an IMS APN and then start Web browsing, or vice versa. Another possible APN could be called "company.com." This APN could mean that the UE wishes to initiate a VPN to company. In this case, the GGSN will forward all the UE's traffic through a tunnel (possibly protected with IPsec) to the company's firewall. In this particular example, the GGSN must be configured manually or must use IKE to establish such tunnel. Of course, the UE can establish the tunnel directly to the company's firewall; however, the 3GPP architecture allows for this case for users who do not wish to use the extra overhead of IPsec over the scarce (and expensive) radio interface.[3]

There are no standard names for APNs; the names are set by each operator and are configured into the UEs manually or via vendor-specific protocols. To allow UEs to roam between different networks, operators must agree on the APN naming or download them to the UEs using nonstandard protocols.

Not all GGSNs within a single operator's network need to support all APNs. The operator configures APN support in the GGSN. When the SGSN receives the PDP context activation request, it searches for the GGSN that best supports the requested APN. The search is done through DNS requests. The SGSN sends a DNS request for the requested APN and can receive a reply containing one or more addresses assigned to one or more GGSNs. The SGSN selects an address and sends the PDP context-creation request to it. If the GGSN accepts the request, it sends a reply indicating success to the SGSN, which in turn sends an acceptance message to the UE. All the messages shown in Figure 10–6 between the SGSN and GGSN are carried by the GTP-C protocol; messages between the RNC (as a relay) and the UE are carried on the Uu interface in the Radio Resource Control (RRC) protocol. Hence, all messages are carried on the control plane and represented by different protocols between different nodes.

The PDP context activation is, in many cases, triggered by an application running on the UE. Applications can set the different parameters in the PDP context activation request based on its needs. For instance, an application might need and specifically request IPv6. An application could also request special filters or specific QoS parameters for special flows. This information is included in the PDP context activation message. A *Traffic Flow Template* (TFT) can be set inside the GGSN to accommodate the UE's QoS demands. The TFT acts as a flow identifier for the GGSN based on various parameters, including the flow label, protocol number (for transport layer protocols), IP addresses, or prefixes for source, destination addresses, port numbers, and the SPI (if IPsec were used).

Previous versions of the GPRS core network standards allowed network-initiated PDP context activation. However, this feature was recently removed

[3] 3GPP standards do not require the GGSN to set up an IPsec tunnel to the Firewall. However, this is expected to be supported in many implementations.

from the standard. Therefore, a UE that performs a GPRS attach procedure but chooses to not activate a PDP context will not be reachable. Hence, UEs should always have a PDP context activated.

### 10.4.2 Mobility Management in the Core Network

In UMTS, UEs are allocated addresses during the PDP context activation process. These addresses remain valid for the lifetime of the PDP context, regardless of the UE's location. For a given PDP context, the UE will continue to use the same GGSN at all times. In other words, the UE's movement within a UMTS coverage area will not involve any IP mobility.

Mobility within an RNS is handled by softer and soft handovers as described earlier. The core network is only invoked when the UE moves between two different RNSs or SGSN domains (routing areas).

An inter-SGSN handover can take place due to two different events: first, soft handover between two different RNSs connected to two different SGSNs, or second, a hard handover between two RNCs connected to two different SGSNs. When soft handover is taking place, the core network (SGSN in this case) is informed that an SRNS relocation procedure must be invoked. An SRNS relocation also means that an inter-SGSN handover will take place if the SRNC and DRNC are connected to two different SGSNs.

In the case of soft handover, the new SGSN is informed about the old SGSN through inter-SGSN signaling, which also uses GTP-C. The new SGSN also knows the IP address of the GGSN that the UE is connected to. During the SRNS relocation procedure, the new SGSN sends GTP-C signaling to the GGSN, informing it that the UE's traffic should be directed to the new SGSN's IP address instead of the old one.

The control planes and user planes between two SGSNs use protocols identical to those used on the Gn interface (i.e., GTP-C and GTP-U). However, the interface between two SGSNs is called the Gp interface to distinguish the SGSN–SGSN communication from the SGSN–GGSN communication.

## 10.5 IPv6 in UMTS

IPv6 was always considered in GPRS core networks as a candidate protocol for UEs. However, when GPRS standards were first released in 1997, IPv6 was not complete in IETF. Initial GPRS releases considered IPv6 only as a PDP type, but no further work was done to see how IPv6 standards could be used in a GPRS core network. Issues like address allocation, coexistence with IPv4, and mobility management were only considered several years later in the context of UMTS networks.

In 2000, 3GPP members voted, based on operators' and vendors' demand, in favor of mandating the use of IPv6 in IMS applications. IMS includes a session control system based on the Session Initiation Protocol (SIP) standardized in IETF. IMS is expected to be used for multimedia applications, including voice. The very large number of anticipated UMTS devices has led 3GPP to decide that supporting peer-to-peer applications, like those based on IMS, is best done with IPv6.

In this section, we describe how IPv6 is used in 3GPP according to both IPv6 and 3GPP standards. As IPv6 is expected to enable peer-to-peer applications that require reachability via globally unique addresses, the main focus of this chapter is IPv6 on the user level. That is, our focus is on introducing IPv6 in UEs and the GGSN. The UMTS architecture allows the transport-level network to use IPv4 while running IPv6 on the user level, as both layers are completely independent of each other. Hence, operators need not upgrade every node in their networks or wait for IPv6 to be supported in all these nodes before they can start deployment; initially, only the UE and the GGSN need to support IPv6. Other nodes can later support it on the transport level.

We now focus on address assignment to the UE, coexistence, and the role of Mobile IPv6.

## 10.5.1 Address Configuration

The current IPv6 address configuration mechanism in UMTS was first documented in 3GPP release 5, based on the IPv6 Working Group recommendation in RFC 3314. The UE is connected to the GGSN via a point-to-point link that falls in the category of connection-oriented link layers described in Chapter 6. The main requirements behind the current address configuration mechanisms are the following:

- The UE should be able to generate multiple addresses to avoid being tracked by traffic analysts that use the UE's IPv6 address to identify the user. These addresses should not be generated by the GGSN, but by the UE, as described in RFC 3041.
- The UE should not be assigned a single address; in the future, the UE might be acting as a mobile router that provides Internet access to other devices, which will need their own addresses.

Based on these main requirements, it was recommended that the GGSN assign a 64-bit prefix that can be used by the UE to generate as many addresses as needed. IPv6 does not include a mechanism that allows a router to assign a 64-bit prefix to another router or host on its link. However, since the link in this case is point-to-point, the UE can be implemented to assume that router advertisements received from the GGSN will include a unique 64-bit prefix assigned to the receiving UE only. This assumption is valid only if the router advertisement is received from the WCDMA interface. The router advertisement is sent to the UE after the PDP context is activated.

Cellular links are known for being error-prone and bandwidth-challenged. It is generally desirable to minimize the amount of signaling over the air interface and reduce the sizes of signals as much as possible. Normally, a host cannot use an address without testing it for duplication using DAD. However, in this case, the host's (UE's) only neighbor is the GGSN. Furthermore, the GGSN assigns a unique 64-bit prefix to the UE. Therefore, it is easy to ensure that address duplication does not happen by ensuring that the GGSN does not use that 64-bit prefix to derive any of its own addresses. Duplication should also be avoided for link-local addresses. The GGSN sends an interface identifier to the UE during the PDP context activation process (in its reply to the SGSN, which is relayed to the UE). The UE must use this interface identifier for its link-local address, which uses a well-known, reserved prefix. This ensures that duplication will not occur for link-local or global addresses. Hence, DAD is not required on the WCDMA interface.

This mechanism meets the two main requirements discussed above. One of the points of discussion was whether a 64-bit prefix was too much. After all, the UE will most likely use one address for all communications. However, the IPv6 Working Group's consensus was that IPv6 addresses are not a scarce resource and that it is better to design a long-lasting mechanism than to design one that will be modified when, for instance, mobile routers are used. The current mechanism allows for greater flexibility and simplicity of design; wasting some IPv6 addresses was not seen as a major issue in this case.

## 10.5.2 Transition and Coexistence

The transition and coexistence problems between IPv6 and IPv4 are the same in UMTS as in any other IP network, with very minor considerations. In this section, we pick the main deployment scenarios and see how the mechanisms shown in the previous chapters can be applied to UMTS networks. As the need for global addresses for UEs presents the strongest motivation for the introduction of IPv6 in UMTS, we focus on transition mechanisms for the user-level IP packets.

Three different scenarios are presented in this section:

- Tunneling over the Gi interface
- End-to-end incompatibility
- IPv4-only GGSNs

Let's see how each problem can be solved.

### 10.5.2.1 TUNNELING OVER THE GI INTERFACE

In early deployment of UMTS, an operator is likely to enable IPv6 in new GGSNs to support IPv6-enabled UEs. However, an operator would not necessarily upgrade all the routers around the GGSN. The GGSN and SGSN will likely be separated by an existing IPv4 backbone. Similarly, GGSNs can be the only IPv6 router present within an IPv4 network for IPv6-enabled UEs to be able to communicate with other IPv6 hosts on the Internet.

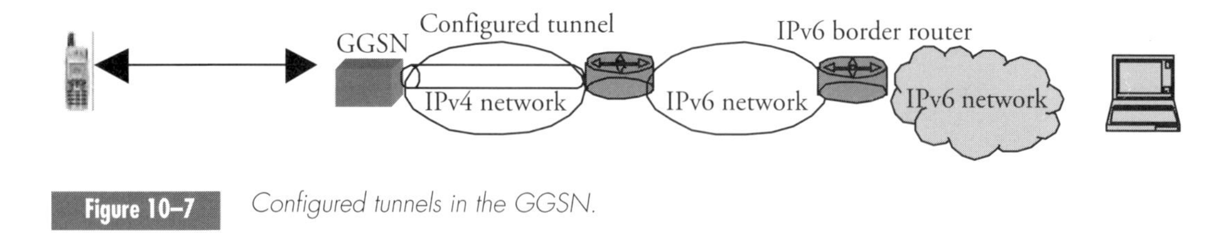


Figure 10–7 *Configured tunnels in the GGSN.*

Configured tunnels can be established between the GGSN and another dual stacked router that can provide native IPv6 connectivity directly or through another tunnel.

Configured tunnels will allow bidirectional communication. Using configured tunnels, IPv6 UEs can send and receive packets through the same path.

Alternatively, if destination addresses (for outgoing packets) are 6-to-4 addresses, the GGSN can dynamically tunnel packets to the IPv4 address embedded inside the destination address. However, 6-to-4 addresses can be inefficient when assigned to IPv6 UEs to allow them to be reachable. The current IPv6 address allocation scheme in UMTS assigns a 64-bit prefix to each UE. Since the 6-to-4 address format reserves 32 bits to be used by the IPv4 address, a 48-bit 6-to-4 prefix can support 65,535 PDP contexts. This translates to fewer UEs, since a UE can have more than one PDP context. Therefore, an increase in the number of IPv6 UEs will require more IPv4 addresses. A GGSN is expected to support hundreds of thousands of PDP contexts which, if 6-to-4 is used, requires a number of IPv4 addresses to be allocated to each GGSN for tunneling purposes. If routing protocols are used instead of 6-to-4 to distribute IPv4 tunnel end points, they would not suffer from the scarcity of IPv4 addresses. Therefore, using routing protocols to distribute tunnel end points might be a preferable option for large operators.

#### 10.5.2.2 END-TO-END INCOMPATIBILITY

IPv6 is a mandatory part of the 3GPP's IMS. When a UE connects to another using IMS to set up the connection (e.g., multimedia sessions, voice over IP), all the signaling involved between the UE and IMS must take place over IPv6. This effectively makes any UE attempting to use IMS seem like an IPv6-only host.

IMS signaling involves requests to set up connections with another host (UE) negotiating the connection's capability and QoS. The signaling is done from the UE to a server in the IMS domain, which relays the session initiation data to the correspondent. However, the actual connection's data flows directly between the two hosts. IMS signaling is not bound to a particular application; it can be used to negotiate any multimedia session.

If a UE initiates an IMS session with another IPv4-only host, some form of translation is required. The need for translators in this case is due to the general-purpose nature of IMS signaling. If all applications were known in advance, you could see that UEs could connect to ALGs, which act as proxies between the two correspondents, using IPv6 toward the IPv6 UE and IPv4 toward the IPv4-only host; however, such assumption may not be feasible in this case.

Another reason for having translators is that a UE might not wish to have multiple PDP contexts active simultaneously. Activating a PDP context consumes resources on the GGSN and the UE. More importantly, due to the delays encountered over the air interface, it can take a significant amount of time to establish a PDP context (hundreds of milliseconds). Therefore, it is possible that some implementers may opt for a single PDP context (with PDP type IPv6) and pay the price of going through translators. Although implementations should always avoid translators if possible, one cannot guarantee that every implementer would agree with this tradeoff.

NAT(P)T-PT is a prime candidate for the application-independent translation function due to its ability to translate packets while using IPv4 addresses in an efficient manner. When used in UMTS networks, NA(P)T-PT is located on the GGSN's egress interface (i.e., on the Internet side of the Gi interface). Normal NA(P)T-PT operation, described in the Chapter 9, allows IPv6-only UEs (including dual stack UEs using the IPv6-IMS) to communicate with other IPv4 hosts on the Internet.

For known and popular applications (e.g., HTTP, email), operators don't need translators; a number of ALGs acting as proxies will suffice in this case. One of the advantages of having a nontransparent proxy (i.e., visible to the end host) is that some form of security can be achieved. Since the host is aware that it is communicating through an ALG (by configuration or automatic discovery), it can establish secure communication to the ALG, which in turn establishes secure communication to the other correspondent. Of course, in this case the ALG can see the plaintext and therefore must be trusted to some extent. However, in this case the traffic is less vulnerable than in the NA(P)T-PT case where no end-to-end security can be guaranteed.

The key requirement for both NA(P)T-PT and ALGs in UMTS networks is scalability. That is, considering that the Gi interface from one GGSN is expected to handle hundreds of thousands of connections, translators and/or ALGs must be able to scale to support this large amount of traffic. This is different from current enterprise and ISP networks where access routers are connected to hundreds—or a few thousand—machines. Allowing these boxes to scale is likely to mean that a number of them will be needed with some load-sharing mechanism. For instance, a DNS can return a different NA(P)T-PT prefix to each DNS query from an IPv6 host (to represent an IPv4 host) in a round-robin fashion. Other more sophisticated load-sharing mechanisms are also available in various products.

#### 10.5.2.3 IPV4-ONLY GGSNS

Pre-IMS product releases of the GPRS core network are already deployed in several parts of the world based on IPv4 handsets. As new host implementations come out with IPv6 enabled by default, operators are likely to gradually migrate their GPRS/UMTS networks toward IPv6. ISATAP provides a simple mechanism that allows operators to support IPv6 UEs while gradually upgrading GGSNs to support IPv6.

In order to support ISATAP, an operator can include one or more ISATAP routers on the GGSN's egress interface (Gi). Normal ISATAP operation will allow UEs to discover the router and establish a bidirectional tunnel to it that allows them to communicate with other IPv6 hosts.

### 10.5.3 IPv6 Mobility

So far we saw three different types of handovers that can take place within the UMTS network: softer handover, soft handover, and hard handover. Each handover involves mobility on a different level, intra-Node B, inter-Node B, or RNS and inter-SGSN mobility. All cases are handled by UMTS-specific signaling and are completely transparent to the UE's IP stack. The only node that is visible to the UE's IP stack is the GGSN, which acts as a default router. Inter-GGSN handovers do not exist in UMTS; the UE's address does not change for any given connection/PDP context, so IP mobility is not invoked.

IP mobility can be invoked in multihomed UEs with interfaces connected to different networks. For instance, movement can take place from a UMTS network to a WLAN network. Such movement requires Mobile IPv6. Furthermore, UEs that wish to be reachable through a permanent address must update their home agent whenever they configure a care-of address based on the prefix assigned by the GGSN.

Another optimization, discussed in Chapter 7, is related to flow movement. A UE may wish to allow some flows to be forwarded on its WCDMA interface and others through its WLAN interface. Again, this optimization can be done with Mobile IPv6 and some extensions similar to those shown in Chapter 7.

The use of Mobile IPv6 need not impact the UMTS architecture. In fact, Mobile IPv6 can be used if the UE implements the mobile and correspondent node functions in the UE and is assigned a home agent anywhere on the Internet. GGSNs need not have any knowledge of Mobile IPv6 and need not support the home agent functions.

In order to allow mobile nodes to move between UMTS and WLAN networks, several issues must be considered. In this section we consider two main issues: AAA and mobility management, both currently being prioritized by 3GPP. AAA has a higher priority, as it is concerned with network access and charging users.

#### 10.5.3.1 AAA ISSUES

In Chapter 8 we discussed the AAA architecture being developed in IETF. It was stated that the protocol used between a host and the AAA *front-end server* currently depends on the access technology being used. In many cases, including UMTS and GSM networks, the authentication protocol runs on the link-layer. For instance, on dial-up links, the Point-to-Point Protocol (PPP) is used between the host and the NAS for authentication and authorization. Currently, the UMTS protocols used for authentication and authorization are different from those used elsewhere on the Internet. Therefore, when a device roams between UMTS and WLAN networks owned by the same operator or another operator, different protocols are used to authenticate and authorize the device for network access. The identifier used in these protocols is also important for charging purposes. If a customer uses a different identifier for each access technology, operators need to either associate each identifier with a separate subscriber in their billing system (although it is really the same customer) or somehow keep track of the various possible identifiers used by each customer and associate them with one subscriber in their billing systems. A much simpler approach is to use the same identifier on all access technologies. This allows the user and operator to identify the subscription by a single identifier and simplify the charging and billing operations from the operator's point of view.

The Extensible Authentication Protocol (EAP) allows for several methods of authentication. In addition, it allows for several identifiers to be used independently of the access technology. EAP is designed to be carried between hosts and the AAA front-end server. In [3], extensions to EAP were proposed to allow it to run the UMTS Authentication and Key Agreement (AKA) protocol, using EAP as a carrier for the AKA messages and algorithm negotiations. AKA runs on the Universal Subscriber Identity Module (USIM), which is typically stored on a smart card used today in mobile phones (SIM card). To authenticate the user for radio network access, the International Mobile Subscriber Identifier (IMSI) is used as an identifier. The IMSI is a 15-digit number that includes a 3-digit Mobile Country Code (MCC), a 2- or 3-digit Mobile Network Code (MNC), and a 10-digit Mobile Subscriber Information Number (MSIN).

Including extensions of EAP to allow it to run AKA means that a UMTS subscriber can use the same identifier (IMSI) and authentication mechanism (AKA) to gain access to other wired and wireless networks independently of the type of access technology being used. The use of this mechanism can be illustrated by considering the AAA architecture in Figure 8–1 and placing the protocols mentioned above on their respective interfaces. Figure 10–8 shows how this mechanism can be used to allow for access-independent authentication and authorization.

There are two ways to encapsulate EAP messages: link-layer encapsulation and application-layer encapsulation. When EAP messages are encapsulated in the link-layer protocols (e.g., 802.1x), the front-end AAA server is essentially the AP that terminates the link layer. Hence, EAP messages are forwarded from the

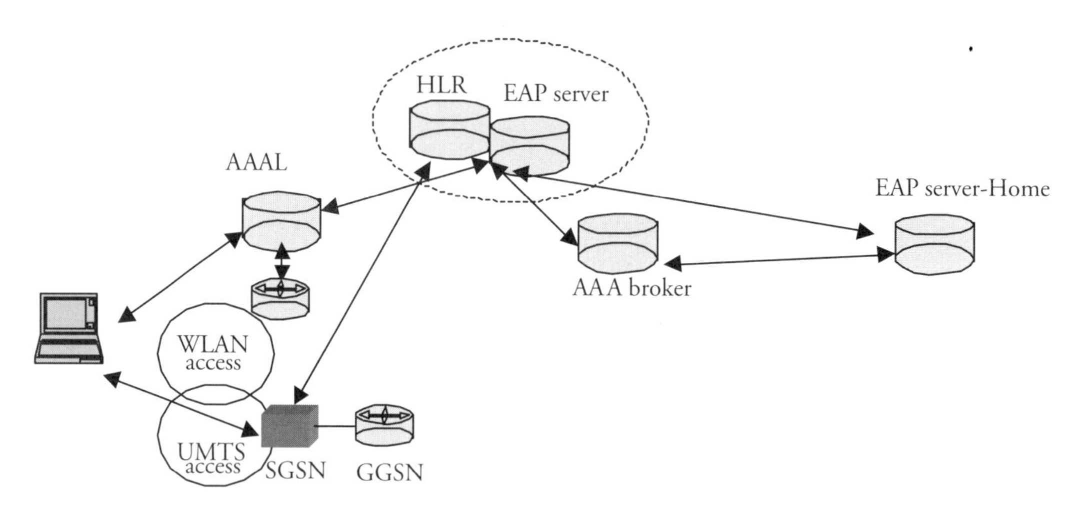


Figure 10–8 *Using EAP/AKA for access-independent authentication and authorization.*

AP (acting as an IP host) to the EAP server, using an AAA protocol (e.g., RADIUS or Diameter). But, if EAP messages were encapsulated by an application-layer protocol (e.g., work in progress in the PANA Working Group in IETF), the front-end AAA server could be located in the AR, which terminates the PANA protocol. In either case, the front-end AAA server would pass the EAP messages to the EAP server, which processes the messages and forwards the AKA parts (in the case where AKA is used) to an Authentication Center (AuC). The AuC generates the AKA parameters required for authentication and can be collocated with the EAP server (and HLR) or implemented in a separate node. In either case, the AuC forwards the authentication parameters to the EAP server, and from there to the front-end AAA server, which passes it back to the host. In this manner, authentication and authorization can take place when a host is connected to a WLAN access.

When the host moves to a UMTS network, normal radio access authentication takes place using AKA (i.e., without EAP). The SGSN relays the host's credentials to the HLR, which provides authentication and authorization to the host.

By collocating the EAP server and the HLR, the UMTS identifier (IMSI) can be reused across different access technologies. Hence, a UMTS operator can allow its users to roam between UMTS and WLAN without having to use different identifiers for authentication.

#### 10.5.3.2 MOBILITY ISSUES

The previous section showed how AAA issues can be solved in order to integrate UMTS and WLAN access technologies and allow users to roam between the different access networks. However, session continuity across different networks requires the use of Mobile IPv6. The use of Mobile IPv6 allows devices to be reachable through a stable home address. In this section, we show one example of such deployment and illustrate how Mobile IPv6 can be used. Figure 10–9 shows an operator's network that contains both UMTS and WLAN access.

In Figure 10–9, we assume that UMTS coverage is almost everywhere, whereas WLAN coverage is concentrated in some hotspots, but they generally cover a much smaller geographical area compared to cellular networks.

We saw in previous sections that Mobile IPv6 is not involved in mobility management within the UMTS network. Therefore, there is no impact on the UMTS architecture when Mobile IPv6 is used. The home agent can be located anywhere on the Internet. In fact, the home agent need not be provided by the cellular operator; it can be provided by another operator, or it can be located at the subscriber's home or office. We include the Mobility Anchor Point (used for HMIPv6, see Chapter 7) within the WLAN access to assist with the handovers between two routers within the WLAN network.

Suppose that a mobile node is currently within the UMTS coverage area and suddenly detects the presence of WLAN coverage. Now the mobile node must make one of three choices:

1. No need to move to WLAN coverage. The mobile node stays in the UMTS network.

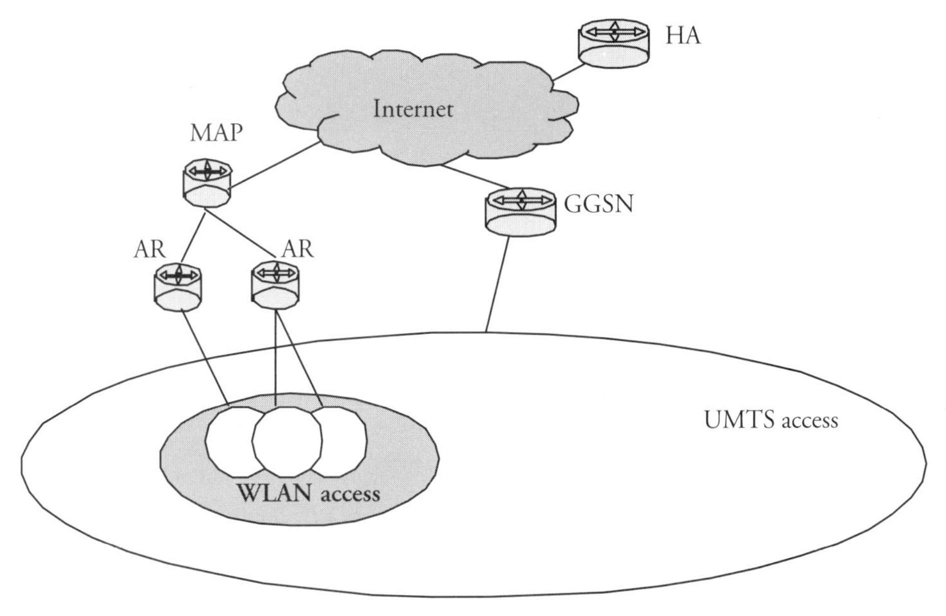


**Figure 10–9** *Using Mobile IPv6 for UMTS-WLAN integration.*

2. The mobile node moves to the WLAN network. In this case, it needs to update its home agent and correspondent nodes if any exist.
3. If flow movement is implemented in the mobile and correspondent node (or home agent if no route optimization is used), the mobile node can choose to receive some flows through the UMTS interface and others through the WLAN interface.

In the first case, the mobile node would not move to the WLAN access. This means that no connection using the home address would be sent through the WLAN access. This does not, however, stop the mobile node from configuring an address on the WLAN interface.

In the second case, the mobile node would configure a care-of address. The mobile node would then update the MAP with its *on-link care-of address* (we assume that it will choose to use the MAP) and update its home agent and correspondent node with its regional care-of address. While moving within the WLAN access, the mobile node would only need to update the MAP. The mobile node need not terminate its UMTS link until it is sure that no new packets will be sent to that interface. This is a typical case for a make-before-break handover. Since we assume that there is a complete overlap between the WLAN access and UMTS access, the mobile node can wait until no new packets are sent to its old care-of address (the UMTS interface). After receiving a binding acknowledgment from the home agent and correspondent node, the mobile node can assume that all new packets will be sent to its WLAN interface. In this scenario, there is no need for anticipation or fast handovers, since the mobile node performs a make-before-break handover. However, fast handovers may be needed to move between two routers within the WLAN network.

In the third case, the mobile node could update the correspondent node or its home agent with its flow movement preferences. The mobile node would keep both interfaces up for the lifetime of the flows or longer, depending on whether both interfaces are needed.

Note that our assumption about a make-before-break handover is based on another assumption: there is a complete overlap between the WLAN and UMTS coverage areas. If we assume that it is a very small overlap, then the mobile node cannot assume that the cellular interface is always available. If the overlap is large enough to ensure that the mobile node can send a binding update and receive the binding acknowledgment before losing the UMTS interface, this is sufficient to ensure that there is no packet loss. However, if the overlap is not that large, then it may be necessary to implement the MAP function in the GGSN or another router connected to the Gi interface; this allows for quicker switching of the packets' path to the new mobile node location. The case for the overlap size is also relevant when the mobile node is moving from the WLAN network to the UMTS network. In this case, the mobile node must decide when the handover should be initiated based on the received power level of the WLAN network. If the received power level drops to below a certain level (determined by the mobile node's implementation), the mobile node must move to the UMTS interface.

## 10.6 Summary

This chapter looked at a practical case for using IPv6 in an existing system, UMTS. We described the network architecture, which is divided into the radio access network (UTRAN) and the core network (GPRS). We discussed the various interfaces defined in 3GPP between the UE and UTRAN as well as that between UTRAN and the core network. We also looked into the operation of UTRAN, including the three different types of handovers supported: softer handover, soft handover, and hard handover.

The operation of the core network was presented. We saw that UMTS brings new concepts, like PDP contexts, APNs, and application-layer tunneling, using the GTP protocols. The pros and cons of those design decisions were also discussed. Following this, the use of IPv6 in terms of address configuration, coexistence with IPv4, and mobility were shown. Point-to-point links actually simplify the address configuration and allow the AR (GGSN) to allocate an entire prefix to the UE, using existing mechanisms. This simplification enables UEs to act as mobile routers in the future without any modifications. In addition, it enables them to generate various addresses without having to test them for their uniqueness.

We applied the transition mechanisms discussed in Chapter 9 to UMTS and saw how operators can gradually introduce IPv6 where it matters (i.e., in the UE) without having to upgrade every node in their networks.

Finally, we saw how Mobile IPv6 can be used in this environment. Although UMTS does not rely on Mobile IPv6 for its mobility, the protocol is still useful when UEs require reachability, moving to other networks, or flow movement. Future releases of UMTS may decide to evolve the GPRS architecture and start using Mobile IPv6 (with IP-in-IP tunneling instead of GTP). The evolution of GPRS is a topic of intense debate in the research and development community.

## Further Reading

[1] 3GPP Technical Specification, 3GPP TS 33.102, "Technical Specification Group Services and System Aspects; 3G Security; Security Architecture (Release 4)," 3rd Generation Partnership Project, December 2001.

[2] Arkko, J., G. Kuipers, H. Soliman, J. Loughney, and J. Wiljaka, "Internet Protocol Version 6 (IPv6) for Some Second and Third Generation Cellular Hosts," RFC 3316, April 2003.

[3] Arkko J., and H. Haverinen, "EAP AKA Authentication," draft-arkko-pppext-eap-aka-09, work in progress, February 2003.

[4] Blunk, L., and J. Vollbrecht, "PPP Extensible Authentication Protocol (EAP)," RFC 2284, March 1998.

[5] Holma, H., and A. Toskala (Eds.), *WCDMA for UMTS: Radio Access For Third Generation Mobile Communications*, 2nd ed. Wiley, 2002.

[6] Inamura, H., G. Montenegro, R. Ludwig, A. Gurtov, and F. Khafizov, "TCP over Second (2.5G) and Third (3G) Generation Wireless Networks," RFC 3481, February 2003.

[7] TS 23.060 version 3.11.0 (release 99), 4.4.0 (release4) and 5.1.0 (release 5).

[8] Wasserman, M. (Ed.), "Recommendations for IPv6 in Third Generation Partnership Project (3GPP) Standards," RFC 3314, September 2002.

# INDEX


## A

*A* flag, mobility header, 89
A records, 9
AAA broker, 266
AAA front-end server, 321
AAA Working Group, 268
AAAA records, 9
Access Control List (ACL), 267
Access networks, 280
Access point (AP), 198, 222-223
Access Point Name (APN), *activate PDP context request* message, 313
Access Router (AR), xix, 198, 222
*activate PDP context request* message, 313
Address resolution, 62
*address[1]...address[n]*, 33
Administration-local scope, multicast addresses, 59
*advertisement interval* option, 95
Advertisement interval value, 96
Aggressive mode, 140
All-nodes multicast address, 63, 70, 98
All-routers multicast address, 60, 63
*alternate-care-of address* option, 85, 90-91, 159
American National Standards Institute T1P1 group (ANSI T1P1), 300
AMPS, 299
Anycast addresses, 7, 60-61
    format for, 61
    scopes, 60
Application independence, and Mobile IPv6, 23
Application layer, 5
Application Level Gateways (ALGs), 145-147, 293, 295, 319
Architecture, UMTS Terrestrial Radio Access Network (UTRAN), 305-11
    hard handovers, 310-311
    power control and handovers, 308-311
    Radio Network Subsystems (RNSs), 305
    soft handovers, 309-310
    softer handovers, 309
    types of handovers in, 308
    Wideband Code Division Multiple Access (WCDMA), 306-308
ARIB, 300, 301
*arpa* domain, 8
Association of Radio Industries and Businesses (ARIB), 300
Asymmetric cryptography, 130, *See also* Public key cryptography
Asymmetric keys, 127
Asynchronous Transfer Mode (ATM), 304
*au* domain, 8
Authentication, 122-123, 130, 146
    mutual, 156
    with secret keys, 128-129
    with public key cryptography, 133
    strong authentication systems, 123
Authentication and Key Agreement (AKA) protocol, 321
Authentication, Authorization, and Accounting (AAA), as enabler for mobility, 265-267
Authentication Center (AuC), 322
*authentication data* field:
    Authentication Header, 41-42
    Encryption Security Payload (ESP) Header, 43
Authentication Header, 39, 40-42, 86, 292
Authentication process, defined, 39-40
Authorization, 122-123, 155-156
    and authentication, 123
Autonomous systems (AS), 16

## B

Back-end AAA server (AAAL), 266-267
Backward compatibility, 110
Bad Guy, defined, xix
Base station, use of term, 198
Basic Service Set identity (BSS ID), 222
Bicasting, 234-235
Binding, 83
Binding acknowledgment message, 91
    format, 92
*binding authorization data* option, 169-170
Binding cache, 83, 85
Binding cache entries, removing in the correspondent node, 173-174

Binding error messages, 112-113, 184
Binding management key, 169
*binding refresh advice* option, 172
Binding refresh requests, 93
  securing, 176
Binding Update (BU) message, 84
Binding update list (BUL), 83, 92
Binding updates, 88-92
  future mechanisms for authenticating, 185-189
  goal of, 109
  Man in the Middle (MITM) attacks on, 152
  securing using public keys and certificates, 182-183
  using to launch attacks, 150-153
Birthday paradox, 138
Bit Error Rates (BER), 213
Bombing attacks, *See* Flooding attacks
Border Gateway Protocol (BGP), 15-16
Break-before-make handovers, 201, 202, 259
Breaking algorithms, 126
Brute-force attack, 126

## C

CA1/CA3/CA5, 136
Caching, 9
Caesar cipher, 124
Candidate Access Router Discovery (CARD), 270, 274
Care-of address, 83
Care-Of address Test Init (COTI), 167
*care-of init* cookie, 168
Care-Of Test (COT), 167
cdma2000, 299
Cerificates, 147
Certificate Authority (CA), 136
Certificate Revocation List (CRL), 136
Certificate Signing Unit (CSU), 136
Certificates, 134-137
Channelization codes, 307
Checksum field, 27-28
*checksum* field:
  ICMPv6, 45
  mobility header, 88, 89
China Wireless Telecommunication Standard group (CWTS), 300
Ciphering, 125
Ciphertext, 125
Circuit-switched core network, 305
Client-server model, 17
Client-server vs. peer-to-peer communication, 17-18
Closed-loop power control, 308-309
Code Division Multiple Access (CDMA), 197
*code* field, ICMPv6, 45
*com* domain, 8
Confidentiality, 122, 123-124, 126, 130
Configured tunnels, 282-283
Congestion avoidance, 212
Congestion window, 212
Connectionless links, 198
Connectionless Network Protocol (CLNP), 18
Connectionless radio links, 199
Connection-oriented links, 198
Conservative approach, to vulnerabilities, 181
Control plane, 312
Control Plane GTP, 312
Cookies, 139
Core network, Universal Mobile Telephony System (UMTS), 311-315
  mobility management in, 315
  PDP context activation, 312-315
Core Network (TSG-CN), 301
Correspondent node (CN), 82
Correspondent nodes, 116
  avoiding DOS attacks in, 156
  binding updates sent to, acknowledging, 112
  failure of, 112-113
  receiving route optimized packets from, 111
  sending route optimized packets to, 109-111
COTI/COT messages, 240, 262
Country domains, 7
Cryptanalysts, 125
Cryptographers, 125
Cryptographically Generated Addresses (CGAs), 25, 105, 142-145, 186-187, 188, 236
Cryptography, 124-146
  Caesar cipher, 124
  cryptographic functions, 125
  defined, 124
  encryption algorithms and keys, 126-127
  hash functions, 125
  public key, 125, 130-137, 147, 162, 177, 182
  secret key, 125, 127-129, 159
Cu interface, 304
*cur hop limit* field, router advertisements, 64
Cwnd, 212
CWTS, 300
Cyclic Redundancy Check (CRC), 309

## D

DAD, *See* Duplicate Address Detection (DAD)
DAD delays, 218

eliminating, 206
reducing, 205-207
Data Encryption Standard (DES), 127
Data origin authentication, 122
Data-link layer, 4
Decision point, 260
Decryption, 130
Deering, Steve, 18
Default router, 62
Default router list, 72
DeMilitarized Zone (DMZ), 146
Demultiplexing, 11-12
Denial of Service (DoS) attacks, 104, 152-153, 178
*destination address* field, 71
IPv6 packet, 27
Destination cache, 72, 75
Destination options header, 43, 44, 51, 85
DHAAD reply message, 100-101
DHAAD request message, 100-101
Diameter, 322
Diameter (new-generation AAA protocol), 268
Digital signatures, 124, 133-134
*dist* field, MAP discovery, 245, 247
Distance-based selection, 247
DNS security protocol (DNSSEC), 104
Domain name, 7
Domain Name System (DNS), 7-9
caching, 9
name servers, 8-9
Resource Records, 9
reverse DNS lookup, 7
structure of, 9
Top Level Domains (TLDs), 7
zones, 8-9
DoS attacks, *See* Denial of Service (DoS) attacks
Dots, in labels, 8
Drift RNC, 310
Dual Stack Transition Mechanism (DSTM), 294
Dual stacked mobile nodes, with both IPv4 and IPv6 enabled, 295
Duplicate Address Detection (DAD), 62, 68-69, 89-90, 188
DAD delays, 218
eliminating, 206
reducing, 205-207
optimistic DAD algorithm, 238
Dynamic Home Agent Address Discovery, 99-103
defined, 100
Dynamic Home Agent Address Discovery (DHAAD), 92, 162
Dynamic Host Configuration Protocol for IPv6 (DHCPv6), 73
Dynamic keying, 162-163
Dynamic MAP discovery, defined, 247


## E


Eager movement detection algorithm, 293
Eager-lazy behavior, 250-251
Echo reply, 33
Echo request, 33
Echo request messages, 247
*edu* domain, 8
Eifel algorithm, 214
Encapsulation, 10-11
Encryption, 124-125, 146
defined, 40, 124
using public key cryptography, 133
Encryption algorithms and keys, 126-127
Encryption Security Payload (ESP) Header, 42-43
format, 42
End-to-end signaling, and Mobile IPv6, 23-24
Enforcement Point (EP), 266
Enhanced Data Rates for GSM Evolution (EDGE), 300
Entity authentication, 122
Ephemeral port numbers, 11
Ethernet, 4
European Telecommunications Standards Institute (ETSI), 300, 301
Exhaustive search, 126
Exponential back-off algorithm, 91-92
Extensible Authentication Protocol (EAP), 321-322
Extension headers, 29-44, 77
defined, 29
Destination options header, 43
encoding options in, 29-30
Fragmentation Header, 35-39
Hop-by-Hop options header, 30-31
IP layer security, 39-43
Authentication Header, 39, 40-42
Encryption Security Payload (ESP) Header, 42-43
ordering of, 43-44
Routing Header, 31-35
*extensions* field, X.509 certificates, 136
Exterior Gateway Protocols (EGPs), 15


## F


Fast handoffs, *See* Fast handovers
Fast handovers, 235-236

Fast handovers *(cont.)*:
alternate approach to, 237-239
  anticipation and handover initiation, 222-224
  attacks on F-BU, 235
  and CGAs, 236
  cost of anticipation, 228-235
  current access router, updating, 224-226
  DoS attacks, 235
  failure cases, 227-228
  HI/Hack messages, 236
  moving to a new link, 226-227
  ping-pong movement, 232-233
  proxy advertisements, 235-236
  security issues, 235-236
    attacks on F-BU, 235
    DoS attacks, 235
    HI/Hack messages, 236
    proxy advertisements, 235-236
Fast Neighbor Advertisement (F-NA) message, 227
Fast path, 29
Fast RAs, 205
Fast retransmission algorithm, 212
Fast-binding update (F-BU), 225
  attacks on, 235
Fields, in X.509 certificates, 135-136
File Transfer Protocol (FTP), 5
Filtering router, 145
Firewalls, 88, 145-146, 147, 185
*flags* field, multicast addresses, 58
Flooding attacks, 152, 183
*flow label* field:
  IPv6 packet, 26, 75
  tunnel header, 50
Flow mobility, 261
Flow movement, in Mobile IPv6, 259-263
*flow movement* option, 261
FMIPv6, *See* Fast handoffs
F-NA message, 237
Foreign link, 83
Forward progress, 70
Forwarding, 5
Forwarding mechanism, 13
*fragment offset* field, Fragmentation Header, 36
Fragmentation Header, 35-39
Fragmentation, in IPv6, 37-38
Francis, Paul, 18
FreeBSD operating system, 164
Frequency division duplex (FDD), 307-308
Frequency Division Multiplexing (FDM), 199
Front-end AAA server, 266
Fully Qualified Domain Name (FQDN), 8

## G

Gateway GPRS Support Node (GGSN), 302-303
General Packet Radio Service (GPRS), 300
Generic domains, 7
Generic Packet Radio Service (GPRS) core network, 302
Global addresses, 54
Global Mobile Suppliers Association, 301
Global routing prefix, 54
Global System for Mobile communication (GSM), 196
*gov* domain, 8
GPRS attach procedure, 312
GPRS Tunneling Protocol (GTP) header, 305
GPRS/EDGE Radio Access Network (TSG-GERAN), 301
GSM association, 301
GSM communication Technical Specifications and Technical Reports, 300
GTP tunnel identifier, 312

## H

*H* flag, mobility header, 89
Half open connection, 208
Handover Acknowledgment (Hack) message, 225
Handover Initiate (HI) message, 224
Handover time, 202-205
Handovers:
  break-before-make and make-before-break handovers, 201
  evaluating, 195-220
  impacts on TCP and UDP traffic, 207-217
  layer 2 vs. layer 3 handovers, 196-201
Hash functions, 125, 133, 137-138, 147
  using for authentication, 138
Hashed MAC, 139
*header ext len* field, 32
*header length* field, mobility header, 88
Hierarchical Mobile IPv6 (HMIPv6), 239-253
  aim of, 242
  combining fast handovers and, 253-259
  deploying, 248-251
  design of, 242
  local mobility without updating correspondent nodes, 252
  location privacy, 251
  overview of, 242-245
  securing binding updates between a mobile node and a MAP, 252-253
HI/Hack messages, 236, 237

HMAC, 139
HMAC MD5, 139
HMAC SHA-1, 139
HMIPv6, *See* Hierarchical Mobile IPv6 (HMIPv6)
Home AAA server (AAAH), 266-267
Home address, 81, 82, 86, 88
*home address* option, 85, 109, 160, 262
    preventing attacks using, 184-185
HOme address Test Init (HOTI), 167
Home agent (HA), 81, 83, 149-150
    anycast address, 100
    binding cache entries in, 90
    failure of, 102-103
    information option, 100
    lifetime, 100
    list, 100
    routing packets through, 107
    securing binding updates to, 158-162
    securing messages to, 157
*home init* cookie, 167
*home keygen* token, 168-169, 240
Home link, 82, 86
Home Location Register (HLR), 302-303
Home networks, 57
HOme Test (HOT), 167
*hop limit* field:
    IPv6 packet, 27
    tunnel header, 50
Hop-by-Hop options header, 30-31
Host-router and router-router incompatibility, overcoming, 281
Host-to-host communication, 9-12
    demultiplexing, 11-12
    encapsulation, 10-11
HOTI/HOT messages, 205, 240, 262
    securing between the mobile node and the home agent, 175-176
Hypertext Transfer Protocol (HTTP), 5

## I

ICMPv6, 4, 44-47, 77
    error messages, 45-45
    informational messages, 47
    and Neighbor Discovery, 62
*identification* field, Fragmentation Header, 36-37
*identifier* field, DHAAD request/reply messages, 101
Identity authentication, 122
IEEE802.1X, 199
IEEE802.11b, 22, 197, 222-223, 259, 272
    purpose of, 200
Immutable fields, 40
Incompatibility, host-router and router-router, 281
Ingress interface, 74
Integrity check, 139
Integrity Check Value (ICV), 40
Integrity checking, 122
Interdomain routing protocols, 15
Interface identifier, 54
Interface-local unicast addresses, 59
Interior Gateway Protocols (IGPs), 15-16
Intermediate System to Intermediate System (IS-IS), 15
International Data Encryption Algorithm (IDEA), 128
International Mobile Subscriber Identifier (IMSI), 321
International Telecommunications Union (ITU), 299
Internet Activities Board (IAB), 18
Internet Architecture Board (IAB), 18
Internet Control Message Protocol for IPv6 (ICMPv6), *See* ICMPv6
Internet Control Message Protocol (ICMP), 4
Internet Key Exchange (IKE) protocol, 104, 140-141, 147
Internet Protocol (IP), 3-24, 18
    application layer, 5
    client-server vs. peer-to-peer communication, 17-18
    data-link layer, 4
    Domain Name System (DNS), 7-9
    host-to-host communication, 9-12
    IP addresses, 7
    network layer, 4
    networking with, 5-6
    physical layer, 3
    routing, 13-17
    sending information using, 10
    transport layer, 4-5
Internet Security Association and Key Management Protocol (ISAKMP), 140
Internet Service Providers (ISPs), 14-15
    peering, 15
Intradomain routing protocols, 15
Intrasite Automatic Addressing Protocol (ISATAP), 286-287
IP addresses, 7
IP in IP encapsulation, 50
IP in IP tunneling, for route optimization, 113-114
IP layer, 4, 6
IP layer security, 62
    extension headers, 39-43
        Authentication Header, 39, 40-42
        Encryption Security Payload (ESP) Header, 42-43

IP mobility, 19-24
 defined, 19
 example, 20
 importance of, 21-22
 mobility management, 21-23
IP Multimedia Subsystem (IMS), 313-314
IP security (IPsec) protocols, 25
IP stack, 268
IPng Directorate, 18-19
IPsec, 86, 102
 security associations:
  manual vs. dynamic, between mobile node and home agent configuration, 162-163
  mobile node's home address as basis of, 160
  using manual configuration to set up, 157-158
IPv4-capable nodes, 288
IPv4-enabled nodes, 288
IPv4-mapped IPv6 addresses, 61
IPv4-translated IPv6 addresses, 61
IPv6, 3, *See also* IPv6 addresses; IPv6 deployment
 addresses, 51-61
 communication example, 74-77
 extension headers, 29-44
 flow label, 28-29
 improvements over IPv4-based Internet, 25
 Neighbor Discovery, 61-72
 primer, 25-78
 protocol, 26-29
 in Universal Mobile Telephony System (UMTS), 315-325
  address configuration, 316-317
  end-to-end incompatibility, 318-319
  IPV4-ONLY GGSNS, 320
  mobility, 320-323
  transition and coexistence, 317-320
  tunneling over the GI interface, 317-318
IPv6 addresses, 51-61
 anycast addresses, 60-61
 containing IPv4 addresses, 61
 interface identifier, 51
 multicast addresses, 58-60
 prefix, 51
 textual representation of, 52-53
 unicast addresses, 53-58
 unspecified address, 61
 written format of, 52-53
IPv6 deployment, 280
 IPv6-only networks, 294
 problem associated with, 280-281
 translation, 288-296
  Network Address Translator and Protocol Translator (NAT-PT), 292-293
  Stateless IP ICMP Translator (SIIT), 289-292
 tunneling, 281-288
IPv6 Forum, 280, 301
IPv6 header, 26
IPv6 in 3GPP networks, case study, 299-326
IPv6 mobility, current/future work on, 265-275
IPv6 Neighbor Discovery protocol, 14
IPv6 packet, 26-27
IPv6 site, 7
IPv6 Working Group, 316, 317
ISATAP addresses, 287-288
"Islands," 280
*issuer* field, X.509 certificates, 136
*issuerUniqueIdentifier* field, 136

## J

Jumbograms, 26

## K

*K* flag, mobility header, 89
Keepalive timer, 210
Keyed hash function, 139
Known plaintext attack, 127

## L

*L* flag, router advertisements, 64-65
Label, nodes, 7-8
Layer 2, 4
 termination of, 198
 trigger, 203
Layers, 3-5
Link layer, 4
Link-layer address option, 68
Link-layer control, 5
Link-layer handover, 259
Link-layer header, 6
Link-local address, 53
Link-local all-routers multicast address, 60
Link-local multicast addresses, 59-60
Link-local unicast addresses, 59
Local Area Networks (LANs), 21, 200
Local binding update, 243
Local Mobility Management (LMM) network, 250
Loopback address, 52, 53
*loose source routing* option, IPv4, 31
Lower layers independence, and Mobile IPv6, 23

## M

*m* flag, Fragmentation Header, 36
Main mode, 140
Make-before-break handovers, 201-202
Man in the Middle (MITM) attacks, 152, 178
  on binding updates, 152
  on MPS./MPA, 154
  on MPS/MPA, 154
MAP, *See* Mobility Anchor Point (MAP)
Market Representation Partners (MRPs), 301
Masquerading, 178
Maximum Transmission Unit (MTU), 35
Media Access Control (MAC), 4
Message Authentication Code (MAC), 139
Message digests, 137-139
Message encryption, using public key cryptography, 133
Message integrity, 124, 156
MITM attacks, *See* Man in the Middle (MITM) attacks
Mobile Authentication Protocol (MAP), 304
Mobile Country Code (MCC), 321
Mobile IP Working Group, 195, 221
Mobile IPv6, 3, 20, 44, 77, 81-120
  binding updates and acknowledgments, 88-92
  binding updates sent to correspondent nodes, acknowledging, 112
  communication scenarios, 149-150
  correspondent nodes, failure of, 112-113
  defined, 149
  Dynamic Home Agent Discovery, 99-103
  fast handovers for, 221-239
  flow movement in, 259-263
  handovers:
    evaluating, 195-220
    optimizations and extensions, 221-264
  main requirements, 23-24
  mobile node's return home, 97-99
  mobility header, 88
  movement detection, 93-97
  overview of, 83-108
  refreshing bindings, 93
  reverse tunneling, 93
  route optimization, 106-114
  route optimized packets:
    receiving from correspondent nodes, 111
    sending to correspondent nodes, 109-111
  security, 121-148, 157-184
    assumptions about, 157
    requirements for, 154-157
  and site-local addresses, 115
  source address selection in mobile nodes, 99
  terminology, 82-83
  tunneling, 48-49
  virtual home links, 106
Mobile IPv6 signaling, securing, 149-192
Mobile Network Code (MNC), 321
Mobile networks, 274
  need for route optimization, 273
Mobile nodes, 82, 91-92, 149-150
  failure of, 114-115
  generation of random home addresses, 105
  home address, discovery of, 104
  MAP selection, 247-248
  multiple home addresses, 105-106
Mobile Prefix Advertisement (MPA), 102-103
  MITM attacks on, 154
  securing, 162
Mobile Prefix Solicitation (MPS), 102-103
  MITM attacks on, 154
  securing, 162
Mobile routers (MRs), 271
Mobile Subscriber Information Number (MSIN), 321
Mobile termination (MT), 302
Mobility Anchor Point (MAP), 242-243
  defined, 242
  MAP discovery, 245-248
    *dist* field, 245, 247
    *pref* field, 245
    *preferred lifetime* field, 246
    *R* flag, 246
    *res* field, 246
    *valid lifetime* field, 246
Mobility header, 44, 84, 88, 163
*mobility header type* field, mobility header, 88
Moderate approach, to vulnerabilities, 181
Modular arithmetic, 130
Modulation, 5
Mobile IPv6, binding update message format, 88
Multicast addresses, 7, 58-60
Multicast Listener Discovery (MLD) protocol, 58, 86
Multihomed hosts, 6, 259
Multihomed mobile network, 271-272
Multihomed node, 6
Multiple access links, 62
Multiple home addresses, 105-106
Multiplicative inverses, 131
Mutable fields, 40
Mutual authentication, 156
  with secret keys, 128-129

## N

Name servers, 8-9
NAT(P)T-PT, 319
Neighbor Advertisement Acknowledgment (NAACK), 227
Neighbor Advertisements (NAs), 68, 69-70, 188-189
Neighbor cache, 72, 75
Neighbor Discovery, 44, 45, 61-72, 77, 95, 102, 188, 218
  address resolution, 62
  conceptual data structures in hosts, 72
  default router, 62
  defined, 61-62
  and ICMPv6, 62
  neighbor advertisements, 69-70
  neighbor solicitations, 66-69
  Redirect messages, 71-72
  reducing, 205-207
  router advertisements, 63-66
  router solicitations, 63
  stateless address autoconfiguration, 73-74
Neighbor Solicitation (NS), 66-69, 188
Neighbor Unreachability Detection (NUD), 62, 66, 69, 72
nemo (network mobility) Working Group, IETF, 274
Nested encapsulation, 50-51
Nested mobile network, 272
*net* domain, 8
Network Access Server (NAS), 267
Network Address Identifier (NAI), 266
Network Address Translator and Protocol Translator (NAT-PT), 292-293
Network architecture, Universal Mobile Telephony System (UMTS), 301-305
  circuit-switched core network, 305
  components, 301
  Core Network (CN), 301-302
  interface, 302
  packet-switched core network, 302-305
  UMTS Terrestrial Radio
  User Equipment (UE), 301-302
Network layer, 4
Network layer independence, 311
Network mobility, 270-274
Networking, defined, 3
*next header* field:
  IPv6 packet, 26-27
  Routing Header, 32, 34
  tunnel header, 50
Next-hop, 13
  determination, 75
NMT, 299
Non-Broadcast Multiple Access (NBMA) link layer, 288
Nonce index, 168
*nonce indices* option, 169-170
Nonces, 139
  maintaining, 174
Nonrepudiation, 122, 124, 139

## O

*O* flag, Neighbor Advertisements, 69
One-way transformations, 125
On-link care-of address (LCoA), 242, 324
Open Shortest Path First (OSPF), 15
Optimistic DAD, 206
Optimistic node, 206
Option length, 90
Option type, 90
*org* domain, 8
Organization scope, unicast addresses, 59
Original header, 50
Orthogonal codes, 307
Outer-loop power control, 309

## P

Packet Data Convergence Protocol (PDCP), 304-305
Packet Data Protocol (PDP), 312
Packet-switched core network, 302-305
PANA Working Group, 268
  IETF, 322
Password-based schemes, and authentication systems, 122
Path MTU (PMTU), 35
Paul's IP (PIP), 18-19
*payload data* field, Encryption Security Payload (ESP) Header, 43
*payload length* field:
  Authentication Header, 41
  IPv6 packet, 26, 28
  tunnel header, 50
*payload proto* field, 88
PDP type, *activate PDP context request* message, 313
Peers, 5
Peer-to-peer communication, defined, 18
Permanent home address, 105
Persist timer, 210
Personal Area Network (PAN), 271
Personal Digital Assistant (PDA), 271-272
Physical layer, 3
ping program, 33
Ping-pong movement, 232-233

Plaintext, 125, 126
PMTU discovery, performance of, 38-39
Point-to-Point Protocol (PPP), 321
Port number space, 10-11
Port numbers, 10-11
Port translation, and NAT-PT, 293
Potential Router List (PRL), 287
*pref* field, MAP discovery, 245
*preference* field, 100
Preference parameter, 100
*preferred lifetime* field:
  MAP discovery, 246
  router advertisements, 66
Prefix, 17
*prefix* field, router advertisements, 66
Prefix information option, 64-65
*prefix length* field, router advertisements, 64
Prefix list, 72
Primary DNS, 9
Prime numbers, 130-131
Privacy Extended Mail (PEM), 135
Proxy advertisements, 235-236
  and fast handovers, 235-236
Proxy Neighbor Advertisements, 70, 85
Proxy router advertisements, 223
Public key cryptography, 125, 127, 130-137, 140, 147, 162, 177, 182
  authentication with, 132
  certificates, 134-137
  digital signatures, 133-134
  encryption using, 132-133
  modular arithmetic, 130
  RSA example, 131-132
Public Key Cryptography Standard (PKCS), 135
Public key, defined, 130
Public Key Infrastructure (PKI), 155

## Q

QoS signaling protocols, 31
QoS/filtering-related parameters, *activate PDP context request* message, 313
Quality of Service (QoS), 200

## R

*R* flag:
  MAP discovery, 246
  Neighbor Advertisements, 69
Radio Access Network Application Protocol (RANAP), 302
Radio Access Network (TSG-RAN), 301
Radio Network Controller (RNC), 305
Radio Network Subsystems (RNSs), 305
Radio Resource Control (RRC) protocol, 314
Radio Resource Management (RRM), 200, 308
RADIUS, 268, 322
Reachability, and Mobile IPv6, 23
Reachability confirmation, 70
*reachable lifetime* field, router advertisements, 64
Real-time applications, 5
Real-time video conferencing over IP, 5
Redirect messages, 71-72, 189
Reflection attacks, 151, 178
Regional care-of address (RCoA), 242
Relatively prime numbers, 130-131
Remote Authentication Dial-In User Service (RADIUS), 268, 322
Replay attacks, 124
*res* field:
  Fragmentation Header, 36
  MAP discovery, 246
*reserved* field, Routing Header, 33
Reserved port numbers, 11
Resolver, 9
Resource Records, 9
Retransmission timer, 210
*retransmission timer* field, router advertisements, 64
Return routability procedure, 176-177
  operation of, 166
  overview of, 166-172
Return routability, securing, 178-182
Reverse DNS lookup, 7
Reverse tunneling, 93
RFC 2408, 140
RFC 2409, 77, 140
RFC 2460, 26, 77, 312
RFC 2461, 62-63
RFC 2462, 55, 73
RFC 2463, 26, 77
RFC 2473, 77
RFC 2526, 60
RFC 2675, 77
RFC 2710, 77
RFC 2765, 291
RFC 3041, 73, 316
RFC 3314, 316
RFC 3484, 77
Root DNS, 7
Round Trip Time (RTT), 33
Roundtrip time (RTT), 211
Route aggregation, 16-17
Route optimization, 106-114, 149-150, 177, 244-245, 273
Route optimization *(cont.)*:
  IP in IP tunneling for, 113-114

Route optimized packets:
  receiving from correspondent nodes, 111
  sending to corresponding nodes, 109-111
Router Advertisements (RAs), 63-66, 95, 189
*router alert* option, Hop-by-Hop header, 31
*router lifetime* field, router advertisements, 64
Router solicitation for proxy, 223
Router Solicitation (RS) message, 203
Router solicitations, 63
Routers, 6, 95
  and the MAP option, 246-247
Routing, 13-17
Routing Area Identifier (RAI), 310
Routing Header, 31-35, 44
  preventing attacks using, 184-185
  security hazards of, 34-35
  using, 33-34
Routing Header and *home address* option, attacks using, 153
*routing header type 2*, 109, 150
Routing Information Protocol (RIP), 15
Routing intelligence, 13
Routing protocols, 13
Routing protocols-based tunnel end point discovery, 286
Routing table, 13
*routing type* field, Routing Header, 32
RSA algorithm, 131-132
RSVP, 31

## S

*S* flag, Neighbor Advertisements, 69
*scope* field, multicast addresses, 59
Scoped addresses, 25
  challenges imposed by, 55-58
Scrambling code, 307
*se* domain, 8
Seamless mobility:
  achieving, 268-270
  context transfer, 269-270
  link-layer agnostic interface to the IP layer, 268-269
Seamoby Working Group, IETF, 270
Secondary DNS, 9
Secret, and authentication systems, 122
Secret key, 126
Secret key cryptography, 125, 127-129, 140, 146, 159
  limitations of, 129
  mutual authentication with secret keys, 128-129
Secure communication, 125
Secure handshakes, 40
Secure HTTP (S-HTTP), 135
SEcure Neighbor Discovery (SEND) Working Group, IETF, 189
Secure Socket Layer (SSL), 135
Security, 121-148
  Application Level Gateways (ALGs), 145-146
  authorization, 122-123, 155-156
  binding refresh requests, 176
  binding updates, securing to correspondent nodes, 165-184
  communication between mobile and correspondent nodes, 155-156
  cookies, 139
  correspondent nodes, avoiding DOS attacks in, 156
  Cryptographically Generated Addresses (CGAs), 142-145
  cryptography, 124-146
  defined, 121
  denial of service attacks (DOS), 152-153
  fast handovers, 235-236
    attacks on F-BU, 235
    DoS attacks, 235
    HI/Hack messages, 236
    proxy advertisements, 235-236
  firewalls, 145-146
  flooding attacks, 152
  hash functions, 137-138, 147
  home agent, securing messages to, 157
  message digests, 137-139
  MITM attacks, 152
  Mobile IPv6 signaling, 149-192
  need for, 121-122
  nonces, 139
  reflection attack, 151
  requirements for, 154-157
  Routing Header and *home address* option, attacks using, 153
  security associations (SAs), establishing, 40, 140-142
Security Association Database (SAD), 41, 158
Security Associations (SAs):
  establishing, 40, 140-142
Security Parameter Index (SPI), 41
Security Policy Database (SPD), 42, 162
Security, stealing traffic, 150-151
*segments left* field, Routing Header, 32, 34
Selective Acknowledgments (SACK), 212-213
Selectors, 42, 116
Sequence number, 124
*sequence number* field:
  Authentication Header, 41
  mobility header, 89

Service and System Aspects (TSG-SA), 301
Serving GPRS Support Node (SGSN), 302-303
Session continuity, Mobile IPv6, 23
Session Initiation Protocol (SIP), 316
Session keys, 267
Shadowing, 196
Signaling, 4
Signal-to-noise ratio (SNR), 269
*signature* field, X.509 certificates, 135-136
SIIT host, 289
SIIT routers, 289-291
Simple IP (SIP), 18-19
Simple Mail Transfer Protocol (SMTP), 5
Site Border Router (SBR), 55-56
Site-local addresses, 54
  and Mobile IPv6, 115
Site-local all-routers multicast address, 60
Site-local scopes, unicast addresses, 59
6-to-4 addresses, 61
6-to-4 tunneling, 283-286
Slow path, 29
Slow start threshold, 212
Socket, 10
Socket Application Program Interface (socket API), 10
Sockets API, 268
Solicited node multicast address, 60, 68, 98
*source address* field, 262
  IPv6 packet, 27
Specific route, 17
SRNS relocation procedure, 310
Stateless address autoconfiguration, 73-74
  ingress filtering, 74
Stateless IP ICMP Translator (SIIT), 289-292
Static keying, 162-163
Stealing traffic, 150-151
Stream Control Transmission Protocol (SCTP), 4-5
Strong authentication systems, 123
*subjectPublicKeyInfo*, 136
*subjectUniqueIdentifier* field, 136, 144, 184
*subnet id* field, 54
Subnet numbers, 16
Subscriber Identity Module (SIM), 304
Symmetric cryptography, 141
Symmetric key encryption, 128
Symmetric reachability, 70

## T

*target address* field:
  Neighbor Solicitations, 67-69
  NUD, 69-70
  proxy neighbor advertisement, 85
  Redirect messages, 71
TCP, 218
  assumptions in wireless networks, 213
  congestion avoidance, 212
  defined, 207
  fast retransmission algorithm, 212
  flow control, 210-211
  how it works, 207-208
  mobility Impacts on, 215-217
  segment, 11
  Selective Acknowledgments (SACK), 212-213
  slow start, 211
  timers, 210
  timeouts, 211-212
TCP/IP protocol suite, 271, *See* Internet protocol suite
Telecommunication Technology Committee (TTC), 300
Telecommunications Industry Association (TIA), 300
TELNET, 5
Temporary home address, MAP, 242
Tentative address, 68
Terminal equipment (TE), 302
3rd Generation Partnership Project 2 (3GPP2), 300
  background, 300-301
  Technical Specification Groups (TSGs), 301
  UMTS network architecture, 301-305
  Working Groups (WGs), 301
3DES, 128, 138
3GPP Release 1999 (R99) specifications, 301
Three-way handshake, 208
Time division duplex (TDD), 300, 307
Time stamp, 124
*Timestamp* option, TCP header, 214
Top Level Domains (TLDs), 7
Top-Level Mobile Router (TLMR), 272-273
Topologically correct addresses, 17
Totient, 131
*traffic class* field, IPv6 packet, 26
Traffic Flow Template (TFT), 314
Transmission Control Protocol (TCP), *See* TCP
Transport layer, 4-5
Transport Layer Security (TLS), 135
Transport level IP layer, 304
Triggers, 268
TSG-RAN WG1/TSG-RAN WG2, 301
TTA, 300
TUBA (TCP and UDP over Big Addresses), 18
Tunnel encapsulation limit, 50-51
Tunnel End Point (TEP), 294
Tunnel endpoint, 50

Tunnel entry point, 48, 50, 86
Tunnel exit point, 48, 50, 86, 244, 253, 261
Tunnel header, 50
Tunnel interface, 118
Tunnel mode, 50, 142
Tunneling, 48-51, 77, 86, 281-288
  configured tunnels, 282-283
  Intrasite Automatic Addressing Protocol (ISATAP), 286-287
  and Mobile IPv6, 48-49
  nested encapsulation, 50
  routing protocols-based tunnel end point discovery, 286
  6-to-4 tunneling, 283-286
  and Virtual Private Networks (VPNs), 49-50
2MSL timer, 210
Two-faced DNS, 57
*type* field:
  ICMPv6, 45
  mobility header, 88
*type of service* field, tunnel header, 50

## U

UDP, 218
  and Mobile IPv6 handover, 217
UDP datagram, 11
*uk* domain, 8
UMTS Terrestrial Radio Access Network (UTRAN):
  architecture, 305-11
    hard handovers, 310-311
    power control and handovers, 308-311
    Radio Network Subsystems (RNSs), 305
    soft handovers, 309-310
    softer handovers, 309
    types of handovers in, 308
    Wideband Code Division Multiple Access (WCDMA), 306-308
Unicast addresses, 7, 53-58
Universal Mobile Telephony System (UMTS), 299, 300
  Authentication and Key Agreement (AKA) protocol, 321
  core network, 311-315
    mobility management in, 315
    PDP context activation, 312-315
  network architecture, 301-305
    circuit-switched core network, 305
    components, 301
    Core Network (CN), 301-302
    interface, 302
    packet-switched core network, 302-305
    UMTS Terrestrial Radio
    User Equipment (UE), 301-302
Universal Subscriber Identity Module (USIM), 321-322
Universal Wireless Communications Consortium (UWCC), 301
Unspecified address, 52, 61
Usable window, 210
User Datagram Protocol (UDP), 4-5
User level IP layer, 304
User plane, 312
User Plane GTP, 312
User space, 3
UTRAN, *See* UMTS Terrestrial Radio Access Network (UTRAN)

## V

*valid lifetime* field:
  MAP discovery, 246
  router advertisements, 66
*validity* field, X.509 certificates, 136
*version* field:
  IPv6 packet, 26
  X.509 certificates, 136
Virtual home links, 106
Virtual Private Networks (VPNs), 142, 312, 314
  and tunneling, 49-50
Voice over IP (VoIP), 195, 259

## W

Wide Area Networks (WANs), 21, 200
Wideband Code Division Multiple Access (WCDMA), 201, 235, 300
Window, 210-211
Wireless LAN (WLAN), 22, 187, 200, 259, 300, 320-325
Wireless links, categorizing, 198-199
WLAN cells, 197
Words, 52

## X

X.25, 304
X.509 certificates, 144-145, 147
  standard, 135

## Z

Zones, 8-9